T0185911

Risk Science

Risk science is becoming increasingly important as businesses, policymakers and public sector leaders are tasked with decision-making and investment using varying levels of knowledge and information. *Risk Science: An Introduction* explores the theory and practice of risk science, providing concepts and tools for understanding and acting under conditions of uncertainty.

The chapters in this work cover the fundamental concepts, principles, approaches, methods and models for how to understand, assess, communicate, manage and govern risk. These topics are presented and examined in a way which details how they relate, for example, how to characterize and communicate risk with particular emphasis on reflecting uncertainties; how to distinguish risk perception and professional risk judgments; how to assess risk and guide decision-makers, especially for cases involving large uncertainties and value differences; and how to integrate risk assessment with resilience-based strategies. The text provides a variety of examples and case studies that relate to highly visible and relevant issues facing risk academics, practitioners and non-risk leaders who must make risk-related decisions.

Presenting both the foundational and most recent advancements in the subject matter, this work particularly suits students of risk science courses at college and university level. The book also provides broader key reading for students and scholars in other domains, including business, engineering and public health.

Terje Aven is Professor of Risk Analysis and Risk Management at the University of Stavanger, Norway.

Shital Thekdi is Associate Professor of Management at the University of Richmond, USA.

Risk Science

An Introduction

Terje Aven and Shital Thekdi

Routledge
Taylor & Francis Group

LONDON AND NEW YORK

First published 2022
by Routledge
2 Park Square, Milton Park, Abingdon, Oxon OX14 4RN

and by Routledge
605 Third Avenue, New York, NY 10158

Routledge is an imprint of the Taylor & Francis Group, an informa business

British Library Cataloguing-in-Publication Data
A catalogue record for this book is available from the British Library

Library of Congress Cataloging-in-Publication Data
Names: Aven, Terje, author. | Thekdi, Shital, author.
Title: Risk science : an introduction / Terje Aven and Shital Thekdi.
Description: 1 Edition. | New York : Routledge, 2022. | Includes bibliographical
    references and index.
Subjects: LCSH: Risk management. | Decision making.
Classification: LCC HD61 .A9463 2022 (print) | LCC HD61 (ebook) | DDC
    338.5—dc23
LC record available at https://lccn.loc.gov/2021011831
LC ebook record available at https://lccn.loc.gov/2021011832

ISBN: 978-0-367-74269-0 (hbk)
ISBN: 978-0-367-74268-3 (pbk)
ISBN: 978-1-003-15686-4 (ebk)

DOI: 10.4324/9781003156864

Typeset in Sabon
by Apex CoVantage, LLC

Contents

About the authors

Terje Aven has been Professor in Risk Analysis and Risk Science at the University of Stavanger, Norway, since 1992. Previously he was also Professor (adjunct) in Risk Analysis at the University of Oslo and the Norwegian University of Science and Technology. He has many years of experience as a risk analyst in industry and as a consultant. He is the author of many books and papers in the field, covering both fundamental issues as well as practical risk analysis methods. He has led several large research programs in the risk area, with strong international participation. He has developed many master programs in the field and has lectured on many courses in risk analysis and risk management. Aven is Editor-in-Chief of the *Journal of Risk and Reliability* and Area Editor of *Risk Analysis in Policy*. He has served as President of the International Society for Risk Analysis (SRA) and Chairman of the European Safety and Reliability Association (ESRA) (2014–2018).

Shital Thekdi has been Associate Professor of Management at the University of Richmond, USA, since 2012. She has co-authored several papers on risk management and decision-making. She has a Ph.D. in Systems & Information Engineering from the University of Virginia; she has a M.S.E. and B.S.E. in Industrial & Operations Engineering from the University of Michigan. She has several years of experience working in industry, with extensive supply chain management and operational analytics experience.

Preface

This is a textbook on risk science – covering the basic concepts, principles, approaches, methods and models for how to understand, assess, communicate, manage and govern risk. It presents the foundational and most recent advancements in the subject matter for a risk science course at the university/college level while applying these to recent high-profile risk-related events and issues. It is relevant for students from all types of domains, including business, engineering and public health, as it highlights the generic and fundamental concepts, principles, approaches, methods and models of this science and field. In order to adequately handle risk, it is essential to understand all the core subjects of risk science and how they relate, for example, what is risk science; how to characterize and communicate risk, with particular emphasis on reflecting uncertainties; how to distinguish risk judgments and risk perception; how to assess risk and guide decision-makers, especially for cases involving large uncertainties and value differences; and how to integrate risk assessment with resilience-based strategies.

We encounter risk in all types of situations, such as in our personal lives and when considering high-level issues like climate change and politics. At a firm level, an organization invests in a project with a goal to be profitable; in reality, this project could result in profitability or failure and losses. At a global health level, the potential occurrence of a health pandemic represents a risk for societies and activities all over the world. Today we do not know what types of epidemics that will happen in coming years, how they will develop or what the consequences will be.

Risk is a concept that is fundamental to our lives, our work and our societies. The essence of the concept is that events may occur in the future that are not as planned or anticipated. Being subject to risk means that there is a potential for some undesirable consequences of these events. The actual consequences could, however, be positive. For example, following a major investment, the firm could be profitable. Without risk, little happens. As the saying goes, *who dares nothing, need hope for nothing*. The saying is just as relevant to individuals as it is to businesses and society. On the other hand, we also want to eliminate and reduce many types of risk. We do not want the food we eat to increase the risk of cancer, the health risk in relation to an epidemic should be as low as possible and jobs should be safe. Taking risks implies that mistakes or accidents can occur. These failures can eventually result in learning and improvement over time, but too high a risk obviously can be disastrous. The challenge is to find the right balance between creating value and taking risk on the one hand and protecting and reducing risk on the other. We strive to find this balance, as individuals, businesses and societies at large.

Risk science can help us find this balance and handle the risk in a good way. This requires that we be able to understand what risks we are facing. We must be able to distinguish between what is worth worrying about and what is not. For example, is it safe to travel by air? Can we ignore the risks related to an emerging epidemic if we implement the measures recommended by health organizations? Does investment in a specific project involve too high of a risk? Risk science will help you and your organization respond prudently. It clarifies how we should conduct risk assessments. It emphasizes that we must be careful not to mix professional risk assessments with 'risk perception' that reflects feelings such as fear and dread. For example, many people do not like to travel by plane. For many of them, the feelings – the risk perception – explain their behavior. Risk science clarifies and discusses what professional risk assessments are and what risk perception is. It may lead to the conclusion that you will confront the risk, as the problem is largely about fear and dread and not high risk.

However, risk science does not prescribe what decision to make. For example, travelers must make the choice whether to fly. They must consider the cost of the trip, the risks and the benefit of the flight and any other factors, such as environmental considerations. Risk science helps structure the thinking and perform the assessments so that the 'right' decision can be made under the current circumstances.

As another example, consider the early stage of an emerging new type of epidemic. No objective scientific answers can be derived because of the large uncertainties. Yet risk science can provide informative risk characterizations and support the development of suitable policies for handling the epidemic. The policies may include weights given to precautionary measures, as there is a potential for severe losses, and the consequences of the epidemic are subject to scientific uncertainties. Risk science explains what a precautionary approach is about and how it relates to different concerns, in particular the need for developments and creating values.

To guide decision-making, economists recommend the use of cost-benefit analysis. Risk science explains how these analyses should be applied in a risk context and combined with risk assessments to provide even a better decision basis. To provide a response to the issue about a project investment involving too high a risk, the risks need to be identified, analyzed and understood. Risk science provides concepts and tools for this purpose. Basic questions addressed in risk science and risk assessments are: What might go wrong? Why and how might it go wrong? And what are the consequences? What do we know, and what do we not know? On the basis of such knowledge, we see better what is required, what measures are needed and how we should proceed. Is it possible to avoid some risks? Can we control the risks if something surprising happens? Are the systems and organizations involved sufficiently robust and resilient in case something happens, such as a disturbance or change of any kind, even not foreseen?

Risk science is very much concerned about such issues, to understand the risks and how the risks relate to robustness and resilience. A main strategy of risk handling is adopting robustness and resilience-based thinking and measures. Risk assessments alone cannot guide us in how to deal with risk in situations of large uncertainties. These assessments have weaknesses and limitations, and some aspects of risk may be overlooked. To meet potential surprises, we need to think resilience. Risk science guides us in how to do this and integrate risk and resilience-based analysis and management.

Even if we all agree on the magnitude of the risks, we may have different views on what to do about the risks, as we have different values and priorities. How much weight to put on the uncertainties, for example, in relation to climate change, is a value judgment. Resilience is a key strategy to meeting uncertainties but so are dialogue and participation. Because of the uncertainties, there is no objective best approach to handle the risks. Different stakeholders weigh the various concerns differently. Risk science explains how dialogue and public involvement processes can reveal existing positions and perspectives and lead to improved understanding among different stakeholders, which in its turn can lead to broad support for a specific risk policy and risk-handling approach.

OBJECTIVES

The main objectives of the book are to present concepts, principles and tools for understanding and acting under conditions of uncertainty and risk. This is becoming increasingly important as businesses, policymakers and public-sector leaders are tasked with decision-making and investment using varying levels of information and knowledge. The major categories of topic interest are *risk assessment, risk perception and communication* and *risk management* (which here also covers *risk governance* and *risk policy*). These terms are fully defined and explored in the text, addressing features such as:

Risk assessment: People and organizations generally begin to understand risk by investigating potential risk events and gauging the severity or urgency of acting on those events and risks. Following the risk assessment process would provide a structure to answering questions such as:

- What are the risk events of concern? Why? What are the consequences of these events? What are the main uncertainties?
- How does the availability of information and knowledge influence the assessment of the risk?

Risk perception and communication: Professionals assess risk, but for proper understanding, communication and handling of risk, it is critical also to study how people perceive the risks. The perceptions reflect the real concerns of people and may include important aspects of risk not fully captured by risk assessments. There is often a need to communicate aspects of a risk with various audiences, such as the general public, the media and other stakeholders. In this setting, a person or organization needs to build on risk science and be intentional in communicating with stakeholders, ensuring an appropriate tone, including the necessary information, providing a clear message and encouraging an appropriate response.

Risk management: Following a risk assessment process, people and organizations must decide *how* to act, such as asking the following questions:

- How does the availability of information and knowledge influence how I (we) should address the risk? Should we strengthen the knowledge base before making a decision?
- Do we need to act now to confront the risk?
- Do we need to invest in anything to protect lives or some other values? If so, what resources can or should we invest?
- Do we need insurance?
- What is an acceptable risk level? How safe is safe enough?
- How much risk should we take in pursuit of values?
- What policies should be implemented in order to best handle the risks?
- What structures for decision-making authority and accountability should be used?

The topics described previously apply to students and of all domains and expertise levels. We acknowledge that some students may pursue entire careers related to risk and safety. Others may be seeking (or be currently working in) careers that are not directly devoted to these areas. This book is aimed at both types of students. The text and the case studies will demonstrate that roles across domain areas involve decision-making under conditions of uncertainty and risk. For example, healthcare professionals must decide on the most appropriate treatment for a patient, recognizing uncertainties related to the effectiveness of the treatment, the accuracy of the diagnosis and variability that exists within biological or human mechanisms dictating the patient's response. As another example, supply chain professionals must decide on the most appropriate response to natural disasters, recognizing uncertainties existing in the timing, geographic location and severity of the disaster itself.

This book is written at an introductory level with a focus on fundamental concepts and application of those concepts without compromising stringency and precision. While this book is designed to be accessible to students with minimal background in mathematics, students will benefit from having some basic understanding of probability theory and statistics. In general, the book is, however, very much self contained. The key terms and concepts will be carefully introduced and explained within the text. To make the book accessible and practical for students across domains, there is particular emphasis on using an interactive and engaging format for comprehension activities. These activities include written exercises and presentations. A number of exercises, with answers, are included.

The main target group for the book is students, but the book should also be of interest for many others, including managers, professionals and scientists working with risk issues related to, for example, engineering, business and public health, who would like to get an overview of the core subjects of risk science.

As risk science covers a wide variety of topics and issues, it is impossible to cover them all. There will always be a discussion about what in fact the state-of-the-art concepts, principles, approaches, methods and models of a science are. The present book builds on recent work conducted by the Society for Risk Analysis (SRA 2015, 2017a, 2017b) and related research, which point to key risk science terminology, core subjects and fundamental principles. These documents have been developed by a group of senior risk scientists with different backgrounds and competences. The subjectivity in topics, issues and examples highlighted in this book is, however, acknowledged. For further studies of the scientific basis of the present book, see Aven (2018a, 2020a) and Aven and Thekdi (2020).

> *Risk science is the most updated and justified knowledge on risk fundamentals (concepts), risk assessment, risk perception and communication and risk management (including risk governance and policies on risk). This book aims at presenting the core of this knowledge. Risk science is also about the process – the practice – that gives us this knowledge. The present book also discusses to some extent this aspect of risk science, but its main scope and focus are on presenting the current knowledge on risk fundamentals (concepts), risk assessment, risk perception and communication and risk management.*

CONTENTS

Chapter 1 introduces three main cases which will be used throughout the book to illustrate the concepts, principles, approaches, methods and models presented. These cases relate to the personal, organizational and societal level, respectively. Case 1 discusses risk issues for a student in an academically rigorous environment; Case 2 looks into a company considering supply chain investments in response to major potential changes in business and technology; and Case 3 addresses a public health crisis, a global pandemic.

The following chapters of the book are organized into five parts. Part I covers basic concepts, Part II risk assessment, Part III risk perception and communication, Part IV risk management and Part V tackling practical risk problems and issues.

The aim of Part I is to motivate and define the basic concepts of risk science. The main focus is on the risk term, but we also discuss related terms, including probability, uncertainty, knowledge, vulnerability, resilience, safety and security. A distinction is made between the concept of risk (Chapter 2) and how risk is measured, described or characterized (Chapter 3). We face risk when we drive a car without consciously measuring or describing the level of the risk. We can also make a similar dichotomy for other terms, like vulnerability and resilience.

Part II covers risk assessment. Chapter 4 reviews basic theory, including scope and aims of risk assessment and stages and processes in a risk assessment: planning, hazard (threat, opportunity) identification, cause analysis and consequence analysis. It also discusses the quality of risk assessments and challenges in relation to their use. Examples of risk assessment methods are outlined and followed up in Chapter 5, which reviews basic methods such as sensitivity and importance analysis, Bayesian networks, Monte Carlo simulation, probabilistic risk assessment (quantitative risk assessment) – PRA/QRA – and specific methods for analyzing vulnerability and resilience. There exist a number of risk assessment methods, some general and relevant for all types of situations, others more tailored to meet specific application needs. We highlight the former category but also present some specific methods and models, in particular related to safety applications.

The aim of Part III is to review basic knowledge about risk perception (Chapter 6) and risk communication (Chapter 7). Topics of Chapter 6 include risk perception and emotions/affects; the meaning of the two modes of thought, System 1 and System 2, and their link to risk perception and risk assessments; differences between expert and non-expert judgments of risk; factors shaping risk perception; social amplification of risk; and the relationship between risk perception and trust.

Risk communication is about how exchange or sharing of risk-related data, information and knowledge between and among different parties (such as regulators, experts, consumers, media, general public) can be provided. Chapter 7 discusses various topics related to risk communication, including the different functions and goals of risk communication, the importance of building risk communication on risk science, key ideas and features of basic models and theories of risk communication, what characterizes good risk communication messages and common methods for developing and testing risk communication messages.

Part IV covers fundamentals of risk management, including risk governance issues and policies on risk. Chapter 8 addresses topics such as the basic strategies and policies for handling risk, the main elements of the risk management process, the use of risk acceptance and tolerability criteria in risk handling, the meaning and use of the cautionary and precautionary principles and the importance of vulnerability and resilience.

Chapter 9 reviews related tools, including utility theory, cost-benefit analyses and multi-attribute analyses. In particular, the chapter discusses the concepts of value of a statistical life (VSL) and implied cost of averting a fatality (ICAF).

Part V discusses the tackling of practical problems and issues by integrating theories and methods from the other categories of topics (parts). Chapter 10 covers general insights and experiences. Three topics are highlighted in this chapter: 1) competences: to understand risk and solve risk problems in practice, broad risk analysis and risk science competences are required. It is not enough to rely on knowledge gained through standards and applied certification arrangements. 2) Pre-analysis and framing issues, which relate to, inter alia, the objectives of the analysis and use of standards. 3) Treatment and communication of uncertainties, which is a main challenge in using risk analysis in practice. Chapter 11 returns to the three cases of Chapter 1. In the student example, we focus on the planning, execution and use of a risk assessment. In the enterprise example, we look more closely into the process of integrating performance, risk and vulnerability (resilience) analysis and management. In the social case, we highlight issues linked to science, policies and politics.

In addition, the book includes three appendices, on terminology (Appendix A), basic risk science (Appendix B) and a summary of answers to problems and exercises throughout the book. Appendix B reviews basic knowledge about risk science; how it is defined;

and what the difference is between generic, fundamental risk science and applied risk science. The former category covers scientific knowledge on generic issues like 'What is risk?' and 'How should risk best be described?', whereas the latter category covers scientific risk knowledge related to specific applications, for example, on safety and security issues for a concrete activity. Appendix B also discusses educational programs in the area.

The title of this book is *Risk Science: An Introduction*. It reflects that the aim is to present an overview of the most updated and justified knowledge of this science on risk fundamentals (concepts), risk assessment, risk perception and communication and risk management and governance (Chapters 1–11). This knowledge can be used to understand, assess, communicate and handle risk in practical situations, such as for climate change risks, business risks and health risks. Risk science is also about the process – the practice – that gives us this knowledge. This book also discusses to some extent this aspect of risk science, but its main scope and focus are on presenting current knowledge on risk fundamentals (concepts), risk assessment, risk perception and communication and risk management. For the risk science practice aspect, several examples are presented. Appendix B provides a general background text for clarifying the meaning of these risk science concepts. However, the following chapters can be read without a deeper understanding of these concepts, keeping in mind that the chapters simply provide state-of-the-art knowledge on risk fundamentals (concepts), risk assessment, risk perception and communication and risk management – and they include some illustrations of the risk science knowledge generation process (practice).

The book covers several features that facilitate the learning and application of risk science. The features include 1) a summary at the start of each chapter explaining what the coming chapter is about, which includes a list of learning objectives; 2) the highlighting of new terms and definitions introduced, as well as key points made; 3) a summary of key symbols; 4) many examples and figures throughout the book to illustrate the theories studied; 5) a number of questions (reflection boxes) throughout the text to stimulate the reader to reflect on the concepts, principles, approaches, methods and models discussed; 6) a review at the end of each chapter highlighting the most important ideas and features of the chapter (key takeaways); and 7) problems at the end of each chapter, with answers (Appendix C). For some of the problems, the answers extend beyond the material presented in previous sections. In this way, additional knowledge is added to the chapter.

The learning objectives highlight the knowledge of central concepts, principles, approaches, methods and models of risk analysis and risk science (covering risk assessment, risk characterization, risk perception, risk communication and risk management). In addition, we can formulate some objectives related to 'know-how' (skills), including

- be able to identify state-of-the-art concepts, principles, approaches, methods and models of risk analysis and risk science
- be able to use basic concepts, principles, approaches, methods and models in risk analysis and risk science settings
- be able to support the tackling of risk problems using contemporary risk analysis and science

As a concrete example, think about the skill of being able to conduct a risk assessment. The book provides basic knowledge ('know-that') for obtaining this skill using theory and illustrative examples, but considerable training beyond studying this book is needed to be a competent risk assessor conducting professional risk assessments in industry, business and so on.

Acknowledgments

We are grateful to Enrico Zio, Seth Guikema, Ortwin Renn, Roger Flage, Eirik B. Abrahamsen, Tom Logan and Lisbet Fjæran for their important input to this book through their collaboration on a number of papers and projects over the last 15 years. We would particularly like to acknowledge Eirik B. Abrahamsen and Roger Flage for the time and effort they spent on commenting on the book manuscript. Many other risk researchers have also contributed to the book by participating in foundational work on risk science, including Sven-Ove Hansson, Katherine McComas, Tony Cox, Michael Greenberg and Wolfgang Kröger. Thanks. We are also grateful to Michael Siegrist for his useful comments on Chapter 6. None of you bear any responsibility for the content of this book with its perspectives, views, limitations and possible shortcomings.

Terje Aven and Shital Thekdi
January 1, 2021

Key symbols

A	event (hazard, threat, opportunity) occurring as a result of an activity
C	consequences of an activity
(A,C)	event and consequences given this event
U	uncertainty (about what will be the consequences of the activity)
(C,U)	risk related to an activity
(A,C,U)	risk related to an activity
(C,U|A)	vulnerability given event A
A'	specified events in a risk assessment
C'	specified consequences in a risk assessment
Q	measure or description of the uncertainties U
P	probability (subjective, knowledge-based)
P_f	frequentist probability
K	knowledge
SoK	strength of knowledge
(C',Q,K)	risk description
(A',C',Q,K)	risk description
(C',Q|A)	vulnerability description given event A
E	expectation
PLL	potential loss of lives
FAR	fatal accident rate
VaR	value at risk
IR	Individual risk
ALARP	as low as reasonably practicable
VSL	value of a statistical life
ICAF	implied cost of averting a fatality

CHAPTER 1

Illustrative examples

CHAPTER SUMMARY

This chapter presents three cases which will be used to illustrate the coming discussions about risk science concepts, principles, approaches, methods and models. These cases relate to the individual, organizational and societal level, respectively. The first concerns a student's personal and academic lifestyle, the second a company considering some investments and the third a pandemic. The cases will be followed up in detail in Chapter 11.

1.1 PERSONAL CASE: MANAGEMENT OF TIME IN AN ACADEMICALLY RIGOROUS ENVIRONMENT

A student in a highly competitive academic program is facing a common issue encountered by almost all students: the 24-hour day. This student has many critical tasks to perform, along with some elective activities that are of value. However, the student must first consider several high-level objectives for college life, including:

- Academic performance: The student must earn high grades in order to qualify for prestigious jobs and graduate programs
- Financial security: The student could elect to pursue high-interest loans or choose to work a part-time job to help offset college expenses
- Physical wellness: The student values time spent exercising and recognizes that physical activity is linked to better academic performance and mental wellness
- Nutrition: The student recognizes the importance of consuming healthy meals, which involve cost and/or time to prepare
- Mental and emotional wellness: The student values activities, such as time spent outdoors or interacting with peers, as a source of wellness
- Social growth: The student values extracurricular activities, such as social clubs, academic fraternities and academic student organizations on campus
- Sleep: Any student or person requires sufficient rest in order to perform all other activities

DOI: 10.4324/9781003156864-1

While balancing these objectives is a challenge for any person, the risk management implications are unique for a student in an academically rigorous environment. Insufficient investment in any of the objectives can be disastrous for students, such as by causing degradation of health and emotional issues. Less severe impacts include poor job prospects and financial losses. There are also opportunity costs related to not taking advantage of growth that generally happens in college environments, such as interactions in student organizations and other social activities.

You as an expert of risk science are tasked with mentoring this college student. You are tasked with exploring the following questions:

- What risks does the student face?
- How should the risks be measured and described for this student?
- What are the relevant issues related to risk perception and communication when consulting with this student?
- What are the relevant risk management issues? What risk strategies and policies are relevant for the student?

In the upcoming chapters, you will explore how to address the issues described previously.

1.2 ORGANIZATIONAL CASE: TOTAL BUSINESS MANAGEMENT INC.: SUPPLY CHAIN INVESTMENTS IN RESPONSE TO MAJOR POTENTIAL CHANGES IN BUSINESS AND TECHNOLOGY

Consider a fictional retailer that sells goods and service through brick-and-mortar stores and also has a rapidly growing online presence. In recent years, the business has changed considerably in both operations and customer habits. Customer traffic at brick-and-mortar locations has decreased considerably, while revenues from within those locations have also decreased. However, online sales have grown, with online sales expected to encompass 90% of revenues within the next five years. This retailer is competing against many global retailers who guarantee fast product delivery, often delivering purchases within a few hours or a few days. As a result, customer expectations are high: They require a quality product, with no shipping damage, that is delivered quickly.

In the past, this retailer has relied on package delivery and supply chain management vendors to handle the shipping. This is often a sophisticated process, involving several methods of warehousing and delivery. For large or bulky items that are considered low volume, Total Business Management Inc. (TBM) typically contracts with manufacturers to ship products directly from the manufacturer's warehousing facility. While TBM maintains a list of preferred shipping vendors, it does not contract shipping rates, nor does it require any particular shipping timeframe or vendor. The second method involves maintaining inventory of high-volume items at several strategically placed warehouses. Over the last five years, TBM has found traditional shipping vendors to be costly, slow and non-transparent in their practices for both moving product between warehouses and also in delivering products from the warehouse to the customer.

In response to the challenges in working with shipping vendors, TBM considers building up its own supply chain capabilities. It considers investing in several features: 1) dedicated shipping channels with TBM's privately owned infrastructure to move product among its regional warehouses, 2) some dedicated shipping channels to deliver products directly to customers located near regional warehouses and 3) capability to outsource delivery to customers by using delivery workers who accept delivery jobs on a case-by-case basis and perform the task using their own vehicle.

The technology landscape has changed considerably in recent years. Recent issues facing TBM include:

- Gig economy: TBM can coordinate with contractors to deliver products easily, without any fixed cost or agreement
- Self-driving vehicles: There is a global technological shift toward autonomous vehicles. The technology is being developed but is not yet fully validated, with several safety features currently being tested. There are also major legislative hurdles, including issues with safety and influences from lobbying groups
- Other evolving technologies: There is rapid development of other related technologies, such as drones
- Sustainability initiatives: There is increased emphasis on sustainability initiatives, calling for improved fuel efficiency, lower miles traveled by the TBM fleet, sustainable packaging material policies, consolidation of items in deliveries and others
- Electronic materials: There is increased popularity in selling product through electronic mediums, such as e-books, digital music, digital video and so on to avoid packaging and shipping costs
- Updatable technologies: There is some interest in developing electronic devices that are designed to be updated using software upgrades, thereby extending the life of the product and delaying obsolescence
- Subscriptions: There is an opportunity to increase sales using subscription methods in an effort to make sales and deliveries more predicable
- Rising cost of packaging for delivery: There is a need to address how to avoid breakage of product during shipping while also acknowledging packaging costs
- Fluctuating shopping habits: Recent economic, health-related and other factors have caused demand patterns to exhibit major shifts over short periods of time

TBM also recognizes that the issues listed here are not comprehensive. There may be many issues that can be surprises in the future and other issues that may be pressing but not considered a high priority by the executive team. The executive team has contracted with your organization to study the following questions:

- What are the relevant risks for TBM?
- How should the risks be measured and characterized?

- What are the relevant risk management issues? What risk strategies and policies are relevant for TBM?

The following chapters will address the issues described previously.

1.3 PUBLIC HEALTH CRISIS: ADDRESSING A GLOBAL PANDEMIC

Suppose you are a risk advisor to the leadership for a nation. In recent weeks, a previously unknown virus has begun spreading globally, with potential to spread to your nation. Because this virus is new and unstudied, there are several questions that have remained unanswered about this virus, including:

- What are the symptoms?
- How can the disease caused by the virus be diagnosed?
- What are the origins (e.g., what is the natural source? is it manmade? is it an act of biological warfare?, etc.)
- How does it spread (e.g., through aerosols, surfaces, etc.)
- How long does it 'live' on surfaces?
- What methods and medications are appropriate for treating the disease caused by the virus?
- Is a vaccine possible? If possible, is the vaccine effective and safe for all populations?
- Would populations be willing to receive a vaccine, if available?
- Is immunity possible (e.g., are certain populations naturally immune? Are recovered patients immune for some time period, etc.)
- What is the effectiveness of personal protective equipment (PPE) in managing the spread of the virus?
- What are the implications for other unforeseen contagions, as climate change and other factors may increase the likelihood of future pandemics?

You foresee several challenges related to addressing this virus. Due to movement of individuals across nations, the impact is not localized. This is a drastically different condition versus your past experiences dealing with localized disasters, such as hurricanes or earthquakes. As a result of this potential global situation, resources would be needed quickly in all areas of the globe. Also, because the virus impacts workforces through direct illness of employees, quarantining of employees if the virus is present and government-issued quarantine orders for regions, manufacturing of resources needed to address this potential pandemic may become scarce. For example, manufacturing of vehicles may be temporarily halted due to social distancing requirements, but those facilities could potentially be reconfigured to produce PPE.

At a socio-economic level, there are several challenges to consider. For example, there is concern over availability of food, healthcare and care for various populations with special needs, such as children and the elderly. There are some industries in particular that can potentially be impacted by declining economic conditions and quarantine policies, such as retail and restaurant industries.

At a healthcare level, there is concern over the availability of healthcare for patients who are not being treated for this particular virus. For example, parents may choose to not expose their children to healthcare facilities out of fear of contracting the virus. As a secondary impact, those children may not receive adequate medical care and vaccinations, possibly increasing likelihood of a secondary epidemic related to illnesses associated with missed vaccinations (e.g., measles, rubella, hepatitis A/B, etc.). Similarly, patients may elect to not seek medical attention for emergencies out of avoidance of healthcare facilities, possibly electing to not treat serious conditions such as heart attacks and strokes. Some regions may also ban the practice of elective health procedures. Finally, there is concern over adequate treatment of the virus itself. For example, it is unclear whether certain drugs or devices, such as ventilators, are effective and/or improve health outcomes for patients.

With these concerns and more, you are tasked with understanding and providing guidance on how risk science, with its concepts, principles, approaches, methods and models, can support the actual assessments, communication and handling of the risks related to the virus. This includes responding to challenges related to, for example:

- How should the risks be measured and characterized?
- What are the relevant issues related to risk perception and communication when consulting with political leaders and the public?
- How should the risks be communicated?
- What are the relevant risk management and government issues?
- How should the risks be handled? What risk strategies and policies should be adopted?

The virus triggers many issues concerning risk and risk analysis, as indicated. For example, consider the COVID-19 pandemic of early March 2020. Is the risk really worth worrying about, and should one, for example, cancel all plane trips because of the risk? Refer to the discussion in the preface. In this example, some experts stated that the risk for the average healthy person was small, and as long as trips to and from some specific areas were avoided, life should proceed as normal. However, the circumstances resulted in an overwhelming reduction in air travel. Risk psychologists explain why: "The coronavirus 'hits all the hot buttons' for how we misjudge risk". These buttons include lack of control, catastrophic potential, delays in effects and being new and unknown, often summarized by the two dimensions dread and newness. Because of such factors, people tend to raise their risk estimates – how likely it is to happen to them. The fact that we are repeatedly hearing about cases and deaths has the same effect. At the same time, it is acknowledged that assessing the risk posed by the virus is very difficult. How can we then conclude that we 'misjudge the risk'? What is the reference for comparison when faced with situations like this characterized by large uncertainties? To what extent can we trust the experts' judgments?

In coming chapters, we will discuss these issues. We will explain the difference between lay people's risk perceptions and professional judgments about risk, and we will provide guidance on how we should best characterize, communicate and handle the virus risks. In this case, we are faced with the potential for serious consequences, and there are scientific uncertainties about these consequences. Models providing accurate predictions of the implications of the virus and possible governmental interventions are not available. How should we then best meet this challenge?

The answer is a combination of precautionary measures and scientific reasoning, to be explained and justified in the coming chapters of the book.

PART I

Basic concepts

CHAPTER 2

What is risk?

CHAPTER SUMMARY

Risk is related to an activity, for example, an investment, the operation of a technical system or life on earth or in a specific country. Risk reflects the potential for undesirable consequences of the activity. When we drive a car, we face risk – an accident could occur, leading to injuries and/or deaths. However, the outcome could also be positive – a successful trip. Thus, risk is about both undesirable and desirable consequences of the activity and uncertainties about what these consequences will be. The consequences are with respect to some values (e.g., human lives, health, the environment, and monetary values). Vulnerability is basically risk given the occurrence of an event. For example, we can speak about the vulnerability of persons to various diseases. The chapter also discusses the meaning of related concepts such as resilience, safety, safe, security and secure.

LEARNING OBJECTIVES

After studying this chapter, you will be able to explain

- the basic ideas of the risk concept
- the meaning of the statement: *uncertainty is a main component of risk*
- the difference between risk and hazards/threats/opportunities

DOI: 10.4324/9781003156864-2

- that risk is about both negative (undesirable) and positive (desirable) outcomes
- what risk means in practical situations
- the relationship with related concepts such as vulnerability, resilience, reliability, safety, security and sustainability
- how risk relates to time

This chapter looks into the risk concept; the coming Chapter 3 will study how to describe the magnitude of the risk. A distinction is made between the risk concept on the one hand and measures and descriptions of the risk on the other. Section 2.1 discusses the definition of risk, Section 2.2 gives some illustrations of the definition and Section 2.3 looks into related concepts such as vulnerability, resilience, safety and reliability. The final Section 2.4 presents some problems. The chapter is partly based on Aven (2020a) and Logan et al. (2021).

2.1 DEFINITIONS OF THE RISK CONCEPT

We will present several definitions of risk, all based on the same ideas and illustrated in Figure 2.1. The setting is this: We consider a future activity, for example, driving a car from one place to another, the operation of a nuclear plant, an investment or the life on earth or in a specific country. This activity will lead to some consequences seen in relation to some values, such as human lives and health, the environment and monetary values. The consequences could, for example, be some injuries and loss of lives, the current state (no events occur), deviations from a planned profit level or failure to meet a defined goal. There is at least one consequence or outcome that is considered negative or undesirable. Looking forward in time, there are uncertainties what the consequences will be.

Think about the example of driving the car. The driver and the passengers face risk, as there is a potential for an accident to occur, which could lead to injuries or fatalities. The accident, injuries and fatalities are undesirable consequences of the activity. This leads us to the first intuitive definition of risk: *the potential for undesirable consequences* of the activity. The term 'potential' relates to the consequences but points also to the uncertainties – an accident with some injuries/fatalities may be the result,

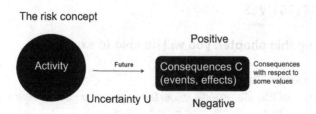

FIGURE 2.1 The basic features of the risk concept (based on Aven and Thekdi 2020)

but we do not know before the activity is realized. We are led to a second definition of risk: *the consequences of the activity and associated uncertainties*. This definition is appealing, as it explicitly incorporates both consequences and uncertainties, which can be seen as the two key components of the risk concept. Returning to the car example, the driver and the passengers face risk, as there are consequences of the activity (of which some are undesirable) and uncertainties about those consequences. Using this second definition of risk, we are encouraged to consider both undesirable and desirable consequences, which is attractive, as risk is also something positive – refer to discussion in the preface – and addressing risk, we need to take into account both positive and negative effects of the activity. What is desirable or not may vary between different stakeholders and be dependent on time. Consider as an example risk related to political tensions between two countries. The tensions can lead to confrontation between the countries and even a war. The immediate consequences related to this activity will typically be negative, but the activity could also lead to positivity in a longer-term horizon for some stakeholders, depending on the outcome of the conflict. Applying the second definition of risk, all types of relevant consequences are considered.

We denote by C the actual consequences of the activity and U the associated uncertainties. Simplified, we write risk as (C,U). For definition 1, where risk is understood as the potential for undesirable consequences, C is restricted to 'undesirable consequences'. Definition 2 extends C also to cover other types of consequences (i.e., non-negative consequences). The consequences C could relate to all types of values, but often restrictions are made, for example, by only considering loss of lives or monetary values.

To ease the understanding of this general definition of risk, it is common to relate the consequences to a reference value (r), for example, the current state, a planned level, or meeting an objective. This leads to a third definition of risk: *the deviation D from a 'reference value' r, and associated uncertainties U*. We write risk = (D,U). If C denotes, for example, the number of fatalities of an activity, r is typically zero and D = C. If r relates to a planned production level, D = C − r, where C is the actual production level. Risk captures the potential for a deviation from the planned level, often with a focus on values below this level.

Instead of writing risk as (C,U), we often use the notation (A,C,U), where A refers to an event (or a set of events) and C then refers to the consequences or effects given that A has occurred. Think again about the car example. Here A could be the occurrence of an accident and C the consequences given the accident. The event A could also represent 'no accident'.

Definition 1:
Risk: The potential for undesirable consequences

Definition 2
Risk: The consequences C of the activity and associated uncertainties U
Risk: (C,U)

Risk: (A,C,U), where A is an event (or a set of events) and C the consequences given the occurrence of A

Definition 3
Risk: The deviation D from a 'reference value' r, and associated uncertainties U
Risk: (D,U)

The event A is referred to as a *hazard, threat* or *opportunity*, as well as a *risk source*, which are terms we will come back to in Section 2.3 and Chapter 3.

When discussing risk, the activity is to varying degrees explicitly defined. For some cases, the activity is tacitly understood, but in others, it is essential to clarify in detail what the activity encompasses, for example, with respect to time. For instance, when considering the risks related to the operation of a nuclear power plant, we need to specify for what time interval, for example, the next ten years. In addition, we need to specify for how long of a time we would consider the consequences of events occurring in this interval. Say that an accident occurs after two years of operation. The effects of this accident could, however, last many years, far beyond the time interval for which the activity is considered. Should we include also these effects when considering risk in the operation interval of ten years?

When considering an activity for a specific period of time, we refer to T as the length of this interval. The maximum time from the occurrence of an event for which the consequences are considered is denoted τ. In the previous example, T is equal to 10 years, and if we define $\tau = 100$, we consider effects of nuclear accidents for a period of a maximum of 100 years if they should occur in the time interval [0,10].

REFLECTION

The previous discussion referred to several definitions of risk. Would it not be better to introduce and use one definition only?

All the definitions capture the same basic idea, but there could be different ways of expressing this idea depending on the situation studied. If you are to refer to one scientific definition of risk only, it is (C,U). If you are to express risk in a more commonly used way, you could say that risk is the potential for undesirable consequences.

REFLECTION

It is common to refer to risk as probability times consequences (expected loss) and as the combination of consequences (scenarios/events with effects) and probabilities. Are these definitions wrong or inadequate?

These probability-based expressions are ways of measuring or describing risk, not risk as such. It is theoretically and practically sound to separate the concept and its measurement (description). Risk as a concept exists without introducing probabilities. We face risk when driving a car – no probabilities are needed to make such a conclusion. How we measure or describe risk will depend on the situation considered. As will be thoroughly discussed in Chapter 3, the probability-based metrics have strong limitations in capturing all relevant aspects of risk. The distinction between the concept of risk and its measurement (description) will stimulate a discussion about to what degree the measurements and descriptions adequately reflect risk in various situations, which can lead to improvements in risk characterizations.

2.2 ILLUSTRATING EXAMPLES

The following examples illustrate the ideas of the risk concept with its main components, A, C, D and U. We will discuss the magnitude of the risks in Chapter 3.

2.2.1 A gamble

John offers you a game, only one game. He has two dice, which you are to throw. If both dice show 6, you have to pay him $36,000; for all other outcomes, you will receive $3,000 from John.

   If you play this game, you face the risk of losing $36,000, but you can also win $3,000. There is a potential for negative consequences: you lose this amount of money. The activity considered is the gamble, and the consequences C is the outcome of the activity, which is either a win of $3,000 or a loss of $36,000. Before throwing the dice, the outcome is unknown; therefore, C is subject to uncertainty U. Hence you face risk.

2.2.2 Isabella's exam

Isabella is to take an exam. For her, any grade worse than *B* is a failure. There is a potential for an undesirable outcome, although she is well prepared. Here the consequences C is the grade of the exam, or simplified, meeting the goal of grade *B* or not. Before the exam, this grade is unknown, subject to uncertainty U. Isabella faces risk in relation to the activity, taking this exam. The grade *B* can be seen as a reference value r. The term D then expresses the deviation from grade *B*.

   It is also possible to extend the consequences C beyond this particular exam. Failure of this exam may result in failure to study a specific program at the university, which requires a grade of *B* or better. When considering the consequences C of this exam, we need to clarify the values that we are relating the consequences. We may choose to focus on the grades themselves or the access to the desired study program. In the latter case, the exam represents a potential for a failure of access to this study program. The failure to get a grade *B* or better at the exam can be viewed as an A event in the (A,C,U) representation of risk. Then C captures the consequences in relation to the study program, where the main focus is on failure to get access to the university program. The uncertainties relate to the occurrence of A and the consequences given that event A has occurred. There could be uncertainties about getting access to this program both in case Isabella gets the desired exam result or she does not.

2.2.3 The consequences of epidemics caused by viruses

Let us go back to summer 2019. We consider the risk related to virus disease outbreaks in the coming five years. Let A denote the occurrence of a virus disease outbreak in that period and C the related consequences. These consequences relate to human health, but also economic, social and security types of impacts, for a short- and long-term horizon. Looking into the future, the occurrences of such outbreaks are not known, nor what outbreaks that occur, what type of viruses and when. And, given an outbreak, the spread and effects are unknown, subject to uncertainties. The consequences will depend on efficiencies of the mitigation measures taken. In this example, T = 5 years, but the consequences are considered for a much longer period, although not yet specified.

2.2.4 Dose-response

We consider the health of a population exposed to some toxin. The people are exposed for a time interval of length T days. We consider the consequences of the exposure for a period of 10 years, that is, $\tau = 10$ years. The event A is defined by the exposure to this toxin of duration T days. The consequences C is defined as the percentage of the population that develops lung cancer within the period of τ years since their exposure. Today we do not know if they will develop lung cancer within the specified time intervals or not – which explains the U of the risk concept (A,C,U).

2.2.5 A community threatened by hazards

We consider a community over a period of one year; thus T = 1 year. The community is potentially impacted by a hazard A (for example, a hurricane or flooding). The consequences considered relate to the residents within the city limits being able to have access to food, up to the time that the conditions are basically as pre-disruption. Hence τ is of an unknown length. The uncertainty U is explained by the fact that we today do not know if a hazard will strike the community within the year, nor do we know what the consequences will be. The community faces a risk, as there is a potential for a hazard to occur with this type of negative effects.

2.2.6 Nuclear waste

We consider the risk related to events occurring which could disrupt a nuclear waste repository, which its turn could lead to releases, severe health issues and environmental damage. Examples of such events include natural hazards (e.g., earthquakes) and security-type incidents (for example, a terrorist attack). The time period considered is $T = \tau = 10{,}000$ years. Two levels of consequences C are considered: disruption of the nuclear waste repository and impacts of potential releases in case of such a disruption. Today we do not know if a disruption will occur within the specified time interval and what the effects will be – which explains the U of the risk concept (A,C,U).

2.2.7 Climate change

What are the effects of climate change, for a short- and long-term horizon? On your business? Your life? Your children and grandchildren? Our country and society? And the earth? When defining risk is relation to climate change, the answers to these questions form the consequences C. Today we do not know what these consequences will be. We face uncertainties and hence risk.

Here a reference level r is often defined by the global temperature at 'pre-industrial' time. A main goal of the 2015 Paris Agreement on climate change is to ensure that the increases in global temperature are less than 2°C above 'pre-industrial' levels. Hence, we can interpret C and D as the global temperature rise with reference to the 'pre-industrial' level. The climate change risk is the size of C (D) with associated uncertainty. The consequences can be related to physical quantities, such as the temperature rise, but also other dimensions, such as societal damage and loss, societal justice, metrics of human well-being and economic indices. Using the (A,C) scheme, it may be more natural to relate the global temperature rise to the event A, but, as will be discussed in Section 2.3, these are relative concepts, and the choices depend on the purpose of the conceptualization and analysis.

Risk can also be associated with specific goals, for example, a goal of becoming a low-emission society. The consequences may then be related to deviations in relation to this objective or deviations and surprises in relation to plans and knowledge we today have about this transition, for example, as defined by the Paris Agreement. The associated risk is often referred to as transition risk.

These are references specified at the authority or government level. For a company, other references will be more relevant, such as expected or assumed profitability. The company will be concerned about whether and how climate change and the transition to a low-emission society will affect profitability. For example, will the authorities introduce climate measures that will have serious consequences for the company's future income? The company faces climate-related risks. It is also common to talk about transition risk in this sense.

REFLECTION

The consequences C relate to the values of interest and could be multidimensional, covering aspects associated with, for example, human health, environment and material assets. Do you think it is a good idea to transform all of these aspects into one dimension to ease the considerations of risk?

There exist several approaches for making such transformations, but as we will see, they are not straightforward to implement. There are technical analysis problems, and there are fundamental problems related to losing information. We will discuss this topic in more detail in Chapter 9.

2.3 RISK AND RELATED CONCEPTS

In this section, we will look into concepts related to risk, starting with vulnerability. Then we discuss the terms resilience, reliability, safety and security.

2.3.1 Vulnerability

The vulnerability concept is closely linked to risk; see Figure 2.2. Starting from the risk representation (A,C,U), we can split risk into two main contributions:

Risk = 'Event risk' (A,U) & Vulnerability (C,U|A)

Let us return to the car example discussed in Section 2.1. Here the event A refers to an accident, and the consequences C relate to the health impacts for the driver and passengers from the accident. The actual event A occurring could also be 'no accident', reflected by writing (A,U), which states that we do not know today before the activity is realized if the accident will occur or not. We refer to (A,U) as the 'event contribution' or 'event risk' – it expresses the risk of an accident to occur or, in more general terms, the risk of event A occurring. This risk is not the same as the probability of the event A occurring. Probability is a way of expressing the uncertainty U, but the existence of the risk concept is not dependent on the uncertainty being quantified. We will explain and discuss this in more detail in Chapter 3.

The accident is an example of a *risk source*, which we define as *an element (action, sub-activity, component, system, event) which alone or in combination with other elements has the potential to give rise to some undesirable consequences.* For the car case, examples of risk sources in addition to the accident are: a driver distracted by a mobile device, health emergencies for the driver and/or passengers, a high stress level

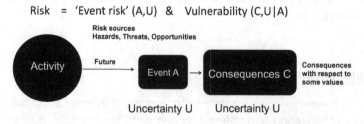

FIGURE 2.2 The basic features of the risk concept split into an 'event risk' contribution and vulnerability

of the driver, high speed, intense discussions among the passengers, active use of car-audio system, slippery roads, brake failure and worn tires. Typically, the risk sources are underlying factors of importance for the event A to occur, but formally we also refer to the event A as a risk source. What is identified as the event A in the (A,C,U) representation of risk is not always obvious. In the car example, we could alternatively have introduced A as 'brake failure', which is seen as an underlying risk source for the event accident. We will return to the process of identifying and selecting the A events in Chapters 3–5.

The terms *hazard, threat* and *opportunity* are examples of risk sources. A hazard is a risk source where the potential consequences relate to harm, that is, physical or psychological injury or damage. In the car example, all the previous risk factors can be referred to as hazards. The term *threat* is commonly used in relation to security applications, but it is also used more widely, for example, when referring to threat of an earthquake. When used in relation to a malicious attack, the term *threat* is understood as a stated or inferred intention to initiate an attack with the intention to inflict harm, fear, pain or misery. When talking about opportunities, there is an aspect of positivity associated with the risk source influencing the consequences. Think about a business context where an opportunity arises as a result of the government increasing the funds for industrial research and development. Or in the Isabella's exam example in Section 2.2.2, where an opportunity arises because of a misprint in the text leading to an extension of the exam by 1 extra hour.

The vulnerability concept (C,U|A) is interpreted as risk given the occurrence of the event A. The sign '|' is read as 'given'. Thus, *vulnerability represents the potential for undesirable consequences given an event* (risk source). The focused values determine what 'undesirable' means. In the car example, the vulnerability is related to what happens given the occurrence of the accident for the driver and the passengers. As for risk, we can extend this definition by specifically addressing the consequences and uncertainties, leading to the vulnerability representations (C,U|A) and (D,U|A).

Definition 1:
Vulnerability: The potential for undesirable consequences given an event (risk source)

Definition 2
Vulnerability: The consequences C of the activity and associated uncertainties U, given an event A (risk source)
Vulnerability: (C,U|A)

Definition 3
Vulnerability: The deviation D from a 'reference value' r, and associated uncertainties U given an event A (risk source)
Vulnerability: (D,U|A)

As mentioned in Section 2.1, A can be one event or a set of events. Which events A to consider is critical for the vulnerability judgments. A system may not be vulnerable to one type of event but very vulnerable to others. If we also allow for unknown types of events, we will clearly have problems in measuring the level of the vulnerability. Yet it is interesting to question if the system is also vulnerable to new types of events. We will return to this discussion in Chapter 3.

We use the term *vulnerability* in relation to the activity considered and the values addressed. Often vulnerability is discussed in relation to a system, like a person or a technical structure. A related activity is defined by the 'operation' of this system. We say, for example, that a system is vulnerable if a failure of one component leads to system failure. However, this leads us to a discussion of how to measure or describe the level and magnitude of the vulnerability, which is a topic of Chapter 3.

Returning to the examples in Section 2.2, risk can be divided into 'event risk' (A,U) and vulnerability (C,U|A). For the epidemic case of Section 2.2.3, a virus outbreak is the event A, and vulnerability is the potential for undesirable consequences given an outbreak. Clearly both components are essential for understanding risk and next handling the risk. Similar interpretations can be made for the other examples of Section 2.2.

2.3.2 Resilience

Consider the health condition of some population. We speak about this population as being resilient, meaning that if they are exposed to various risk sources, for example, some infections, they are able to quickly return to the normal state. Resilience is also commonly defined as the ability of the system to sustain or restore its basic functionality (alternatively and more general: achieving desirable functionality) following a risk source or an event (even unknown). Often resilience is concerned about recovery given a major event, for example, a community hit by a strong hurricane. We see that resilience provides key input to the vulnerability concept, as the resilience influences the degree to which something undesirable is to happen given the event or risk source.

Definition 1:
Resilience: The ability to quickly return to the normal state given an event (risk source)

Definition 2
Resilience: The ability of the system to sustain or restore its basic functionality following an event (risk source)

The resilient system

As for vulnerability, the event or risk source referred to could be a set of events or risk sources. It is tempting to say that lack of resilience is the same as vulnerability. However, as we use the terms here, there is a difference. Vulnerability is a broader concept. In the previous example, lack of resilience means that the people struggle with returning to a normal health state given the risk source. The vulnerability concept, on the other hand, highlights what the actual consequences could be of this lack of resilience. Referring to the representation (C,U|A), the vulnerability concept addresses all consequences defined by the values considered, for example, implications of deaths for other family members. This type of consequences would, however, not be relevant for the resilience concept. What matters here is the ability of the system (here these persons) to return to the pre-event state following the disruptive event (the infection). Resilience is thus an input to and an aspect of vulnerability. See Problem 2.11.

Vulnerability and resilience are not the same. Vulnerability, represented by (C,U|A), is a broader concept. Resilience – how fast a system recovers – is an aspect of the vulnerability concept.

To further illustrate the resilience concept, consider the following system model.

A system has four states, functioning as normal (3), intermediate state (2), intermediate state (1) and failure state (0). The state process is denoted X_t and is a function of time t, t = 0,1,2, ... Different stressors (events of type A) lead the system to jump to state 2, 1 or 0. We may, for example, let state 2 relate to a known type of stressor and state 1 to an unknown type. Or state 2 may relate to a stressor from which we know that the system will always recover quickly, whereas 1 is a state in which there are considerable uncertainties as to what will happen. Being in state 2 or 1, the system may recover and jump to state 3, but it could also jump directly to state 0. We may think of resilience as being related to the degree to which the system is able to return to state 3 if a stressor causes it to go to state 2 or 1. For example, if it is a fact that the system returns to state 3 in the next period of time if a stressor has made it jump to state 2 or 1, the system is indeed resilient. However, in practice, the system may not return to state 3, or it may return but take an extremely long time to do so. And it may return from state 2 but not state 1.

We can also think about a system which frequently jumps from state 3 to 0 (failures occur) but hardly visits states 1 or 2. The system could be resilient in these two intermediate states, but if the system seldom or never visits these states, it cannot be viewed as very resilient. Hence the system resilience is also linked to the risk of direct system failures (jumping from state 3 to state 0) relative to the risk associated with visits to the intermediate states (jumping from state 3 to state 2 or 1). We will discuss this further in Section 5.6.

As the system in Figure 2.3 is resilient, it quickly recovers when in states 1 and 2. Thus it cannot be vulnerable when in one of these states. If the system is not resilient for these states, the system can be considered vulnerable: it stays for a long time in states 1 or 2 and could jump to the worst state, 0. Hence the concepts in this example capture the same ideas. Consider then these two concepts for state zero. For the resilience of the system, we are focused on the time it takes to recover. However, for vulnerability, we may also include considerations of effects of being in state 0 relative to the values of interest, which is of importance for the risk and vulnerability concepts but not for the issue of recovery and resilience, repeating the arguments provided previously for why there is a difference between these two concepts.

2.3.3 Reliability

The reliability of a system, for example, a water pump or a power plant, is defined as the ability of the system to work as intended. The reliability – or rather the unreliability – concept can be viewed as a special case of risk by considering the activity

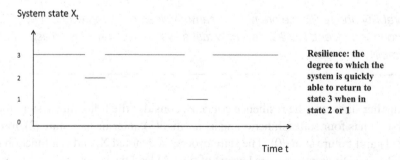

FIGURE 2.3 Illustration of a resilient system: the system quickly returns to the normal functioning state (3) when in one of the intermediate states (1 or 2) (based on Aven 2017a)

generated by the operation of the system and limiting the consequences C to failures or reduced performance relative to the system functions. Unreliability captures, then, the potential for undesirable consequences, namely system failure or reduced performance, and is represented by (C,U) with this understanding of C. The (C,U) definition of risk covers both negative and positive consequences; hence, it is a matter of preference whether we refer to risk in this setting as unreliability or reliability.

2.3.4 Safe and safety, secure and security

Safety is often referred to as an absence of accidents and losses, but such a definition is not very useful in practice, as it refers to an unknown event when looking into the future. However, adding the uncertainty dimension, we see that we are led to the risk concept (D,U), (C,U) and (A,C,U). The event 'absence of accidents and losses' is considered a reference value r, and risk is understood as deviation from this reference with associated uncertainties. Thus, *safety can be viewed as the antonym of risk* (as a result of accidents). Consider a situation in which you walk on a frozen lake with a thick layer of ice. The risk is low, and the safety level is high. We may conclude that walking on the lake is safe, but then we have made a judgment about the risk: we find it acceptable or tolerable. Thus, the term *safe refers to acceptable or tolerable risk*. We also often use the safety term in this way, for example, when saying that safety is achieved.

Analogously, we define secure as acceptable or tolerable risk when restricting the concept of risk to intentional malicious acts by intelligent actors. Security is understood in the same way as secure (for example, when saying that security is achieved) and as the antonym of risk when restricting the concept of risk to intentional malicious acts by intelligent actors.

> **REFLECTION**
>
> A risk manager of a company states that the company has secure deliveries of essential equipment for their production. How would you interpret this statement in relation to the previous analysis?
>
> *The risk manager is referencing the risk related to the deliveries as acceptable or tolerable. The reasons for adverse consequences could include intentional malicious acts, but other reasons could also be possible, including accidents. The term secure (and also security) is thus used in a broader way than defined previously.*

2.4 PROBLEMS

Problem 2.1

A person, Tom, walks underneath a boulder that may or may not dislodge from a ledge and hit him.

a   Define and explain the risk concept for this situation (activity, event A and consequences C).

b   Define and explain the vulnerability concept for this situation. Consider two different events A.

c   Use the example to illustrate both a safety and a security situation.

d   What do you call the boulder using risk terminology?

e   Which of the following risk statements do you consider consistent with the definition of risk provided in Section 2.1?

    i   There is a risk that the boulder will hit Tom.

    ii   Tom takes a risk by walking under the boulder.

    iii Tom is in a risky situation.

    iv   Tom risks getting hit by the boulder.

    v   Tom faces risk – the boulder may dislodge from the ledge and hit him.

    vi   The boulder represents a threat to Tom.

    vii The boulder represents a hazard.

Problem 2.2

A person, Sara, invests $1 million in a project. Define risk in relation to this investment. Clarify what type of time considerations that are needed.

Problem 2.3

Discuss the term *resilience* in relation to the human body and potential infections and trauma. Do you consider the human body resilient? Why/why not?

Problem 2.4

We all would like the food we eat to be safe. Based on the discussion in this chapter, what does it mean that food is safe?

Problem 2.5

In security contexts, when talking about risk, it is common to refer to the triplet: value, threat and vulnerability. Explain how this perspective is covered by the general framework presented in this chapter.

Problem 2.6

The concept of sustainability relates to meeting the needs of the present without compromising the ability of future generations to meet their needs. Define and explain related risks, in line with the ideas of this chapter (hint: relate sustainability to the definition of safety).

Problem 2.7

The *risk* term is defined in many different ways in the scientific and popular literature. Identify one or more of these definitions. Are they consistent with the definitions provided in this book?

Problem 2.8

Conceptualize risks for the example presented in Section 1.1.

Problem 2.9

Conceptualize risks for the example presented in Section 1.2.

Problem 2.10

Conceptualize risks for the example presented in Section 1.3.

Problem 2.11

Pam is a resilient athlete in the sense that she is rather quickly back following injuries. At the same time, she can be vulnerable. Explain.

Problem 2.12

Consider the occurrence of a major incident A in a community (for example, an earthquake, a hurricane or a flooding event). Resilience is about the recovery phase, the ability of the community to return to the normal state. Explain how risk can be defined in relation to this recovery.

KEY CHAPTER TAKEAWAYS

- Risk is the potential for undesirable consequences of the activity considered and can be formalized as (C,U), where C is the consequences of the activity considered and U the associated uncertainties. Some features of the consequences are undesirable. Alternatively, risk can be represented by (A,C,U), where A is an event and C the consequences given the occurrence of A. Two aspects of time apply, the time interval for which the activity is considered and the time for which the consequences are considered given the occurrence of an event.
- Vulnerability is essentially risk conditioned on the occurrence of an event A. Risk can be written Risk = (A,U) & (C,U|A), where the former term is the event risk and the latter is vulnerability.
- The terms *safe* and *secure* imply acceptable or tolerable risks, typically accident risk when talking about safe and intentional malicious acts in a security context.
- Resilience is the ability of a system to sustain or restore its basic functionality following a risk source (known or unknown). It is an aspect of vulnerability.

CHAPTER **3**

Measuring and describing risk

CHAPTER SUMMARY

Anne and Amy are university students who are considering a trip from Washington, DC, to San Francisco. Anne prefers to travel by plane, but Amy is hesitant because she views this as being risky. Anne argues that traveling by commercial airliners is very safe because there have been very few plane crashes over the recent years. Amy is not convinced. She calls for more information and a more careful explanation of what 'risky' means.

Anne and Amy are debating several issues. First, they are debating the likelihood of a plane crash. What is a probability, and how can it be used to evaluate this situation? Second, they are debating the concept of risk. What aspects are included in a characterization of risk? Third, they are debating the concept of safety. Both are making value judgments about how safe air travel is, using their own values of personal welfare and uncertainty (probability) judgments.

This chapter will discuss the issues described previously, including a thorough discussion of probability and risk. It is explained that probability is used to both represent variation and to express uncertainties. In the former case, we speak about frequentist probabilities and probability models, whereas in the latter case, we refer to subjective (knowledge-based, judgmental) probabilities, or just probabilities. It is shown that the probability concept is an essential tool for measuring and describing risk, in particular how large or small the risks are. The limitations of the probability concept for this purpose are also discussed. Using the (C,U) and (A,C,U) definitions of risk, we derive at general risk characterizations (C',Q,K) and (A',C',Q,K), where A' is a set of specified events, C' some specified consequences, Q a measurement or description of uncertainties and K the knowledge that Q and (A',C') are based on. The main aim of this chapter is to motivate and explain these risk characterizations. A number of examples will be presented to illustrate the characterizations and their use in practice. Common risk metrics, such as expected values and value-at-risk, are presented and discussed. A related vulnerability characterization will also be looked into: (C',Q,K|A'). A common approach is to use probability P as the uncertainty description Q, but it should always be supplemented with judgments of the strength of the knowledge (SoK) supporting the probabilities. The chapter also discusses the potential for surprises relative

DOI: 10.4324/9781003156864-3

to the available knowledge. The actual event A occurring may be overlooked by the specification of events in A'. The event A is what we refer to as a black swan, which is a popular metaphor in risk science. Its definition and use will be carefully discussed, as well as a related metaphor – the 'perfect storm'.

LEARNING OBJECTIVES

After studying this chapter you will be able to explain
- the basic differences between the risk concept and its measurement/ description
- the meaning of a frequentist probability
- the meaning of a probability model
- the meaning of a subjective (knowledge-based, judgmental) probability
- the meaning of the risk characterizations (C',Q,K) and (A',C',Q,K)
- the meaning of the vulnerability characterizations (C',Q,K|A')
- what type of risk metrics are commonly used, and what their strengths and weaknesses are
- why strength of knowledge judgments are needed to supplement probabilities
- why potential surprises and the unforeseen are important issues in relation to risk characterizations
- the meaning of the black swan metaphor
- the meaning of the perfect storm metaphor

First in this chapter, we discuss the probability concept, which gives a basis for introducing and discussing the risk characterizations (C',Q,K) and (A',C',Q,K). Then some illustrating examples are provided before we end the section with some reflections about potential surprises and black swans. Background literature for the chapter includes Lindley (2006), Aven and Reniers (2013), Aven and Renn (2009) and Aven (2012a, 2014, 2020a).

3.1 THE PROBABILITY CONCEPT

Leo plays a game, throwing a die. If the outcome is 1, he loses $36. If the outcome is 2, 3, 4, 5 or 6, he wins $6. What is the probability that he loses money, that the die shows 1?

   Leo thinks about this and answers quickly: it is 1/6, as the possible number of outcomes is six, and each outcome is equally probable. The sum over all outcomes is 1, and hence the sought probability must be 1/6. Leo has used the *classical interpretation of probability*. He has assumed that the die is fair, meaning that the probability is the same for all six potential outcomes. But how can we ensure that the die is fair? He can study the die's physical properties, its symmetries and weight, and from that conclude that each outcome should have the same probability of being the result when throwing the die.

   This study of the die's physical properties may instead reveal that the die is not symmetrical, is weighted or there is some other minor physical difference, but Leo does not believe that would change the probability much. He is thinking more deeply into what he is actually saying: "would not change the probability much". He understands that he is now interpreting the probability in a different way than according to the classical interpretation. The probability depends on the properties of the die and is referred to as a *frequentist probability*. He performs an experiment, throwing the die 60 times. The frequency of outcomes $1, 2, \ldots, 6$ is 7, 12, 8, 9, 14 and 10, respectively, resulting in observed relative frequencies 7/60, 12/60, 8/60, 9/60, 14/60 and 10/60. If the die were fair, as in the classical interpretation of probability, he would have expected a relative frequency of 10/60 for each outcome. Clearly these frequencies do not confirm that the probability of getting the outcome 1 is 10/60. However, when Leo throws the die many more times, he observes that the fraction of times the outcome is 1 becomes closer and closer to 1/6. We say that the observed frequency converges to 1/6. A similar process happens for the other outcomes. The die can be considered fair. The probability of outcome 1 is interpreted as the long-run limiting frequency of outcome 1 if we could perform an infinite number of similar throws of the die. In practice, we cannot perform an infinite number of throws; thus, the frequentist probability is a mind-constructed quantity. In this example, the frequentist probability is equal to the classical interpreted probability. However, in general, the frequentist probability is unknown. Let us think about a case where the die is not fair. What is then the frequentist probability of getting the outcome 1?

The answer is that we do not know. It needs to be estimated. Yet this underlying frequentist probability exists. We therefore need to be careful in writing and distinguishing between the frequentist probability and estimates of this probability. In the example previously, 7/60 is an estimate of the frequentist probability of getting outcome 1. We happen to know the true frequentist probability in this case (1/6), but in general that is not the case. In writing, we refer in general to a frequentist probability as P_f and its estimate as $(P_f)^*$.

Thus, we need to take into account that the estimates could deviate from the frequentist probabilities. There is estimation uncertainty. Textbooks in statistics analyze these uncertainties using measures such as variance (standard deviation) and confidence intervals; see Section 3.2.3.

The prominent mathematician Jacob Bernoulli (1654–1705) discussed such issues more than 300 years ago (Aven 2010). He tried to determine probabilities of different types of events, with an accuracy of 1/1000. He referred to the term 'moral certainty' about this accuracy. He concluded that a total number of 25,500 trials were required to establish this accuracy. Jacob Bernoulli found this number very high, and it made his analysis difficult. Working on a book project on the topic, Ars Conjectandi, he wrote to Leibniz on October 3, 1703: "I have completed the larger part of my book, but the most important part is missing, where I show how the basics of Ars Conjectandi can be applied to civil, moral and economic matters" (Polasek 2000). Jacob Bernoulli did not finalize the book. Researchers have indicated that Jacob was not really convinced by his examples. He could not specify the probabilities with the necessary precision for situations of practical importance.

The problem with this type of reasoning, the search for accurate determination of an underlying true probability (i.e., frequentist probability), is the need for repetitions of similar situations, as in the die example previously. In many real-life events, such repetition is not possible. As an example, think about the frequentist probability that

Jacob Bernoulli would finish the book within one year following the letter to Leibniz. Clearly, such a probability cannot meaningfully be defined. It is not possible to establish repeated experiments. Either he will finish the book or not. It is a unique event.

A frequentist probability represents the fraction of time the event considered occurs if the situation can be repeated under similar conditions infinitely. It represents the variation – event occurs and event does not occur. When such a population of similar situations can be meaningfully defined, is a judgment call. Thus, the frequentist probability is a model concept, which needs to be justified. It does not exist in all situations.

Think again about the die example. If the die is fair, the probability that the outcome is any of the numbers 1, 2, 3, 4, 5 or 6 is 1/6. These probabilities are classical probabilities but also frequentist probabilities representing the variation in outcomes: throwing the die many times, the fraction of times it shows a particular number is about 1/6. In the limit, it is exactly 1/6. If the die is not fair (and also if the die *is* fair), trials can be used to accurately estimate the true underlying frequentist probabilities, that is, the variation in outcomes, when conducting an infinite number of trials.

The third and final interpretation of probability we will discuss here is *subjective probability*, also referred to as *knowledge-based* or *judgmental probability*. These are of a different type than classical and frequentist probabilities and can always be specified, as they express the assessor's uncertainty and degree of belief for an event to occur or a statement to be true – therefore the term 'subjective'. Let A denote the event that Jacob Bernoulli will finish the book within one year following the letter to Leibniz. We can think about a friend of Jacob on October 3, 1703, making a subjective probability of A to occur equal to 0.10, that is, $P(A) = 0.10$. It means that this friend has the same uncertainty and the same degree of belief for A to occur as randomly drawing a red ball out of an urn containing ten balls, of which one is red. Hence this probability is a judgment of the assessor, not a property of 'the world', that is, the situation considered.

If the statement were that the probability is maximum 0.10 (10%), it means that this friend has the same uncertainty and the same degree of belief for A to occur as randomly drawing a red ball out of an urn containing ten balls, of which one or zero is red. The probability is an *imprecise probability*; the assessor is not willing to be more precise.

Let us return to the Leo example introduced in the beginning of Section 3.1 and suppose he notices that the die is not fair. And suppose he is not allowed to throw the die before the game. What is then the probability of getting 1? It is possible to think about an unknown frequentist probability of this event, but we can also consider this probability as a subjective probability. Leo studies the die carefully and assigns a probability of 0.1. This number represents Leo's judgment; it cannot be said to be objectively right or wrong, as there is no reference for such a statement. The probability is given or conditional on some knowledge, which we refer to as K. We write $P = P(A|K)$, where the sign '|' is read as 'given'. Thus, Leo assigns a probability equal to 0.10 given the knowledge K. He has the same uncertainty and degree of belief for the die to occur as randomly drawing a red ball out of an urn containing ten balls, of which one is red. A subjective probability always has to be seen in relation to the knowledge K. We often refer to K as the background knowledge for the probability assignment. The term 'knowledge-based probability' is motivated by this construction $P(A|K)$.

Probability interpretations

Classical probability of event A, $P_c(A)$

$P_c(A) = m/n$, where n is the total number of outcomes and m the number of outcomes resulting in event A.

It is assumed that all outcomes have the same probability.

Frequentist probability of event A, $P_f(A)$

$P_f(A) = \lim x_n/n$, as n goes to infinity, where x_n is the number of A events in n trials similar to the one studied.

Subjective (knowledge-based, judgmental) probability of event A, P(A)

P(A): the assigner has the same uncertainty and degree of belief for the event A to occur as randomly selecting a red ball out of an urn containing P(A) · 100 % red balls.

REFLECTION

Ronny states that he tells the truth, but Lisa is not convinced. She believes he is lying and specifies a probability of at least 90% that he is doing so. How would you interpret this probability?

It is an imprecise subjective probability. The probability means that Lisa's uncertainty and degree of belief that Ronny is lying is comparable to randomly drawing a red ball out of an urn containing 100 balls, of which at least 90 are red. Lisa is not willing to be more precise.

A frequentist or classical probability interpretation would not make sense in this case.

3.1.1 Expected value

Return again to the die example. Suppose the die is fair. If you throw the die over and over again, what would the average outcome be? The answer is 3.5. We refer to this value as the *expected value*, or the statistical expected value. We note that this value in this case can never be obtained; the possible outcomes are 1, 2, 3, 4, 5 and 6. Hence we cannot interpret the expected value as a probable outcome or something like that.

Probability theory shows that the expected value, referred to as E[X], where X is the outcome of the die trial, can for this example be computed in this way:

$$E[X] = 1 \cdot P(X = 1) + 2 \cdot P(X = 2) + 3 \cdot P(X = 3) + 4 \cdot P(X = 4) + 5 \cdot P(X = 5)$$
$$+ 6 \cdot P(X = 6). \tag{3.1}$$

Therefore, if $P(X = x) = 1/6$ for x = 1, 2, . . . , 6, the expected value is indeed 3.5. The probability P here may refer to any of the three interpretations discussed previously.

When the expected value is founded on frequentist probabilities, it can be interpreted as an average value when the situation considered is repeated over and over again infinitely. However, when the probabilities are subjective (or classical), an interpretation based on averages cannot be used, as such repeatability is not always possible. If X is the number of fatalities as a result of a specific pandemic (e.g., COVID-19), frequentist probabilities cannot meaningfully be defined, and hence a related frequentist interpreted expected value would not exist. However, subjective probabilities can always be defined, and an expected value based on a formula similar to (3.1) can be introduced:

$$E[X] = 1 \cdot P(X = 1) + 2 \cdot P(X = 2) + 3 \cdot P(X = 3) + \ldots \qquad (3.2)$$

We can interpret E[X] according to this type of formula (3.1 and 3.2) as a *center of gravity* of the probabilities $P(X = x)$. This interpretation of an expected value is always applicable, independent of the way probability is understood.

3.1.2 Examples of probabilistic statements

Probability of getting a disease

In a large population group of people, about 1 out of 1 million get a specific disease on a yearly basis. Frank is a member of this population. What is the probability that Frank gets this disease next year? How should this probability be interpreted?

It is possible to provide two types of interpretations, using frequentist probabilities and subjective probabilities. In the former case, a frequentist probability P_f is introduced representing the proportion of people in the population who get this disease. The P_f is unknown, but based on the information provided previously, $1 \cdot 10^{-6}$ is an accurate estimate of this frequentist probability.

Alternatively, $1 \cdot 10^{-6}$ can be viewed as a subjective probability P for the event that "Frank gets this disease next year" (A), given the available knowledge K, that is, $P(A|K) = 1 \cdot 10^{-6}$. The assessor has the same uncertainty and degree of belief for the event A to occur as randomly drawing a red ball out of an urn containing 1 million balls, of which one is red. The knowledge is strong in the sense that it is based on evidence that shows that about 1 out of 1 million get this disease on a yearly basis, and Frank is considered a typical member of this population.

Now say that we have information that Frank is somewhat more exposed than average for getting this disease. How could we then formulate the probabilities?

We could use an imprecise probability statement, expressing that $P(A|K) \geq 1 \cdot 10^{-6}$; that is, the probability for Frank to get this disease the next year is at least $1 \cdot 10^{-6}$.

Using a frequentist interpretation is not straightforward, as the frequentist probability P_f is a property of the population and not Frank. Hence, we need to change the population in order to take into account this new information. We must define a new population with members who are 'similar' to Frank. Then a new, unknown frequentist probability is constructed. The information available provides an estimate of this frequentist probability which is higher than $1 \cdot 10^{-6}$.

The example demonstrates the importance of being precise on what the event considered is and how the probabilities are interpreted.

The probability of a defect unit in a production process

Consider a production process of units of a specific type. Let $p = P_f(A)$ be the frequentist probability that an arbitrary unit from the production the coming week is defective, where A refers to the event that the unit is defective. This probability is interpreted as the fraction of units being defective in this production process the coming week. To estimate p, a sample of 100 units is collected, showing 2 defects. From this, an estimate p* equal to 2/100 is derived.

Textbooks in statistics show how to express uncertainties in relation to p and p* using concepts like variance (standard deviation), confidence intervals and prediction intervals. Two main lines of thought are used, traditional statistical analysis and Bayesian analysis. For details, see Problem 4.12 and Section 5.2.

Alternatively, we can define a subjective probability P(A|K), expressing the assessor's uncertainty or degree of belief for the event A to occur, using the urn interpretation discussed previously. Based on the observations, the assessor may conclude that P(A|K) = 2/100, seeing these observations as the background knowledge K.

3.1.3 Probability models

A probability model is a model built on frequentist probabilities, that is, a model of variation. Consider again the previous die example, where the die is not necessarily fair. A probability model is defined by frequentist probabilities $P_f(X = x)$ of the six outcomes. To simplify the notation, we introduce $p_x = P_f(X = x)$ for x = 1, 2, . . . , 6. The probability model is specified by p_x, for x = 1, 2, . . . , 6. The frequentist probabilities are in most cases unknown and need to be estimated.

Probability models are commonly used in probability theory and statistics. Well-known types of probability models are binominal, Poisson and normal (Gaussian).

3.2 BASIC IDEAS ON HOW TO MEASURE OR DESCRIBE RISK

Let us return to the example of Section 2.2.1, where you face the risk of losing $36,000, but you can also win $3,000, depending on the outcome of the two dice. Now the issue is the magnitude of the risk. How large is the risk? How should it be described? Formally, we defined the risk by the pair (C,U), where C is the consequences (outcome) of the activity (the game) and U the associated uncertainties. The consequences C is either a loss of $36,000 or a win of $3,000; hence, it remains to express the uncertainties U. Assuming that the dice are fair, we are led to the probabilities 1/36 and 35/36, respectively, as the loss occurs only if both dice show 6. These probabilities can be interpreted as classical, frequentist or subjective. All apply in this case.

We note that the associated expected win for John is $3,000 \cdot (35/36) - 36,000 \cdot (1/36) = 6,900/36 \approx 1,917$.

This expected value is not really providing much information for you, as you are allowed to play the game only once. However, for John, it is informative. The expected value shows that on average, if the game is offered to many people, he would lose about $1,917 per game. Clearly this would be a poorly designed game if the purpose was to earn money from it. However, we do not know the background for this particular game. It is also possible that John is cheating, using dice which are not fair. Clearly you as a player also need to reflect on that risk. How should the risk characterization be updated to take into account this risk aspect?

There are different approaches; see Table 3.1. One is to consider the previous probabilities as subjective probabilities conditional on the assumption that the dice are fair and provide a judgment of the reasonability of this assumption. Based on the observation that John will lose money in the long run if the dice are fair, this assumption can indeed be questioned. The knowledge supporting the probabilities 1/36 and 35/36 is weak.

An alternative and related approach is to add to the previous subjective probabilities a judgment about the risk related to this assumption being wrong. This risk is

TABLE 3.1 Alternative approaches for expressing your risk

Approach	Event/ consequences	Probabilities	Knowledge	Comments
The dice are fair	3,000 −36,000	35/36 1/36	Condition on the assumption	Based on the assumption that the dice are fair
Assigned probabilities, strength of knowledge judgments	3,000 −36,000	35/36 1/36	Weak knowledge supporting these probabilities	
Assigned probabilities, judgment of risk related to assumption deviation	3,000 −36,000	35/36 1/36	Weak knowledge supporting these probabilities (35/36 and 1/36)	An imprecise probability expressing that it is likely – at least a probability of 50% – that the dice are not fair, strong knowledge supporting this judgment
Assigned probabilities, probabilistic analysis John is cheating	3,000 −36,000	P(loss =36,000|K) ≈ 0.45	Weak supporting knowledge	P(John cheating |K) = P(not fair dice |K) = 0.90, P(losing 36,000|John cheating, K) = 0.50

Approach	Event/ consequences	Probabilities	Knowledge	Comments
Assigned probabilities, probabilistic analysis John is cheating, imprecise probabilities	3,000 −36,000	P(loss = 36,000|K) ≥ 0.05	Stronger supporting knowledge	P(John cheating |K) ≥ 0.50 P(losing 36,000| John cheating, K) ≥ 0.10
Assigned probabilities, strength of knowledge judgments, focusing on event (John cheating) and consequences given event	3,000 −36,000 3,000 −36,000 John is cheating	35/36 1/36 P(loss = 36,000| John is cheating, K) is large Rather high	Strong knowledge Relatively strong Relatively strong	Assuming John is not cheating Assuming John is cheating, the dice are not fair

expressed as an imprecise probability with associated knowledge strength judgments, for example, an imprecise probability expressing that it is likely – there is at least a probability of 50% – that the dice are not fair, with knowledge judgments considered rather strong arguing that the expected value of the game is favoring the player: John will lose money in the long run if the dice are fair.

As a third approach, we seek to add a probabilistic analysis of John cheating in this game. Suppose you assign a probability of John cheating equal to 90% given your background knowledge, that is

$$P(\text{John cheating }|K) = P(\text{not fair dice }|K) = 0.90,$$

and a probability of 50% of losing \$36,000 if the dice are not fair; that is, John is cheating. Then using probability calculus (the law of total probability), you can calculate a probability of losing \$36,000 equal to

$$P(C = 36,000|K) = P(C = 36,000|\text{ fair dice, K}) \cdot 0.10 + P(C = 36,000|\text{ not fair dice, K}) \cdot 0.90$$
$$= (1/36) \cdot 0.10 + 0.50 \cdot 0.90 \approx 0.45.$$

The background knowledge for this probability would be weak, as it is difficult to assign the probability that John is cheating and the probability of a loss of \$36,000 if the dice are not fair. The probability number 0.45 seems poorly justified and thus somewhat arbitrary. The arbitrariness can be reduced somewhat by using imprecise probabilities, for example, saying that there is at least a 50% probability that he is cheating, and P(C = 36,000| not fair dice, K) is at least 10%, which leads to the calculations

P(C = 36,000|K) = (1/36) · P(fair dice | K) + P(C = 36,000| not fair dice, K) · P(not fair
dice | K)
≥ P(C = 36,000| not fair dice, K) · P(not fair dice | K) ≥ 0.10 · 0.50 ≥ 0.05 = 5%.

Thus, there is at least a probability of 5% of losing $36,000, and this may be sufficient
to conclude that you would not like to play the game. You judge the knowledge supporting the probability as medium strong.

As a final approach, risk is described by two main contributions, as indicated previously. The first is the same as in the quantitative approach, presuming the dice are fair.
The second is simply a qualitative description of the risk in the case that John is cheating. Let A be the event that he is cheating and C the consequences for you as previously.
Then you describe risk by reporting

Risk description when A does not occur (is not true): loss $36,000 with probability
   1/36 and win of $3,000 with probability 35/36
Risk description when A is true: A loss of $36,000 is rather likely. The supporting
   knowledge is considered relatively strong.
It is likely that A is true, as John would lose money in the long run if the dice are fair.
   The background knowledge is rather strong.

Overall, risk is considered high, given that it is rather likely that you will lose $36,000,
and the knowledge supporting that judgment is rather strong.

This description and judgment of the knowledge strength would of course be different if you knew John and were informed about how he had designed this game. You
would then ignore the possibility that he is cheating.

Subjective probabilities are needed if we are to incorporate the cheating risk aspect
into the analysis. Classical or frequentist probabilities cannot be meaningfully defined.

In all of the previous approaches for describing risk, there are three main elements:

• The identified or specified consequences (events, effects/consequences).
• A way of describing the uncertainties.
• The knowledge supporting the previous judgments.

In general terms, we denote these three elements C', Q and K, respectively, leading to
a risk description or characterization (C',Q,K). When events are focused, we are led to
(A',C',Q,K). See Table 3.2.

The identified or specified events and consequences (A' and C') are not the same
as A and C in the risk definition. The A and C are the actual event and consequences
occurring, whereas A' and C' are those considered in the analysis. We may, for example,
think about a situation in which you play the game and win but do not get the money
from John: he just disappears. Hence the actual C is equal to 0, which is an outcome
not covered in your risk description. We will discuss this issue further in Section 3.4
when addressing surprises and black swans.

The description Q covers probability (precise or imprecise) and judgments of the
strength of the knowledge (SoK); see Problem 3.12. We have discussed probabilities

TABLE 3.2 General elements of a full risk description or characterization (A',C',Q,K) for a risk (A,C,U) related to an activity

Risk description elements	Meaning	Examples
A'	Specified events	John is cheating.
C'	Specified consequences	Outcome of the game.
(C',Q)	A description of the risk (C,U): The uncertainties U expressed by Q: Probability (imprecise probability) and Strength of knowledge (SoK) judgments	Risk description when A' does not occur (is not true): loss $36,000 with probability 1/36 and win of $3,000 with probability 35/36. Risk description when A' is true: A loss of $36,000 is rather likely, as John would lose money in the long run if the dice are fair. The supporting knowledge is considered relatively strong.
K	The knowledge that Q (and A' and C') are based on	A belief that John will give you the $3,000 if you win.

in Section 3.1 and illustrated the SoK concept previously, but further discussions are needed. How should we conclude that the knowledge is strong or weak? Is it is possible to quantify the SoK?

Thinking about the previous dice example, it can quickly be concluded that SoK judgments cannot be meaningfully quantified on a continuous, interval measurement scale such as, for example, weight or height. Rather, we are restricted to a nominal with order scale, for example, weak knowledge, medium strong knowledge and strong knowledge. Think about your knowledge concerning John cheating. To transform that knowledge into one unique number is problematic, as discussed previously, even for an interval. Clearly the knowledge would be strong if we had a lot of relevant data supporting the probability judgments or a strong understanding of the phenomena studied. If we had performed tests or weight and symmetry analysis of the dice, the supporting knowledge would be strong. In the absence of such data and analysis as in this case, the knowledge is weak. Common issues to consider when making judgments about the SoK are (Aven and Flage 2018):

- The reasonability of the assumptions.
- The amount and relevancy of data/information.
- The degree of agreement among experts.
- The degree to which the phenomena involved are understood and accurate models exist.
- The degree to which the knowledge K has been thoroughly examined (for example, with respect to unknown knowns; i.e., others have the knowledge but not the analysis group).

In the dice example, the assumption of fair dice is critical. Clearly if this assumption is used as a basis for the risk description, the knowledge would be judged as weak. Care has to be shown when it comes to the third criterion, as experts may have the same type of background and training. Hence, this criterion should highlight agreement across different 'schools' and perspectives. The forth criterion could, for example, relate to the justification of a specific probability model. The fifth and last criterion questions whether the knowledge has been scrutinized, to identify potential surprises relative to the current knowledge. For example, in the dice example, we could question whether processes have been conducted to look for events that could lead to different outcomes than 36,000 and 3,000.

See Problem 3.19 for examples of scoring systems based on such issues.

The knowledge K is of different types, but for the present discussion, it is best explained as *justified beliefs*. You have knowledge about the game and John. In general, the knowledge is founded on data, information, modeling, testing, argumentation and so on. Often the knowledge adopted can be formulated as assumptions, as in the previous case, when stating the dice are fair.

It is useful to distinguish between *general knowledge* (GK) and *specific knowledge* (SK) for the activity considered (Aven and Kristensen 2019). To illustrate, consider the epidemic example of Section 1.3. The GK includes all the generic knowledge available on epidemics and pandemics, whereas the SK includes knowledge concerning the specific situation related to the particular epidemic discussed.

In addition, we need to clarify whose knowledge we refer to. We distinguish here between your knowledge and John's knowledge and the total knowledge available when adding also other people's knowledge.

Evidence as a concept is closely related to knowledge. It captures the basis for a belief or statement in the form of data, information, modeling insights, test results, analysis results and so on. A risk assessment can produce evidence in the form of a risk characterizations of relevance for making a judgment about a statement being true. In a court, risk assessment results can be used as evidence, acknowledging that it is not representing the truth but a judgment about the truth subject to uncertainties.

In situations of very weak knowledge, a broad imprecision interval would typically be used. The extreme case is a probability [0,1], which means that the assessor is not willing to use the probability scale at all. The analyst assesses the knowledge and leaves it up to others to conclude whether the relevant event is likely to occur. However, in most cases, there is some data or information available that allows for a judgment like 'unlikely', for example, indicating a probability below 10%. This judgment is then supplemented with a description of the knowledge supporting it, as well as a judgment of the strength of knowledge.

New knowledge, such as in the form of new data, can make the SoK stronger. The knowledge could potentially reduce the uncertainty about the unknown quantities (A' and C'). However, it could also lead to a higher level of uncertainty about these quantities, for instance, if the new knowledge challenges existing beliefs and assumptions. We are also cautious in noting that more data has the potential to mislead and promote over-confidence in the assessment: The data could to a varying degree be informative (accurate and relevant) for the issues discussed.

3.2.1 Vulnerability

If John is cheating, it is likely that your loss will be $36,000. You are vulnerable in the case of John cheating. We remember that vulnerability was defined as (C,U|A), basically reflecting the risk – the potential for negative consequences – given the occurrence of an event A', that John is cheating. Thus, we can describe the vulnerabilities given that John is cheating as follows:

A likely loss of $36,000, as John would lose money in the long run if the dice are fair. The supporting knowledge is considered relatively strong.

In general terms, the vulnerability is described by (C',Q,K|A'), that is, risk conditional on the occurrence of the event A'. We remember from Section 2.3.1 that risk can be written as (A,U) and vulnerability (C,U|A). Similarly, risk can be described by

Risk description = (A',Q,K) & (C',Q,K|A') =
Event risk description and Vulnerability description

The last row of Table 3.1 provides a risk description in line with this separation between event risk description and vulnerability description.

3.2.2 How to conclude that the risk level is high or low

Risk is described by (A',C',Q,K), which is multidimensional, and it is not straightforward to conclude that the risk is high or low. Fortunately, there are ways of simplifying the judgments. The first one relates to small risks.

A risk is judged as low if the probability of the related undesirable events/consequences is small and the associated strength of knowledge is strong. The activity can be seen as safe. For example, food is safe if the probability of getting an illness as a result of the food is small and the supporting knowledge is strong. What is sufficiently low depends on regulatory standards established for food and also depends on the value judgment of the person. We will discuss the issue in Chapter 8 when addressing issues related to risk acceptance. In the present section, our focus is on how to characterize the risk to be able to make appropriate judgments about the acceptability.

Moreover, we can conclude that the risk is high if the probability of the related undesirable events/consequences is high and the associated SoK is strong. Consider the risk related to the number of fatalities in relation to a pandemic in a period of one year, described by a probability distribution as shown in Table 3.3. Is the risk description (C',P) showing that the risk is high? It is of course impossible to say without having a reference for what is high. We can relate the numbers to the typical death rates in that country to the numbers in other countries, as well as to the intervention costs to

TABLE 3.3 Number of fatalities due to a pandemic per 100,000

Number of fatalities	Less than 1	1–10	10–50	50–100	>100
Probability	0.05	0.25	0.40	0.25	0.05

mitigate the pandemic. However, as previously, the discussion in the present chapter is more about how to present the risk than what the conclusions would be. The probability distribution of Table 3.3 can be informative, but, as stressed repeatedly, we also need to add judgments of the strength of knowledge supporting these probabilities.

REFLECTION

Suppose the probability of some undesirable event/consequence is high, and the associated SoK is weak. Would you then categorize the risk as high?

Yes, remembering that risk is the potential for some undesirable consequences, or the combination of undesirable consequences and uncertainties, and in this case, there are considerable uncertainties related to the occurrence of some undesirable event/consequence.

It is challenging to compare risk levels when using probability distributions and SoK judgments. Different metrics are often used to simplify the comparisons; see Section 3.3. In general, such metrics should be used with care, as they often provide a rather poor characterization of the risk. In the following, we discuss additional qualitative aspects to consider when making judgments about the risk being high or low.

For a risk to be classified as high, the following criteria need to be considered (a minimum of one of these criteria applies) (Aven and Kristensen 2019):

a) The risk is judged to be high when considering consequences and probability
b) The risk is judged as high when there is a potential for severe consequences and considerable uncertainties (weak knowledge)
c) A high degree of vulnerability, lack of resilience
d) Weak general knowledge about the consequences C of the activity
e) Weak specific knowledge about the consequences C of the activity
f) Strong general knowledge about undesirable features of the consequences C of the activity
g) Strong specific knowledge about undesirable features of the consequences C of the activity

For example, the risk is classified as high if a critical failure of a safety system has been revealed through testing, by reference to a), c) and g). We can also classify the risk as high if the safety system has not been tested, with reference to b), c) and e). In the former example, we argued according to what we know and, in the latter, to what we did not know.

The undesirable features of the consequences could relate to risk sources and events, as well as barriers implemented to avoid the occurrence of undesirable events and reduce the effects if such events should occur. The category 'high' is typically associated with an unacceptable risk, which requires additional measures in order to proceed.

Analogously, we can specify criteria for low (all relevant criteria apply):

h) No potential for serious consequences
i) The risk is judged to be low when considering consequences and probability, and supporting knowledge is strong
j) Solid robustness/resilience
k) Strong general knowledge about the consequences C of the activity
l) Strong specific knowledge about the consequences C of the activity

Medium applies to all other cases.

3.2.3 Some examples

We first return to the plane example introduced in the beginning of this chapter, where Anne and Amy plan a plane trip from Washington, DC, to San Francisco. Then we discuss the examples in Section 3.1.2, the risk of getting a disease and the risk related to defective units in a production process. Finally, we include a radon example, following up the discussion of Section 3.2.2.

Traveling by plane

Amy argues that the trip is safe: the risk is very low, she argues. But Anne is not convinced.

The natural point of departure for such a discussion is accident statistics. A number of documents present historical data regarding fatalities in aviation for commercial airliners and in general. There are different ways of presenting the statistical data, but a common metric is to use fatalities per 1 million flights. The data show a positive trend over the years. Typical numbers presented for recent years show about 0.1–0.5 fatalities per million flights for commercial airliners. For general aviation, a typical number is 1 fatality per 100,000 exposed hours (FAA 2018), thus considerably higher than for commercial aviation. Extensive amounts of data are available, showing differences with respect to a number of factors, for example, flight type, flight phase (for example, approach, landing, cruise) and continent. Other types of references are also used, for example, fatalities per km travel. Do these data allow us conclude whether it is safe to travel by plane?

To be able to do this, we first need to clarify what *safe* means. As noted in Section 2.3.4, people often understand being safe as being associated with an absence of failure or loss: here, that the flight does not result in a crash with associated fatalities and injuries. However, looking into the future, we do not know if such an event will occur or not; we face uncertainties. Thus, we are led to considerations of risk. We say the flight is safe if the risk is sufficiently low. This forces us to clarify what 'sufficiently low' risk means. The historical accident numbers referred to previously clearly relate to risk, but risk is about the future, and, hence, we cannot simply use the data we have without any type of reflections on the relevancy of these data. However, taking into account the comprehensiveness and solidness of the accident data related to flying, and the structure and features of the aviation industry, we quickly conclude that the accident data are relevant for flights today, tomorrow and later. The industry will basically be the same, although events like COVID-19 could lead to some changes. Looking at the data, overall positive accident trends are observed over the years. The reasonable explanation is better planes and improved safety and risk management systems.

Thus, for Anne and Amy's travel, these historical metrics are informative for the judgment about the risk being sufficiently low. What is sufficiently low is in principle a subjective judgment. Amy may find the risk sufficiently low, but not Anne. Amy concludes that the risk is sufficiently low and acceptable for the trip, as the accident risk metrics are so small and the basis for these metrics is strong. Clearly, if the data supporting the metric of, for example, 0.1–0.5 fatalities per million flights had been weak, the metric would not had been given the same weight. To conclude on an activity being safe, it is not sufficient that the derived probability of a negative outcome be low; the knowledge supporting this probability also needs to be strong.

A safe activity:
  The probability of undesirable consequences is low and the supporting knowledge strong

---

The judgment of the risk being sufficiently low and acceptable is subjective. However, in cases like this, with such strongly founded metrics showing such small accident rates, the judgments quickly become intersubjective: There is broad agreement in the judgments. A rationale is provided which compares the metric with other types of activities, for example, other transportation means (e.g., car driving). Although there are some measurement problems in producing comparable results for different activities, the overall message of such comparisons is clear: Flying shows low fatality rates compared to other transportation methods. It is much safer to fly than to drive, if we take into account the traveling time.

A risk and vulnerability description is shown in Table 3.4. The probabilities have a strong knowledge base. Given a crash, the vulnerabilities are high.

Finally, in this example, some comments concerning security issues. Anne and Amy have decided to travel by plane from Washington, DC, to San Francisco. First, they have to go through security to enter the gates. We all know why these security arrangements have been developed. Many travelers view these security precautions as unavoidable but irritating, increasing travel times. Some question security efforts made by the aviation authorities, referring to statistics that the number of deaths as a result of such attacks at airports is basically negligible compared to other activities (Aven 2015a). However, this type of reasoning is unjustified. The low number of incidents can

TABLE 3.4 Risk description for plane example

Risk description element	Meaning	Examples
A'	Specified events	Plane crash
C'	Specified consequences	Injuries and loss of lives
(C',Q)	A description of the risk (C,U): The uncertainties U expressed by Q: Probability (imprecise probability) and Strength of Knowledge (SoK) judgments	Fatality probability less than $1 \cdot 10^{-6}$ for a flight Strong knowledge basis
K	The knowledge that Q (and A' and C') are based on	Historical data are the only source used
(C',Q,K|A)	Vulnerability description given a plane crash	Given a plane crash, the fatality probability is very high. The knowledge is strong. Hence the vulnerability is large

rather be explained by effective security management. Attackers have not been able to conduct events like September 11 since 2001. If we had not implemented such arrangements, it is not unreasonable to conclude that many types of serious incidents would have occurred. There is no certainty, but any serious judgment of the risk facing the aviation industry due to terrorist attacks would state that the risk would be great and clearly unacceptable if we had not implemented strong security measures.

In the absence of security arrangements, Anne and Amy would not have made a plane trip. The reason is simply that they then judge the risk to be too high. However, given the current security arrangements, they judge the risk to be low and acceptable.

It is also possible to travel from Washington, DC, to San Francisco by train, but it would have been a much longer trip. For this transportation means, there are few security arrangements. Yet we consider the related risk acceptable. Is the risk so low? If we look at the statistical data, we would indeed find that the frequencies of security incidents are low, and the knowledge basis is strong. We consider it a safe means of transportation. Now let us perform a thought construction. Suppose a series of terrorist events occurred on trains in one particular country. The number of events is still rather limited, but they raise concern in the country and in other countries, as there is considerable uncertainty about the intentions of the group responsible for the attacks. A security expert claims that there is nothing to be worried about. Using the incident data, the death risk is claimed to be tiny compared to, for example, driving. However, people are concerned, and the expert indicates that this is because of risk perceptional aspects like fear and dread. It is not rational to not travel by train is the claim.

This reasoning fails. The issue is more complicated than this. First, risk is not given by the historical data. Uncertainty is a main component of risk, and, when considering how to deal with the risk, the issue of uncertainties is critical: What are the strategies of the attackers? Will we experience an increase in the frequency and/or intensity of such attacks in the near future? How will these attackers react to the measures taken by the authorities? See further discussion in Chapter 6 (Section 6.1.1 in particular).

The risk of getting a disease

In Section 3.1.2, we considered the probability that Frank would get a specific disease next year. Now we extend the discussion by asking: How big is the related risk?

One way of describing the risk is to use subjective (knowledge-based) probabilities together with judgment of the strength of knowledge. A probability equal to $1 \cdot 10^{-6}$ is assigned, that is, $P(A|K) = 1 \cdot 10^{-6}$, where A is the event that Frank gets this disease next year. The knowledge K is strong, as it is based on data showing that about 1 out of 1 million get this disease on a yearly basis, and Frank is considered a typical member of this population.

Another approach is to define the consequences C' by the frequentist probability P_f representing the proportion of people in the population who get this disease. An estimate of P_f equal to $1 \cdot 10^{-6}$ is derived. Uncertainties about P_f are ignored, as $1 \cdot 10^{-6}$ is considered an accurate estimate of P_f.

In the general case, we also need to consider uncertainties about P_f. Subjective probabilities can be used for that purpose, with associated strength of knowledge judgments. Suppose Frank has shown some symptoms of getting the disease, but the evidence is difficult to interpret. The doctor uses the general statistics together with the specific knowledge (the symptoms) and derives at a 90% uncertainty interval for P_f, [0.001–0.5], meaning that the doctor's subjective probability for P_f to be in this interval is 90%, that is

$$P(0.01 \leq P_f \leq 0.5) = 0.90.$$

The doctor considers the knowledge supporting this judgment relatively weak.

Traditional statistics provide different types of uncertainty characterizations for P_f, including confidence intervals. These intervals do not reflect the assessor's uncertainty about what the value of the unknown P_f is but express variation in observations if we could repeat the type of situation considered over and over again. To establish a confidence interval, we need to have available a set of relevant observations, that is, a population of n persons similar to Frank having the same type of symptoms. If X denotes the number of persons among n who have gotten the disease, we can use probability theory to establish two quantities Y and Z depending on X so that $P_f(Y \leq P_f \leq Z) = 0.90$. The interval [Y,Z] is a confidence interval of level 90%. If we could do the 'experiment' over and over again, the interval would cover the unknown P_f in 90% of the cases. We refer to textbooks in statistics.

The risk related to defective units in a production process

Consider the production process defined in Section 3.1.2, with A referring to the event that a unit is defective. The consequences of the activity are defined by the fraction of units that are defective, and the uncertainties are described by a probability distribution or prediction interval for that fraction, with associated strength of knowledge judgments. Let X be the number of defective units among the 100 sampled units. The assessor predicts X to be two. In addition, a 90% prediction interval for X (or X/100) is presented: [0,5] ([0, 0.05]), meaning that it is 90% probable that X will be in the interval [0,5] based on the current knowledge. The knowledge basis is a previous sample of 100 units.

Radon example

Consider a house development project during spring and summer, with an issue of radon risk (Aven and Kristensen 2019). The owner has currently not been able to make proper radon measurements in the house, as it needs to be measured in the winter. Based on a radon map, the owner observes that there are high radon concentrations in the area where the house is located. The general knowledge concerning radon risk is strong, and the risk is judged high according to the scheme presented in Section 3.2.2, for example, by referring to criterion 6. The general knowledge recommends some protection barriers be installed. The specific knowledge is judged to be rather weak, and to strengthen this knowledge, the owner needs to defer the project some six to nine months to make detailed measurements. The results of such measurements could potentially show that the radon exposure for the house is small, and no special protection is required. However, a delay of six to nine months is considered too costly for the owner, and the conclusion is that the protection barriers be installed on the basis of the general knowledge, in line with the cautionary principle and the implementation of robust/resilient measures; refer to Chapter 8.

3.3 COMMON RISK METRICS

In its most general form, risk is described by (C',Q,K) or (A',C,Q,K). From these expressions, specific metrics are often used summarizing or highlighting aspects of risk. Here we will discuss some of them, including

- Expected value E[C']
- Distributions P(C' ≥ c) and risk matrices based on P(A') and E[C'|A']
- Value at Risk (VaR), defined by a quantile of the probability distribution of C'

Related frequentist metrics can also be defined. The P and E then need to be replaced by estimates P_f^* and E_f^*.

3.3.1 Expected values

Let us return to the die gamble of Section 2.2.1 where John offers you one game, where you will receive $3,000 from him if the outcome is a success for you – the two dice do not both show 6 – and you have to pay him $36,000 if both dice show 6.

Suppose the dice are fair. Then the expected value (win) for you is, as noted in Section 3.2:

$$3000 \cdot (35/36) - 36000 \cdot (1/36) = 6900/36 \approx 1{,}917.$$

Say that you are informed that the expected value (win) of the game is equal to $1,917 with no further information. Would that information be sufficient for you to make a decision whether to play the game? Certainly not – you need to see beyond the expected value. The possible outcomes of the game, here $3,000 and –$36,000, and the associated probabilities are required. A loss of $36,000 could have severe consequences for you. Seemingly the game is attractive for you, as the expected value is positive and large, but you may lose a considerable amount of money, which may cause you think very carefully whether to play. To accept the game, you will reflect on the *value/utility* of the different outcomes.

Hence a risk description based on expected value alone would not be very informative. It would, however, be so for John if he offers this game to many people. Then $1,917 would be an accurate estimate of the actual loss for him per game. This follows from the law of large numbers; see discussion subsequently and Ross (2009). Clearly, if he offered this game to many people, he would lose a considerable amount of money.

The observation that we need to see beyond expected values in decision-making situations goes back to Daniel Bernoulli (1700–1782) more than 250 years ago. In 1738, the Papers of the Imperial Academy of Sciences in St Petersburg presented an essay with this theme: 'the value of an item must not be based on its price, but rather on the utility that it yields' (Bernstein 1996). The 'price' here refers to the expected value.

To reflect the utility, an adjusted risk metric can be derived, $-E[u(C')]$, where u is a utility function. See Problem 3.14.

The previous probabilistic risk numbers are founded on a strong knowledge basis (given that the dice are confirmed fair). If, however, the dice are not necessarily fair, the expected value will be based on a weak knowledge base. Probabilities can be assigned (see Section 3.2 and Problem 3.14), leading to an updated expected value, but this value cannot be given much weight because of the weak supporting knowledge.

It can be argued that the expected value is an attractive risk metric, as it is based on one number – or potentially a few numbers if the metric is split into different consequence attributes (loss of lives, environmental damage, etc.). However, given that the aim is to describe risk, the metric has some strong limitations to a large extent motivated by the previous discussion (Aven 2020e):

1   It does not show the potential for extreme consequences or outcomes.
2   It does not show the uncertainties of the estimates of $E[C']$ (frequentist case).
3   It does not show the strength of the knowledge on which the $E[C']$ is based (knowledge-based case).

The first point represents a serious weakness of this metric and is discussed previously using the die example. A portfolio type of argument can be used to justify the use of expected values in some cases, as for John if he plays the game many times. Another example is an insurance company with a number of activities or projects covered. This company is mainly interested in the average value and not the individual ones. By the law of large numbers, we know that under certain conditions, the average value converges with probability one to the expected value of each quantity. However, for this argument to be valid, the number of projects must be large and there must be some stability to justify the existence of frequentist probabilities. With the potential for extreme and surprising observations, the approximation of replacing the average with the expected value could be rather poor – the uncertainties in the estimates would be large.

The question now is the extent to which we can apply the portfolio type of argumentation for real-life risk, for example, global risk or national risk (e.g., pandemic, extreme weather events, cyber attacks). In the case of global or national risks, we have a number of projects or activities, but, unfortunately, these cannot, in general, be seen as averaging themselves out. There are two main problems. The first is that, in such risk studies, a main focus and interest is extreme types of events, which occur relatively rarely. Second, the world is rapidly changing, and the stability required to think in relation to frequentist probabilities is challenged. A key aspect to consider, when looking into the future and the related risks, concerns incorporating potential surprises and what today is not foreseen. The conclusion is that an expected value computed today could be a poor predictor of the future average value, even when taking a global or national perspective.

As mentioned previously, in practice, $E[C']$ characterizations are often refined by presenting the pair $P(A')$ and $E[C'|A']$, where A' is an event (hazard, threat, opportunity) and $E[C'|A']$ is the conditional expected value, given the occurrence of A'. The previous discussion concerning $E[C']$ also applies to $E[C'|A']$, although this split into these two terms reduces some of the issues discussed previously for $E[C']$, in particular the problem of multiplying a small probability of the event occurring with a large consequence number. However, $E[C'|A']$ is still an expected value and 1–3 apply. Important differences in potentials for extreme outcomes are not revealed using the approach.

Another example to illustrate this discussion is the following (Aven 2017c). A country has about 100 facilities, of which all have the potential for a major accident, leading to a high number of fatalities. A risk assessment is conducted, and the total probability of such an event in the next 10 years is computed as 0.010. From the assessment, one such major event is expected in this period. A safety measure is considered for implementation. It is, however, not justified by reference to expected value calculations also reflecting the benefit of the measure, as well as the costs. The expected benefit of the measure is calculated to be rather small. The costs are considered too large in comparison with this expected value. The rationale is that we should expect one such event in the period, and the measure considered would not really change this conclusion.

However, the perspective taken is close to being deterministic and destiny oriented. One such event does not need to happen. Safety and risk management aim to avoid such accidents, and, if we succeed, the benefits are high – saving many lives. The value of the safety measure is not fully described by the expected number. The measure's value is mainly about confidence and beliefs that the measure can contribute to avoiding an occurrence of the accident.

Probabilities are commonly used for this purpose. Implementing the safety measure can, for example, result in a reduction in the accident probability from 0.010 to 0.009, which shows that it is less likely that a major accident will occur. However, the difference is small and will not really make any difference for the decision-making problem. One major accident is still foreseen.

The full effect of a risk-reducing measure is not adequately described by reference to a probability number alone. A broader concept of 'confidence' is better able to reflect the total effect. This concept is based on probability judgments, as well as assessments of the knowledge supporting these judgments. For example, it matters a great deal whether the probability judgments are based on strong knowledge or weak knowledge. In the case of the safety measure discussed previously, its implementation could reduce risk by reference to criteria h), i) . . . , l) defined in Section 3.2.2.

3.3.2   Distributions $P(C' \geq c)$ and risk matrices based on $P(A')$ and $E[C'|A']$

Next, we consider the risk description based on the pair of consequences and probability, (C',P), partly following the analysis of Aven (2020e). This metric meets critique 1 raised against $E[C']$ discussed in the previous section but not 2 and 3. In practice, standard risk matrices are often used to present the risk. The matrix typically shows the risk expressed for different events (hazards/threats) using the probability of the event $P(A')$ and the conditional expected value $E[C'|A']$ or a typical C' value for different categories of consequences. There are often ambiguity problems in defining the events A' and the related consequence value. The event A' could, for example, be defined as 'flooding', but its consequences vary from small to extreme. The definition used is critical for the probabilities assigned and the magnitude of the consequence dimension. If, for example, a broad event A' is defined, like 'flooding', a rather low consequence value follows. However, the probability of A' is relatively high. If, on the other hand, a more specific A' event is defined, the probability is lower, but the consequence value is higher. Clearly, if the study is not clear on the definitions, different people would make different choices, and the overall results, for example, represented by average judgments, would be meaningless.

For this reason, adjustment should be made by fixing events B, having some defined large consequences – for example, at least 1,000 or 10,000 fatalities, and then focusing on $P(B)$. A set of B events may be considered. The ambiguity problem is then solved.

In the frequentist case, uncertainties of the estimates need to be assessed. If only hard data are used to estimate the frequentist probabilities, confidence intervals can be produced, but these only show variation in data and do not say much about uncertainties regarding the 'true' values of P_f. Similar types of problems may occur in the case that the estimates are based on expert opinions, but, using Bayesian analysis (see Problem 4.12 and Section 5.2), it is possible to make formalized expressions of uncertainties about P_f, which reflect different sources of uncertainties. The Bayesian approach leads to many probability assignment tasks, and, normally, some types of uncertainty intervals are preferred, showing, for example, a 90% uncertainty interval for P_f. The interval can be generated by removing the most extreme expert estimates – for example, below the 5% percentile and above the 95% percentile. Such intervals are most informative when used for the $P_f(B)$ case, as these are then the only probabilistic quantities used.

The case of knowledge-based probabilities is similar to the frequentist case in many respects, with obvious changes in terminology. Here the probabilities are not uncertain, but the knowledge supporting the probabilities can be more or less strong. Different probabilities produced by different assessors are seen as reflecting different knowledge bases and degrees of belief of the relevant events. The analysts may use the total set of expert judgments to generate their best judgments, or the study could present an interval showing the variation in judgments among the experts. A 90% interval could be derived, as for the frequentist case, but the interpretation would thus be different. The interval in the knowledge-based case does not reflect uncertainties, as there is no presumed true value to relate the assignments to. All the assignments in the knowledge-based case could be based on more or less strong knowledge, but that is not revealed in the metric.

Risk matrices presenting scores for impact/consequences of specified events, with associated probabilities, are commonly used in practice. Two main problems with using such risk matrices are as partly reflected by the previous discussion:

- The consequences of the events are in many cases not properly represented by one point in the matrix but by several with different probabilities. If we restrict attention to one point, this value would typically be interpreted as the 'expected value' (conditional expected value given the initiating event), which is the center of gravity of the probability distribution for the appropriate consequences. In most cases, this value is not very informative in showing the range of possible (or even the most likely) consequences. Take the event 'pandemic'. It is possible to foresee many scenarios with severe negative impact, ranging from a rather limited number of affected persons to situations where millions are suffering. Grouping all such scenarios into one, and highlighting only the expected (mean) value, can obviously lose essential information needed to characterize the risk to usefully inform decisions.
- Two events can have the same location in the risk matrix, but the knowledge supporting these judgments could be completely different, as discussed previously.

To meet these challenges, adjusted risk matrices have been developed. See Figures 3.1 and 3.2. Figure 3.1 shows an example of an 'extended risk matrix', which presents risk using consequence categories given the events, probabilities of the events and SoK judgments. Four A' events are shown. As an example, consider the event in the lower-right corner. This event has a catastrophic consequence, a low probability (≤ 0.01) and weak SoK.

In Figure 3.2, the consequences are fixed, as referred to previously when introducing the events B. This eases the interpretation of the diagram. With the consequences specified, the risk characterizations would cover two dimensions for the events considered: the probability P of the event and the strength of knowledge SoK supporting the

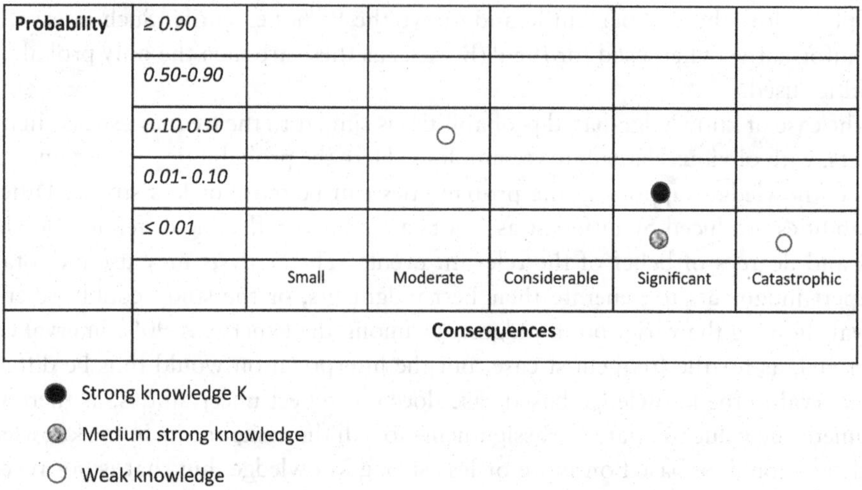

Probability	≥ 0.90					
	0.50-0.90					
	0.10-0.50		○			
	0.01- 0.10				●	
	≤ 0.01				◉	○
		Small	Moderate	Considerable	Significant	Catastrophic
				Consequences		

● Strong knowledge K

◉ Medium strong knowledge

○ Weak knowledge

FIGURE 3.1 Example of an extended risk matrix also including the strength of knowledge (based on Aven and Thekdi 2020, p. 15)

Probability	≥ 0.90			
	0.50-0.90		Event 2	
	0.10-0.50			
	0.01-0.10	Event 3		Event 1
	≤ 0.01			
		Strong	Medium	Weak
			Strength of Knowledge	

FIGURE 3.2 A risk matrix showing risk scores for probability and strength of knowledge, with fixed consequences (based on Aven 2020e)

probability judgments, as shown in Figure 3.2. The scores are judgments made by the assessors. In the example of Figure 3.2, the biggest risks are those in the upper-right corner, as these have high scores on probability and weak knowledge strengths.

3.3.3 Value at Risk

In some applications, in particular business and the financial service industry, it is common to use the risk metric value-at-risk (VaR), x_p, which equals the 100p% quantile of the probability distribution of the potential loss X (Aven 2010). That means that x_p is given by the formula $P(X \leq x_p) = p$. Typical values used for p are 0.99 and

0.999. Thus, if the VaR at probability level 99% is $1 million, there is only a 1% probability of a loss exceeding $1 million. VaR has an intuitive appeal as a risk measure. However, it suffers from some severe problems. The main one is that it does not reflect the size of potential outcomes higher than the VaR value. We may have two situations: in one, X is limited to values close to the VarR value, and in the other, there is a potential for extreme outcomes. The metric does not reveal the difference. In addition, the metric does not take into account the strength of the knowledge supporting the probabilities.

3.4 POTENTIAL SURPRISES AND BLACK SWANS

RISK LAB

"According to the theory, it is highly improbable that anything should ever happen anytime, anywhere"

Let us return to the game introduced in Section 2.2.1 where John offers you a game; you have to pay him $36,000 if both dice show 6, and you will receive $3,000 from John otherwise. Suppose you make your assessment based on the assumption – your belief – that both dice are fair. Then it is revealed at a later stage that the dice were not fair. You are surprised by this new information. You did not foresee this scenario.

Using the general risk setup of (A,C,U) and (A',C',Q,K), the knowledge K was based on the assumption that the dice were fair. However, knowledge and assumptions can be wrong, as in this case. It is obviously difficult to include this aspect in the risk characterization, as it extends beyond the knowledge that forms the characterization. However, following the guidance provided by describing risk through (A',C',Q,K), the knowledge aspects are highlighted, in particular the strength of the knowledge. Judgments of this strength will stimulate reflections and analysis which could reveal potential surprises. A key point in these judgments are considerations of the validity of assumptions. There is no guarantee that all erroneous assumptions and knowledge are identified, but the issue is focused on, and some potential surprises can be revealed.

As discussed in previous sections, knowledge can also be characterized as weak or strong, not only wrong or correct. To explain, think about the situation where you assign a rather high probability of John cheating; the supporting knowledge is considered relatively strong, as John would lose money in the long run if the dice were fair. You make no assumption that the dice are fair or not. If it turns out that the dice are in fact fair, you may be surprised, based on your knowledge. By also presenting the judgment of the supporting knowledge, the overall risk results will shift focus from the probabilities to the arguments and evidence provided.

Here is another example of the issue of surprises and how it relates to risk characterizations.

Terry is a professor and teaches a course in risk analysis. It is time for examinations. Terry walks around among the students, who work hard solving the problems. Several students point to a particular problem and question if a specific formulation is correct. Terry reviews the comments and has to admit that the formulation represents an error. Terry is surprised. He has prepared such examinations for 30 years and never experienced any mistake of this type. In addition, he has implemented strong quality assurance processes, which include providing accurate answers to the problems, checking the examination text carefully several times and having a colleague reviewing it before accepting the test. Terry was convinced that the test was fine. He considered the probability of having errors in the test very small and based on strong knowledge. The error risk was considered very small. Yet, the error occurred – a surprise for Terry relative to his knowledge.

Hence Terry produced a risk characterization according to the book by specifying the consequences (here focusing on errors) and using probability and strength of knowledge judgments (Q as P and SoK). The third component K captures the knowledge supporting the judgments made, and includes, for example, beliefs that that his colleague actually did a serious check of the text. It turns out that Terry, some few days before the examination, actually observed a signal of complacency – he felt that making the examination test was a rather routine work – it always goes fine. However, he ignored this signal. This feeling of complacency could have resulted in a lack of awareness and precision when reviewing the text. The signal should have been added to the knowledge basis. By acknowledging it, Terry could have reconsidered the P and SoK judgments, and he probably would had made an extra check of the text.

The example demonstrates the importance of highlighting all aspects of the risk characterizations (C',Q,K). The knowledge K is not static. By careful analysis, examining, for example, assumptions and signals, new knowledge can be gained and the analysis will be based on a stronger, more justified knowledge basis.

The black swan metaphor is commonly used for surprising events. We define a *black swan* here as *a surprising extreme event relative to one's knowledge* (Aven 2014, 2015c). 'Extreme' means that events have large/severe consequences.

The history of the black swan metaphor goes back a long time. The most common interpretation is the one provided by Taleb (2007). About 300 years ago, all observed swans in the Old World had been white. But then, in 1697, a Dutch expedition to Western Australia, led by Willem de Vlamingh, discovered black swans on the Swan River –a surprising event for people in the Old World. In his famous book from 2007, Taleb refers to a black swan as an event with the following three attributes. First, it is an outlier, as it lies outside the realm of regular expectations, because nothing in the past can convincingly point to its possibility. Second, it carries an extreme impact. Third, in spite of its outlier status, human nature makes us concoct explanations for its occurrence after the fact, making it explainable and predictable. See Problem 3.16 for a discussion of the differences and similarities between two definitions of a black swan.

Taleb (2007) present a number of examples of black swans:

> Just imagine how little your understanding of the world on the eve of the events of 1914 would have helped you guess what was to happen next. (Don't cheat by using the explanations drilled into your cranium by your dull high school teacher). How about the rise of Hitler and the subsequent war? How about the precipitous demise of the Soviet bloc? How about the rise of Islamic fundamentalism? How about the spread of the internet? How about the market crash of 1987 (and the more unexpected recovery)? Fads, epidemics, fashion, ideas, the emergence of art genres and

schools. All follow these Black Swan dynamics. Literally, just about everything of significance around you might qualify.

(Taleb 2007, p. x)

Taleb comments specifically on the September 11 event:

Think of the terrorist attack of September 11, 2001: had the risk been reasonably *conceivable* on September 10, it would not have happened. If such a possibility were deemed worthy of attention, fighter planes would have circled the sky above the twin towers, airplanes would have had locked bulletproof doors, and the attack would not have taken place, period. Something else might have taken place. What? I don't know. Isn't it strange to see an event happening precisely because it was not supposed to happen? What kind of defense do we have against that? Whatever you come to know (that New York is an easy terrorist target, for instance) may become inconsequential if your enemy knows that you know it. It may be odd to realize that, in such a strategic game, what you know can be truly inconsequential.

(Taleb 2007, p. x)

Suppose a national risk assessment in the United States was conducted in 2000. A risk characterization (A',C',Q,K) is presented. Then it could be the case that A' does not include such an attack as September 11. That means that the actual A is not covered by the assessment A'; a surprise occurs relative to the knowledge of the assessors. In this case, we refer to the event as an unknown known, meaning that it was unknown for the analysts but known to some (here the terrorists). We also talk about unknown unknowns (events not known by anybody).

For the September 11 event, a different type of argument can be used to explain why it came as a surprise and is a black swan. Suppose the risk assessment has identified this type of event as a threat. It is included in (A',C',Q,K). However, the likelihood of the event may be judged to be so low (also with strong knowledge support) that it is considered that it will not happen. As discussed in the previous quote by Taleb, if that would not have been the case, some measures would had been implemented to avoid it happening. Figure 3.3 illustrates the different types of black swans.

Clearly these events represent a challenge, as they come as a surprise relative to our knowledge. It is not possible to show or present the related risks as a part of the risk descriptions, but we can and should highlight the knowledge on which the judgments are based and, in particular, the assumptions made. Addressing and discussing this knowledge and these assumptions are, in many cases, equally, if not more, important than highlighting the probabilities derived. The way we have conceptualized and described risk underlines the importance of analysis processes to scrutinize the knowledge K. Many of these approaches, methods and models are based on monitoring and tracking critical assumptions. Irrespective of how thorough the assessment is, potential surprises is an important analysis issue. The focus on such surprises ensures that this type of risk is highlighted and acknowledged. Decision-makers need to cope with the issue, and analysts should guide them on how to do this.

Types of Black swans

Extreme consequences

FIGURE 3.3 Types of black swans (based on Aven and Krohn 2014)

REFLECTION

The black swan has become a very popular metaphor. Why do you think it has?

Because of the black swan metaphor, we have noticed an increased interest and enthusiasm for discussing risk analysis. The metaphor has made a complex issue comprehensible and practical. Risk assessments give us insights about risk. However, these assessments have limitations. They do not show how the world performs but instead provide judgments about the world. Surprises occur relative to these judgments, despite often seemingly strong evidence and expert agreement. That is the essence of the metaphor. It applies not only to risk analysis but any type of seemingly strong justified beliefs. The history and our frameworks of thought did not give us any ideas of a different perspective, a different truth.

3.5 PROBLEMS

Problem 3.1

Let C be the outcome from the throw of a die and let $p = P_f(C = 1)$. Explain what p expresses. The die is of a special type, and p is not necessarily 1/6. Let us assume that p is either 1/6 or 1/3. Explain how you can express your uncertainty about p using knowledge-based probabilities. Assign concrete numbers to illustrate how it is done.

Problem 3.2

A person states that he predicts B to occur 3 times in a 10-week period. Give two possible interpretations of this expression by reference to frequentist probabilities and knowledge-based (subjective) probabilities.

Problem 3.3

In a risk assessment, a risk analyst, Per, has introduced a probability model f(y), where f(y) equals the frequentist probability that Y equals y, y = 0,1,2, 3,4. Explain what f(y) expresses. How will you interpret the expected value $E_f[Y]$? Assume that the following estimates of f(y) have been established: 0.80, 0.10, 0.05, 0.03 and 0.02, respectively. From these numbers, estimate the expected value $E_f[Y]$.

Problem 3.4

Variation is commonly represented using probability models. The Poisson model is used to model the variation in the occurrences of events per unit of time (for example, days). Express the Poisson model and interpret the probabilities.

Let D denote the difference between the Poisson frequentist probabilities $f(x) = P_f$ (X= x) and the true fractions g(x) with x number of events occurring. Explain why it is reasonable to refer to D as model error and uncertainty about D as model uncertainty.

Problem 3.5

Consider again the Poisson model with parameter λ. Assume that λ is unknown. Specify knowledge-based (subjective) probabilities to express your uncertainty (degree of belief) of λ when you would like to reflect that λ could be either 1 or 2 with equal probability.

Problem 3.6

A risk analyst, Kari, is to make a judgment about the value of an unknown future observable quantity N. She knows that N is either 0, 1, 2, 3 or 4. To express her uncertainty about N, she assigns probabilities P(N = 0), P(N = 1), . . . , P(N = 4). Say that these probabilities are 0.80, 0.10, 0.05, 0.03 and 0.02, respectively. Explain the meaning of these probabilities. What is the expected value? Explain the meaning of this value. Is this value uncertain? Explain.

Problem 3.7

Present a 95% prediction interval (i.e., a prediction set) for N in Problem 3.6. Explain what the interval (set) means.

Problem 3.8

Consider a probability model $f(x) = P_f(X = x)$, x = 1, 2. Explain the meaning of f(x). Let p = f(1) (hence f(2) = 1 − p). Assume that p is unknown. Specify a subjective probability density to express your uncertainty (degree of belief) of p when you would like to reflect that any interval of the same length has the same probability.

Problem 3.9

In a safety context, two common risk metrics are PLL (potential loss of lives) and FAR (fatal accident rate). PLL is defined as the expected number of fatalities in a period of one year, whereas FAR is defined as the expected number of fatalities per 100 million exposed hours.

Suppose the PLL is 1/10,000. The number of exposed hours in a year is 1,000. Calculate a related FAR value. What is the 'individual risk' (IR) assuming 10 people being equally exposed? The individual risk is defined as the probability that a specific person is killed in a year.

Problem 3.10

A risk assessment presents risk using a risk matrix, with the two dimensions P(A') and E[C'|A'], where A' = event and C' = consequence/loss. Argue why a risk description based on such a matrix could be rather misleading.

Problem 3.11

A risk assessment presents risk using a risk matrix. Two scenarios are given the same loss-probability score, where probability is a knowledge-based probability. In the risk assessment, it is concluded that the risks related to these scenarios are of the same magnitude. What is the problem with this conclusion?

Problem 3.12

Two risk analysts discuss whether strength of knowledge (SoK) judgments are related to the knowledge supporting the probabilities P or the combination of consequences C' and P. Can you help them clarify the issue?

Problem 3.13

Tim plans a car trip from Paris to Rome. Introduce a frequentist probability p for an accident to occur during this trip. Express what p means. Is it a meaningful concept to use in risk analysis?

Problem 3.14

It is suggested to define risk by the expected disutility $-E[u(C')]$, where u is a utility function. Discuss the suitability of such a definition.

Problem 3.15

Explain what a black swan of the unknown known type means using the (A,C,U) – (A',C',Q,K) conceptualization.

Problem 3.16

Does the definition of a black swan used in this book capture the same ideas as the definition by Taleb?

Problem 3.17

The black swan metaphor is the most common metaphor for reflecting rare, surprising events. Another metaphor often referred to is 'perfect storms'. This storm was a result of the combination of a storm that started over the United States, a cold front coming from the north, and the tail of a tropical storm originating in the south. All three meteorological features were known before and occur regularly, but the combination is very rare. The crew of a fishing boat decided to take the risk and face the storm, but its strength was not foreseen. The storm struck the boat, it capsized and sank; nobody survived (Paté-Cornell 2012). This extreme storm is now used as a metaphor for a rare confluence of well-known phenomena creating an amplifying interplay leading to an extreme event (Glette-Iversen and Aven 2021).

Would you classify this event as a black swan?

Problem 3.18

Was the occurrence of the 2020 COVID-19 pandemic a black swan? Why?

Problem 3.19

Based on the criteria suggested in Section 3.2 to assess the strength of knowledge (SoK), suggest some possible ways that scoring systems can be defined for what is strong knowledge, medium strong knowledge and weak knowledge.

Problem 3.20

Is knowledge expressing the truth?

Problem 3.21

Look for some historical data comparing fatality rates for travel by plane compared to driving. Suppose a specific risk analysis based on some data showing that the FAR value for flying is the double of driving. Would you conclude that it is riskier to fly than driving? The FAR value is defined in Problem 3.9.

Problem 3.22

Give an example of a *known unknown* (events which we know will occur, but we do not know the form of the events and when they will occur).

Problem 3.23

In a class, let the students assign probabilities and SoK judgments for a specific event and present the scores using a scatter plot (see Figure 3.2).

Problem 3.24

Study current reports from the Intergovernmental Panel on Climate Change (IPCC) on climate change risk. How is risk described? Make comparisons with the recommendations of the present chapter. Comment on the meaning of the statement that it is extremely likely (at least 95% probability) that most of the global warming trend is a result of human activities (IPCC 2014).

Problem 3.25

Prepare a five-minute video presentation that explains the differences among classical, frequentist and knowledge-based probability. Post this video on social media with #whatisrisk and ask respondents to comment on the video. Prepare a written response that explains:

- Was it difficult to explain this concept? Why?
- Did the respondents understand your video? Were there parts that were poorly understood?
- What questions do you still have about these probabilities after making the video and receiving feedback?

Problem 3.26

A traditional risk matrix is based on presenting $(P(A_i), E[C|A_i])$, for events A_i, i = 1, 2, . . . , n. Present an alternative approach based on specifying a set of categories of consequences associated with the events. Why do you think many analysts prefer using this alternative approach?

KEY CHAPTER TAKEAWAYS

- Probability is used to reflect uncertainty, imprecision and variation.
- Knowledge-based (subjective) probabilities express uncertainties.
- A knowledge-based probability $P(A|K)$ is equal to 0.15 (say) if the assigner has the same uncertainty and degree of belief for A to occur as randomly drawing a red ball out of urn containing 100 balls, of which 15 are red, given the knowledge K.
- Imprecise probabilities express imprecision. They are interpreted similarly.
- An imprecise knowledge-based probability $P(A|K)$ is equal to an interval [0.10, 0.20] (say) if the assigner has the same uncertainty and degree of belief for A to occur as randomly drawing a red ball out of urn containing 100 balls, of which 10, 11, . . . or 20 are red – given K. The assigner is not willing to be more precise.
- Frequentist probabilities represent variation in infinite populations of similar situations.
- A frequentist probability $P_f(A)$ is interpreted as the limiting fraction of times A would occur when repeating the situations considered over and over again.

- Risk is described by (C',Q,K) and (A',C',Q,K), where A' is a set of specified events, C' some specified consequences, Q a measurement or description of uncertainties and K is the knowledge that Q and (A',C') are based on.
- Q is commonly represented by (P,SoK) where SoK are judgments of the strength of the knowledge supporting the probabilities P.
- Common risk metrics are expected values E[C'], probability distributions $P(C' \leq c)$, value at risk and combinations of P(A') and E[C'|A'] as in risk matrices. SoK judgments should be added to these metrics.
- To present (A',C',Q,K), it is useful in many cases to fix C' and show P(A') and SoK judgments.
- Vulnerability is described as (C',Q,K|A').
- Common vulnerability metrics are conditional expected values E[C'|A'] and probability distributions $P(C' \leq c|A)$. SoK judgments should be added to these metrics.
- A black swan is a surprising extreme event relative to one's knowledge. Different types of black swans are shown in Figure 3.3.

Risk assessment

CHAPTER 4

Basic theory of risk assessment

CHAPTER SUMMARY

This chapter discusses concepts and tools to conduct a risk assessment, which consists of methods that aim at improving our understanding of the risks considered and in this way supporting decision-makers when deciding what actions to take. A comprehensive risk assessment addresses several key aspects:

What can happen ('go wrong')? Leverage past data, expert opinions, trends, modeling and so on to identify potential events that could cause negative (and positive) consequences (outcomes). The task of the risk assessment team is to understand these risk events by distinguishing between risk sources, threats, hazards and opportunities. While this team cannot imagine or list out every possible outcome, a preliminary list is essential for the later steps. How to address potential surprises needs also to be addressed but represents a challenge.

Link risk events to consequences: Why and how the potential events (discussed previously) could result in negative or positive consequences. These consequences can be good or bad or have a mix of good and bad aspects.

Assess uncertainty: Acknowledge uncertainty and use a characterization of uncertainty to frame an understanding of what can and will happen, covering judgments about likelihood and strength of knowledge as explained in Chapter 3.

Evaluate the risk: Determine the significance of the risk (for example, expressing that the risk derived is exceeding what is typically accepted in society) and rank alternatives using relevant criteria. There is a leap from the risk assessment to the decision, referred to as the *management review and judgment*. It is defined as the process of summarizing, interpreting and deliberating over the results of risk assessments and other assessments, as well as of other relevant issues (not covered by the assessments), in order to make a decision. As the risk assessment has limitations in capturing all aspects of risk and there are other aspects than risk that are important for the decision-making, a management review and judgment is always needed.

DOI: 10.4324/9781003156864-4

As an example of a risk assessment, think of a study of a technical system like a nuclear power plant. The aim is to identify what can go wrong and the most critical risk contributors. This information can later be used to improve the system and reduce risks. As another example, think of a business owner who is considering some investments. A risk assessment is conducted in order to enhance the understanding of the risks associated with these investments. Based on the assessment, the business owner will have a better basis for what decision to make.

The main stages of a risk assessment include using the terminology introduced in Chapter 3, identification of events (threats/hazards/opportunities and risk sources A'), cause analysis to investigate how these events can occur, consequence analysis to study the effects (C') of these events, characterization of the risk (A',C',Q,K), study of alternatives and measures to modify risk and evaluation of the risk. A risk assessment comprises two main parts: risk analysis and risk evaluation. The risk evaluation is typically conducted by risk analysts. The risk assessment provides input to the decision-makers, who will conduct a management review and judgment and make a decision.

This chapter also discusses how to characterize a good risk assessment. We point to the importance of the identification stage and that the process of understanding risk is more about knowledge and lack of knowledge than producing probability numbers, in line with the insights provided in previous chapters. A risk assessment which does not include processes scrutinizing the knowledge and in particular the assumptions on which the quantitative analysis are based is not a high-quality risk assessment. Another critical point is the emphasis on vulnerabilities and resilience.

Models, including probability models, play an important role in risk assessments. Examples are provided on how to develop and use such models in risk assessment contexts.

LEARNING OBJECTIVES

After studying this chapter, you will be able to explain

- the basic ideas, purpose and stages of a risk assessment
- the difference between risk assessment, risk analysis and risk evaluation
- what characterizes a high-quality risk assessment
- the concepts of validity and reliability of a risk assessment
- how to treat uncertainties in risk assessments
- the importance of analyzing vulnerabilities and resilience to properly assess risk
- how to relate the risk characterizations of the risk assessments to knowledge
- how to address potential surprises
- how models are used in risk assessments
- the concept of model uncertainty in risk assessments

4.1 BASIC IDEAS

Figure 4.1 summarizes the main features of a risk assessment, covering events, causes and consequences. Risk assessment is a tool used to understand the risk related to an activity, evaluate the significance of the risk and rank or rate relevant options based on established criteria. A simple example will be used to illustrate the basic ideas.

Consider the following example: Tommy is a graduate student in a very competitive academic program. His professor has invited him to present a project to his class, professors in his department and industry professionals. Tommy is honored by this invitation but is hesitant to accept because he does not like to make oral presentations. He asks the professor if he could think about it for a couple of days. He would like to perform a risk assessment for the oral presentation to provide a basis for whether to take on this challenge.

Tommy first reflects on what it is that is important for him in relation to this talk. What are the values of concern for him? He quickly concludes that he would like to perform well and get positive feedback from the audience. Then he starts to think about the usefulness of doing an exercise like this to learn and be a better speaker. Maybe the presentation will not be 'perfect', but he could still benefit from it from a longer perspective. Based on these reflections, he decides not to have an overly ambitious goal for the talk, such as 'a highly successful presentation' and 'a brilliant talk'.

Next Tommy performs a high-level simulation of the presentation, to clarify what it will entail. Using risk terminology, the presentation is the *activity* to be assessed. From this simulation, he starts to think about what can go wrong – what events could occur that could negatively affect the presentation. He writes down a number of such events, including

1   He is so nervous that it affects the talk, and the audience is more concerned about his state than what he is saying

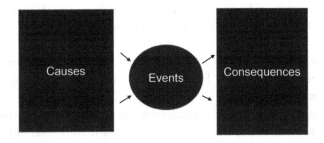

Risk assessment: Process to identify risk sources, threats, hazards and opportunities; understanding how these can occur and what their consequences can be; representing and expressing uncertainties and risk; characterize the risk; and determining the significance of the risk and rank alternatives using relevant criteria

FIGURE 4.1 The main features of a risk assessment, with its focus on events, causes and consequences

2   He forgets a key argument for a statement
3   He is not able to meaningfully answer a question from the audience

He acknowledges that he may have overlooked some important events in this list and considers using a systematic method to reveal others. He knows from the risk assessment course he has taken that there are many methods that can be used to identify events that could negatively affect his presentation. He decides to apply the *structured what-if technique* (SWIFT).

This method is based on using some guidewords, which need to be tailor made to the context considered. Some of the common guidewords are: time, place, amount, things, persons, ideas and process. Using these guidewords, and phrases like 'What if' and 'How could', Tommy is led to questions like

- What if the presentation is too long (or too short)? (wrong time)
- How could the presentation be too long (or too short)? (wrong time)
- What if the presentation is considered boring? (wrong ideas)
- How could the presentation be considered boring? (wrong ideas)
- What if I say something that offends one of the professors? (wrong ideas or persons)
- How could I offend one of the professors? (wrong ideas or persons)

Based on these questions, some undesirable events can be formulated. The process can also be used to generate positive events. For example, how could the presentation be considered interesting and exciting?

Tommy reviews these events and make a judgment about the severity of the consequences. The severity is considerably higher for some events than others, but it is difficult to make a judgment because the effects of the events could range from something rather minor to a quite serious consequence. Think, for example, about the event where Tommy is not able to provide a meaningful answer to a question. Depending on the type of question and what the response is, we can picture a spectrum of effects, some more serious than others. Tommy's focus is on events he thinks would be embarrassing, which is the largest or most severe consequence. He performs similar delineations for other events on his list, ensuring that for all events considered, the consequences are judged as large/severe.

Then Tommy considers the likelihood for these events to occur. He uses score categories: highly likely (higher than 50%), probable (10–50%), low probability (1–10%) and unlikely (less than 1%). Table 4.1 presents the results for some selected events, with associated strength of knowledge judgments. His judgments are based on the assumptions that he will carefully plan the talk and practice a lot. The assessment shows that he is most concerned about events 1, 5 and 6. For event 1, his judgment is that the nervousness should not be that strong if he prepares the talk very well. There is, however, an element of uncertainty. Sometimes he becomes very nervous for reasons he does not really understand, and he therefore assigns a weak score on the strength of knowledge for the low likelihood (unlikely) judgment. For event 5, he considers it rather likely that the talk will be perceived as very boring; he is confident about it, as he plans to use notes, and he is not normally able to improvise. For event 6, the problem is that one of the department professors is said to be very sensitive to what is communicated about some risk science topics, but that professor seldom attends presentations of this type. Thus, it is unlikely that this will be a problem, but the knowledge supporting this

TABLE 4.1 Risk characterization for Tommy presentation example

Events	Severity	Likelihood	Strength of knowledge	Comments	Risk
1 He is so nervous that it affects the talk, and the audience is more concerned about his state than what he is saying	High	Low probability– Unlikely	Weak	He will practice a lot	High
2 He forgets a key argument for a statement	High	Low probability	Medium strong	He will practice a lot	Moderate
3 He is not able to meaningfully answer a simple question from the audience	High	Low probability	Medium strong	He will practice a lot	Moderate
4 The presentation is far too short or long	High	Unlikely	Strong	He will carefully plan the talk and practice a lot	Low
5 The presentation is considered very boring	Moderate– High	Probable– Highly likely	Strong	He reads from a manuscript	Moderate– High
6 A professor is offended or provoked by what he says	High	Unlikely	Weak	One of the professors is very sensitive to what is communicated on specific risk science issues	Moderate– High

judgment is considered weak, as his talk may trigger this professor's interest – it is difficult to know.

The analysis shows the key features of a risk assessment commonly referred to as a coarse or preliminary risk assessment. Often the assessment is divided into sections covering different parts of the activity or subject studied. In this example, Tommy could distinguish between the opening part of the talk, the main part, the conclusions and the discussion part and then perform the analysis for each of these four stages.

Based on the given risk assessment, Tommy has obtained an improved basis for making a judgment about whether to accept the invitation to have this presentation. He reflects on the results of Table 4.1 – is the risk acceptable? The risk assessment does not provide a clear answer on that question, but it provides some input (refer to discussion in Section 4.2.3). And using the assessment, Tommy has an instrument for systematically examining measures that he could implement to reduce the risks, thereby giving him some control over these aspects. For example, he could simply accept the risk related to event 5, but he could also try to do something about it – practice so much that he could talk without notes. Then this particular event would no longer pose a severe risk. However, this risk could affect some of the others; presenting without notes could lead to him being more nervous, thereby affecting event 1.

For event 6, Tommy made a simple event tree model; see Figure 4.2. This type of tree is commonly used in risk assessments to analyze what can happen given the occurrence of an initiating event – or the activity – here that Tommy conducts his talk. Then two branches are introduced, responding to two questions. Depending on whether the answer is yes or no, three scenarios and consequences are derived. The worst case is that the sensitive professor attends the meeting and is provoked. If this professor attends the meeting but is not provoked, the consequences are not so serious for Tommy, but he would still perceive the situation as stressful because there would be uncertainty about whether this professor will be provoked. Tommy uses this model to structure his thoughts in relation to this event. The tree allows for probability calculations, but Tommy chooses not to conduct a detailed precise probabilistic analysis. Based on the rather weak knowledge he has about this professor, it is difficult to assign specific probabilities. However, he still uses probabilistic reasoning, as explained in the following.

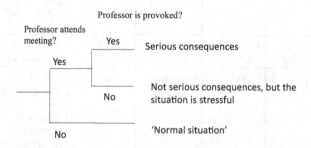

FIGURE 4.2 An event tree model of situations involving the sensitive professor

Let B_1 denote the event that the sensitive professor attends the meeting, B_2 that the professor is provoked and C_1 that the consequences become severe. Then probability calculus shows us that

$$P(C_1) = P(B_1) \, P(B_2 \mid B_1), \tag{4.1}$$

that is, the probability of C_1 is the product of the probability of B_1 and the conditional probability that the professor is provoked given that he is attending the meeting. It follows that, for example, if both $P(B_1)$ and $P(B_2 \mid B_1)$ are less than 10%, the probability of C_1 will be less than 1%. This number does not make Tommy relaxed, as he knows he has weak knowledge supporting such an analysis. Rather he seeks to improve the knowledge about what it is that provokes this professor. Tommy talks to several of his co-students, and it does not take very long to identify one major issue: This professor gets frustrated when hearing students and others referring to risk as expected values – probabilities multiplied by losses, which for this professor is a completely inadequate way of describing risk in most cases. Tommy is in line with this thinking, although he is surprised that someone can be provoked just because of that. With these new insights, the supporting knowledge of event 6 in Table 4.1 is adjusted to medium strong, still acknowledging that there could be other sensitive topics for this professor. The risk is judged as moderately high. Clearly, this professor can be considered a risk source for Tommy's talk.

Tommy then reflects on vulnerabilities and resilience. There could potentially be a poor response in case some type of disturbance, either the attendance of the sensitive professor or a difficult question. Tommy acknowledges that such disturbances represent a main risk contributor. He makes some plans on how he can practice to reduce the vulnerabilities and strengthen the resilience. If these plans can be properly implemented, the risks associated with events 3 and 6 would become small. A main element in this plan is to simulate a talk with his friends and practice answering surprising type of questions.

Despite all of the analyses conducted, Tommy is concerned that he has overlooked something of importance. He decides to also make an assessment specifically addressing potential surprises. He remember from the curriculum on risk assessment the method referred to as *red teaming*. It serves as a 'devil's advocate' by seeking to produce alternative interpretations and challenge established ideas and thinking. A friend of Tommy, Roger, is invited to represent the other perspective. Roger addresses many issues, for example, related to the sensitive professor. He questions if this professor is so terrible as the 'rumors' indicate and if there are examples showing that the professor is in fact generally a very nice and friendly person. Is there really truth in these rumors? Roger specifically challenges Tommy on the events for which a low risk has been assigned. He argues that the risk for having too long a talk is significant, as Tommy has planned to use quite a number of slides, and Roger claims that Tommy often uses far too much time on some detailed topics.

This analysis has a focus on what can go wrong. Tommy also makes an assessment reversing the focus, for example, considering the event that the presentation is perceived as very interesting and exciting. His initial judgment is that this event is unlikely, with a strong strength of knowledge basis, unless he makes some fundamental changes in the plans. He needs to refrain from using a manuscript or notes and also practice a lot to be relaxed and be an engaging speaker. Again, this may influence the assessment associated with other events. We see how the assessment can be used to systemize the available knowledge and judgments and help Tommy make the best decision for himself under the current circumstances.

As mentioned, the aim of a risk assessment is to improve the understanding of risk related to the activity considered, here the talk, and use this understanding to evaluate the significance of the risk and rank alternatives with respect to risk. A risk assessment can be seen as the sum of a risk analysis and a risk evaluation; we write

Risk Assessment = Risk Analysis + Risk Evaluation

The risk analysis covers the first part of the risk assessment, concerned with understanding risk. It covers the identification of risk sources, threats, hazards and opportunities; understanding how these can occur and what their consequences can be; representing and expressing uncertainties and risk; and characterizing risk. The previous example is mainly a risk analysis in this sense.

The risk evaluation is closely linked to judgments about the acceptability of the risk and the following risk treatment. However, the risk evaluation is conducted by risk analysts, and normally they are not the decision-makers. In the previous case, the analyst is also the decision-maker, which makes the difference between risk evaluation and the risk acceptability and treatment somewhat blurred here. In other cases, it is critical to separate the professional risk analysts' risk evaluation and the decision-makers' evaluation and decisions. Think, for example, about a city which contracts a consultant to conduct a risk assessment for its many activities and functions. The analysts perform a risk evaluation to compare the risk levels in this city with other cities and what are generally considered high or low levels for the type of risks considered. The analysts do not, however, conclude what are acceptable or not acceptable risks – that is for the bureaucrats and politicians to decide. We will return to this issue later; see Section 4.2.

Note that the term *risk analysis* is also sometimes used in a broader sense, as the totality of risk assessment, risk characterization, risk communication, risk management, risk governance and policy relating to risk in the context of risks that are a concern for individuals, public- and private-sector organizations and society at a local, regional, national or global level (SRA 2015); see also Appendices A and B. It will be clear from the context what the proper interpretation is.

REFLECTION

Is it important for Tommy to be strict on separating his professional risk evaluation and the risk acceptability and treatment (handling)?

Yes, as in this way, Tommy is able to separate what his professional judgment is about risk and what other concerns are, for example, perceptional factors like fear and dread. The issue will be discussed in more detail in Chapter 6.

4.1.1 Different types of risk assessments

There are many ways of categorizing risk assessments. One common classification is to distinguish between quantitative risk assessment, qualitative risk assessments and semi-quantitative risk assessments. In quantitative assessments, risk is quantified using probabilities and expected values. However, as we thoroughly discussed in Chapter 3, risk quantification should always be supplemented with qualitative strength of knowledge judgments, leading to a semi-quantitative assessment. A full risk characterization (A',C',Q,K) is by definition semi-quantitative or qualitative. In a qualitative risk assessment, risk is expressed qualitatively, without numbers. Tommy's assessment is mainly qualitative. The likelihood judgments are based on imprecise probabilities, and hence the assessment can be interpreted as semi-quantitative.

Another way of classifying risk assessment is to distinguish between data-driven risk assessment methods and model-based methods. The former category uses probability models, but the main element is data and statistics. The model-based methods are used when few data are available. Two examples will be presented in the following to illustrate the approaches.

A data-driven risk assessment: is smoking dangerous (risky)?

Today, there is broad agreement in society and among scientists that smoking is risky; however, it was only a few decades ago that the statement that smoking is dangerous was very much contested. In 1960, a survey by the American Cancer Society found that not more than a third of all US doctors agreed that cigarette smoking was to be considered "a major cause of lung cancer" (Proctor 2011). As late as 2011, research work conducted by the International Tobacco Control Policy Evaluation Project in The Netherlands showed that only 61% of Dutch adults agreed that cigarette smoke endangered non-smokers (Proctor 2011; ITC 2011).

The main sciences dealing with this issue are the medical and health sciences. Risk science and statistics have supporting roles, providing knowledge on what it means that smoking is dangerous or risky, that smoking causes lung cancer and how risk assessments should be conducted to conclude on such questions, taking into account all types of uncertainties. Risk science and statistics provide guidance on how to balance the two main concerns: the need to show confidence by drawing some clear conclusions (expressing that smoking is dangerous) and to be humble by reflecting uncertainties.

Standard risk assessment frameworks are used for these purposes, established by statistics and risk science. For example, a probability model may be introduced based on frequentist probabilities, expressing proportions of persons belonging to specific populations (for example, men of a specific age group) who get lung cancer. By comparing the probability estimates for non-smokers and for smokers, conclusions can be made. For example, let p_1 be the frequentist probability that an arbitrarily selected person in this population who is smoking (more than x number of cigarettes per day) gets lung cancer within a specific period of time, and let p_2 be the corresponding frequentist probability for a person who is not smoking. Based on random samples of persons within this population, estimates $(p_1)^*$ and $(p_2)^*$ are derived for p_1 and p_2, and comparisons can be made. Following statistical reasoning, it is concluded that smoking increases the cancer risk: p_1 is greater than p_2 if $(p_1)^*$ is sufficiently larger than $(p_2)^*$. What is sufficiently larger depends on the 'level' of the test. If the level is 1%, it means that if this type of test is performed over and over again, then in only 1 out of 100 cases on average, it is erroneously concluded that smoking increases the cancer risk; that is, it is concluded that $p_1 > p_2$, when in fact $p_1 = p_2$. We refer to textbooks in statistics dealing with hypothesis testing.

More refined analysis can be carried out introducing parameters of the probability models, representing, for example, the number of cigarettes per day and the duration of smoking; see Flanders et al. (2003) and Yamaguchi et al. (2000).

The statistical analysis may demonstrate that there is a correlation between two factors (here smoking and lung cancer), but that does not prove *causality*. A commonly referenced example is related to ice cream sales in a big city. The sales numbers correlate with the drowning rate in the city swimming pools, but this does not mean that there is a causality link between the two factors. The heat or temperature can explain the correlation. The temperature (heat) is an example of an unseen or hidden variable, also called a confounding variable. In general, we can say that for B to cause A (smoking to cause lung cancer), at a minimum, B must precede A, the two must covary (vary together) and there must be no competing explanation that can better explain the correlation between A and B. See Problem 4.11 for further discussion of the causality concept.

Another common framework for the statistical analysis is the Bayesian one, as briefly discussed in Section 3.3.2, in which epistemic uncertainties are represented by knowledge-based or subjective probabilities expressing degrees of beliefs. When new evidence becomes available, the probabilities are updated, using Bayes' formula. See Problem 4.12. A key quantity computed in this setup is the change in the probability that a person will get lung cancer given some new information about this person's health condition or other issues.

Model-based risk assessment: space exploration (partly based on Aven 2020d)

Consider the problem of assessing the risk for a spacecraft with a specific mission. To be concrete, think about the Apollo or Shuttle projects and plans for sending astronauts to Mars. When preparing for such flights, risk considerations play an important role. Risk science offers guidance on how to think in relation to risk and how to best assess the various risks. The problems are fundamentally different from those discussed previously for smoking, as relevant data and statistics are not available. Alternative analysis approaches and methods are needed. Basically, as mentioned previously, risk science offers three types of perspectives: quantitative, qualitative and a mixture (semi-quantitative), all based on models to represent the system and related processes. Models are needed, as experience in the form of observations of the performance of the spacecraft is not available in the planning phase.

In the Apollo program, probabilistic risk assessment (PRA) was used. This type of risk assessment is also referred to as quantitative risk assessment (QRA). It is based on answering the following three questions (the triplet of Kaplan and Garrick 1981):

- What can happen (i.e., what can go wrong)?
- If it does happen, what are the consequences?
- How likely is it that these events/scenarios will occur?

Scenarios are developed using events trees of the type shown in Figure 4.2. Normally the number of branches is considerably larger than two. For example, to analyze potential leakage from a liquid tank in a spacecraft, the tree could cover branches reflecting: Leak not detected? Leak not isolated? Damage to flight critical avionics? Damage to scientific equipment? Depending on the branch questions, different scenarios are developed. For each of these branches, normally a *fault tree analysis* is conducted aimed at identifying what combinations of failures could lead to this particular branch event occurring. Fault trees are discussed in Problem 4.6. Then, assigning probabilities to each failure event and branch, calculations of the probability for specific types of consequences can be carried out extending the computation shown in formula (4.1). For the Apollo project, focus was placed on the event of having success in landing a man on the moon and returning him safely to earth. PRAs are discussed in more detail in Chapter 5.

It is interesting to note that the use of PRA in the Apollo program was not continued. The Shuttle was designed without PRA; instead, qualitative approaches like *failure mode and failure effect analysis* (FMEA) were used. In an FMEA, we investigate what happens if the component fails for each component of the system studied. A probability judgment is made, and a classification is obtained by combining effect and likelihood categories; see Problems 4.4 and 4.5.

In relation to the Apollo PRA, considerable focus was on the numbers calculated. A probability of success in landing an astronaut on the moon and returning safely to earth at below 5% was indicated (Jones 2019; Bell and Esch 2018). For the NASA management, this number was considered dramatic and harmful for the project: It would be impossible to communicate to society a risk of that magnitude. The result was that, in relation to the Shuttle project, they later stayed away from PRA as a design tool. The judgment was that PRA overestimated the real risk. The result of the high judged risk numbers in relation to Apollo was that risk in that project was acknowledged as a serious problem, and measures were implemented to make improvements in all aspects of the project and design.

NASA management believed and testified to Congress that the Shuttle was very safe, referring to a 1 in 100,000 probability of an accident (Jones 2019). The justification for this number was, however, weak. NASA engineers argued for 1 in 100 and, following the loss of Challenger and more detailed assessments, the latter number was used.

Risk science at that time provided guidance on how to conduct the PRA. These analyses are quantitative, with probabilities computed for different types of failure events and effects using event trees and fault trees. A value of the PRA as important as the quantification is the improved understanding of the system and its vulnerabilities. The systematic processes of a PRA require that the analysts study the interactions of subsystems and components and reveal common-cause failures, that is, multiple component failures to a common source.

This case demonstrates the challenges of using numbers to characterize risk. At the time of Apollo, risk analysis was very much about PRA and quantification of risk using probabilities. Although the importance of gaining system insights was highlighted as mentioned previously, the numbers were considered the main product of the analysis, estimating the real risk level. The main goal of the risk analysis was to accurately estimate risk. If a failure probability of 0.95 was computed, it was interpreted as expressing the frequency of failures occurring when making a thought construction of many similar systems. Clearly, if such a frequency represented the true failure fraction, the project would not have been able to continue – it would have been too risky. Risk science explains, however, that this number does not express the truth, what will happen in the future, but is a judgment based on modeling and analysis, which could be supported by more or less strong knowledge. The actual frequency could deviate strongly from the one estimated or predicted. In this case, the knowledge basis was obviously weak, and the numbers should therefore not be given much weight. The fact that the analysis was also based on many 'conservative assumptions', leading to higher risk numbers than the 'best estimates', provided additional arguments for not founding the risk management only on the numbers.

At the time of these projects, a main thesis of risk science was that risk can be adequately described by probability numbers. More precisely, risk could be well characterized by the risk triplet, as defined by Kaplan and Garrick (1981) and referred to previously: events/scenarios, their consequences and associated probability. This perspective on risk is also commonly used today, but new knowledge has been derived since the 1980s. According to contemporary risk science, it is essential that the risk characterizations also cover the knowledge supporting these probabilities and judgments of the strength of this knowledge, as thoroughly discussed in Chapter 3. Of special importance here is the need to examine the assumptions that the probabilities are based on, as they could conceal aspects of risk and uncertainties and reveal potential surprises relative to the knowledge that the assessment is based on (see case in Bjerga and Aven 2016). The main aim of the risk assessment is not to accurately estimate the 'true' risk but to understand and characterize the risk.

NASA (Jones 2019) makes some interesting statements concerning the importance of risk analysis:

> Shuttle was designed without using risk analysis, under the assumption that good engineering would make it very safe. This approach led to an unnecessarily risky design, which directly led to the Shuttle tragedies. Although the Challenger disaster was directly due to a mistaken launch decision, it might have been avoided by a safer design. The ultimate cause of the Shuttle tragedies was the Apollo era decision to abandon risk analysis. . . . The amazingly favorable safety record of Apollo led to overconfidence, ignoring risk, and inevitable disasters in Shuttle. . . . The Shuttle was cancelled after the space station was completed because of its high risk. NASA's latest Apollo like designs directly reverse the risky choices of Shuttle. The crew capsule with heat shield is placed above the rockets and a launch abort system will be provided.
>
> (Jones 2019)

According to NASA, the experience with the Apollo and Shuttle projects suggests two observations:

> First, the most important thing is the organization's attention to risk. To achieve high reliability and safety, risk must always be a prime concern. Second, the risk to safety must be considered and minimized as far as possible at every step of a program, through mission planning, systems design, testing, and operations.
>
> (Jones 2019)

The message is clearly that what is needed is proper risk management and a good safety and risk culture. The investigations following the Shuttle disasters found a bad safety culture, leading to poor decisions. Risk assessments, like PRAs, are useful tools but alone will not help much if the culture and the leaders are not encouraging scrutiny and follow-up of all types of issues to enhance reliability and safety.

Jones (2019) gives also an illustrative simple example, showing the importance of proper risk assessment and management. A mission is often thought of as a chain of links, and success is believed to be ensured by giving priority to the weakest links, and improving others is considered wasted effort. However, such reasoning could be disastrous, as the overall probability of failure is basically determined by the sum of all the linked failure probabilities (see Problem 4.13). With many links, the overall failure probability could be high, even if each of the linked failure probabilities is small. Risk management needs to take this into account when seeking to control and reduce risk. Risk analysis and risk science provide this type of knowledge. They specifically help decision-makers use organization resources in the best possible way. If a big risk for one link is difficult and expensive to reduce, the same total risk effect could be achieved by improving a set of other links.

4.2 RISK ASSESSMENT STAGES

Figure 4.3 shows the main stages of a risk assessment, covering the planning of the risk assessment (determining the context), the risk analysis and risk evaluation and the use of the risk assessment.

4.2.1 The planning of the risk assessment – establish the context

To conduct a risk assessment, it needs to be properly planned. The key is to clarify the problem/issue and define the objectives of the assessment. In general, the main aim of risk assessment is to improve the understanding of risk concerning risk sources, hazards/threats/opportunities and related consequences, leading to a better basis for

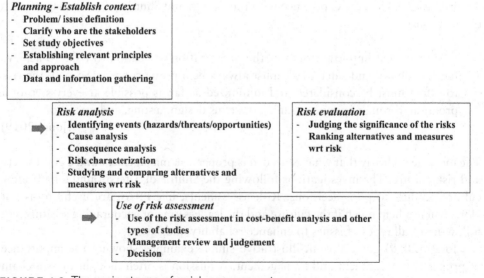

Planning - Establish context
- **Problem/ issue definition**
- **Clarify who are the stakeholders**
- **Set study objectives**
- **Establishing relevant principles and approach**
- **Data and information gathering**

Risk analysis
- **Identifying events (hazards/threats/opportunities)**
- **Cause analysis**
- **Consequence analysis**
- **Risk characterization**
- **Studying and comparing alternatives and measures wrt risk**

Risk evaluation
- **Judging the significance of the risks**
- **Ranking alternatives and measures wrt risk**

Use of risk assessment
- **Use of the risk assessment in cost-benefit analysis and other types of studies**
- **Management review and judgement**
- **Decision**

FIGURE 4.3 The main stages of a risk assessment process

answering questions like: What can go wrong? What are the main risk contributors? What is the effect of implementing a specific measure? This understanding is used to generate measures to modify (commonly reduce) risk and support decision-making on the choice of arrangements and measures. Risk assessments provide input to other types of analysis such as the cost-benefit type of studies, as well as judgments about what are acceptable and unacceptable risks.

Let us return to the smoking and spacecraft examples of Section 4.1.1. For the smoking example, the issue is about demonstrating that smoking is dangerous or risky. Earlier studies, observations and experience have clearly indicated that this is the case, but further studies are sought to strengthen the insights on the issue, for example, by showing how the risk depends on the level of smoking. Hypotheses are formulated and a statistical approach is chosen to test these hypotheses.

For the spacecraft example, the issue and related objectives are not so clear, as indicated by the discussion in Section 4.1.1. Very much of the focus was on accurately estimating a success probability, but it turned out that this type of objective was problematic, and a change was made: the aim of the assessment was to increase the understanding of the risks in order to support the design and operation of the spacecrafts. Different principles and approaches were considered and used for the assessment. For Apollo, PRA formed a pillar, whereas qualitative approaches were used for the Shuttle. The NASA example also shows the importance of clarifying who the relevant stakeholders are. It makes a big difference if risk numbers are used in-house as an instrument for identifying critical risk contributors or they are to be presented to US Congress demonstrating that the spacecraft is safe.

Data and information gathering is partly conducted in the planning phase and partly as an integrated part of the risk analysis. It is not always possible to see what data and information are needed at the early stages of the risk assessment process.

Risk assessments are conducted in different phases of a project. The differences in aims in scope and goals would typically be considerable. Think, for example, about a risk assessment in a planning or design phase of a technical system, like a spacecraft, compared to a risk assessment in the operational phase. When designing a system, there is normally considerable flexibility, and it is possible to choose among many different arrangements and measures. In the operational phase, however, the main system elements are fixed, and the possible changes relate mainly to operational, human and organizational factors. The risk assessment methods vary accordingly. Consider again the NASA case. The risk assessments discussed in Section 4.1.1 were all conducted for the design phase. As an example of risk assessment conducted in the operational phase, think about a risk assessment to be conducted for the operations of the international space station, highlighting workplace risks as a result of chemicals, gases, products, radiation and other types of exposures. Here actual testing and measurements can be carried out to support the assessments, leading to an improved understanding of what types of risk the astronauts face and the severity of these risks. While the differences in scope between risk assessment could be very large, the basic ideas are the same.

REFLECTION

When the stakes are high – there is a potential for severe consequences – should we always try to apply a model-based analysis of the form PRA (QRA)?

No, not in general. If the uncertainties about the system or activity studied are very large, such an approach would not be meaningful, as realistic models cannot be derived. In this case, a crude or preliminary risk assessment is more appropriate. However, if the uncertainties are not too large, a model-based analysis is attractive, as it allows analysts to study effects of changes and identify what the most important risk contributors are, even when no data are available for the activity considered. What type of approach or method to use of course also depends on the amount of resources one is able and willing to use on the assessment.

4.2.2 The risk assessment process

The risk assessment process covers two main stages, the risk analysis and the risk evaluation, as shown in Figure 4.3. We first consider the risk analysis. The bow-tie is commonly used to illustrate the key features of the risk analysis; see Figure 4.4.

The first step of the risk analysis is the *identification of events* A' (hazards, threats, opportunities). We remember the list produced by Tommy in the talk example of Section 4.1. Many analysts consider this step the most important one, as if we overlook an event, it cannot be further assessed with respect to risk. As mentioned in Section 4.1, many methods exist for this purpose. A common characteristic of all the methods is that they are founded on a type of structured brainstorming in which one uses guidewords, checklists and so on adapted to the issue or problem being studied; see also Section 5.7.

In the *cause analysis*, we study how the events A' can occur, what underlying events, risk sources and risk influencing factors may result in A'. Again returning to the Tommy talk example and starting from an event A' equal to 'poor response to a question', a cause analysis could point to factors such as a weak knowledge basis, stress and lack

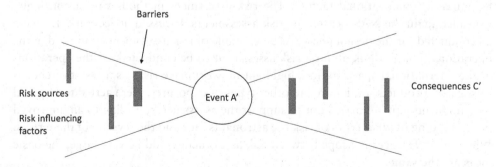

FIGURE 4.4 A schematic example of a bow-tie used in a risk analysis context

of skill in speaking. The barriers, as shown in Figure 4.4, may hamper the influence of these factors and potentially prevent the occurrence of event A'. Training and rehearsals of the talk are examples of such barriers. The quality of the barriers in preventing A' is a key element of the cause analysis.

To conduct a quantitative cause analysis, the common approach is to make a model of the event A' showing the relationship between A' and more underlying factors. A simple example is shown in Figure 4.5 for the Tommy talk example. According to this model, the event A' occurs if one of the (sub)events B1, B2 or B3 occurs.

Tommy assigns probabilities equal to 0.25, 0.10 and 0.01 for the three events B1, B2 and B3, respectively. From these probabilities, we see that the probability of event A' is about 0.36, derived by summing these three input probabilities. Alternatively, we can compute P(A') by the formula

$$P(A') = 1 - P(\text{Not } A') = 1 - [(1 - P(\text{not } B1))(1 - P(\text{Not } B2))(1 - P(\text{Not } B3))]$$
$$= 1 - (0.75 \cdot 0.90 \cdot 0.99)] = 0.33.$$

This formula is based on the assumption that the three events B1, B2 or B3 are independent. The formula expresses that for A not to occur, B1 must not occur, B2 must not occur and B3 must not occur.

The probabilistic analysis also needs to clarify the knowledge supporting these judgments, in particular when it comes to the assumptions made. Using the model, Tommy can analyze changes as a result of measures introduced, for example, on training and rehearsals. The model can be further developed by asking what is needed for each of events B1, B2 or B3 to occur. The method used is similar to a fault tree analysis; see Problem 4.6. Fault tree analysis is one of the most used methods in PRAs. Another common method used for this purpose is Bayesian networks; see Section 5.2.

The next stage is the *consequence analysis*, for which the effects C' of events A' are studied. The consequences can relate to different values, including lives, the environment or economic assets. Event tree analysis, as illustrated in Figure 4.2, is a common method used. The number of stages in the event sequence depends on the number of barriers in the system. These barriers aim at preventing the events from resulting in serious consequences. For each of these barriers, it is common to perform a barrier failure analysis, using, for example, a fault tree in line with the ideas of Figure 4.5.

Studies of system and barrier dependencies are an important part of the analysis. Think about a case where sensitive information is stored in a computer system behind

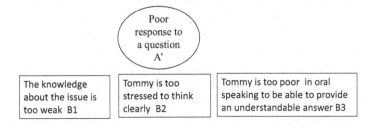

FIGURE 4.5 Simple model linking event A' and the subevents B1, B2 and B3

two password-protected security levels. The idea is, of course, that an unauthorized person shall not gain access to the information even if this person gets through one layer of protection. However, if the user applies the same password for both security levels, this extra protection is lost. There is dependency between the two barriers (the same password). An obvious measure would be not to allow the user to use the same password for both security levels.

Physicians often refer to *dose-response relationship* – it is a type of consequences analysis. Formulas are derived showing the relationship between varying dose levels and the average response. The dose here means, for example, the amount of drug that is introduced into the body or the amount of exposure from a risk source (e.g., radiation). Commonly, probability distributions are used showing the average responses as a function of specified dose levels. More refined analysis will also provide probabilities for the response for a fixed dose level, with strength of knowledge judgments supporting these probabilities. We may, for example, assign a probability of 20% that the effect or response will be a factor twice as high as the typical response value. Also, uncertainty intervals may be suitable for this purpose.

The consequence analysis may capture detailed studies of real-life phenomena, for example, how a fire would spread in a building or how a pandemic affects the economy. Risk scientists support this type of studies by providing suitable risk-based concepts, principles and methods.

For *risk characterizations*, we refer to Chapter 3. The risk characterization allows for studying differences between alternatives and the effects of changes, for example, potential measures to modify (reduce) risk. It also allows for the identification of important risk contributors. A common approach is to assess the difference in the risk characterization when removing a risk source or event. In this way, an *improvement potential* is identified. See Problem 4.8 and Section 5.1. Another similar approach is to adjust the assumptions that the analysis is based on to some extreme level to reveal the contribution from these assumptions.

It is also common to perform other types of sensitivity analyses on the basis of the risk characterization to see how sensitive the results are with respect to changes in the input quantities; see Section 5.1.

Finally, there is a need to discuss *risk evaluation*. It covers judgments of the significance of the risks, ranking of risk priority and measures with respect to risk. The risk evaluation is conducted by risk analysts; decision-makers are typically not involved in the risk evaluation process. When referring to *judgments determining the significance of the risk, comparisons are made with criteria or references made for the type of analysis*. For example, in industry, it is common to formulate some reference values for what is considered high-risk and low-risk numbers. These reference values are commonly referred to as risk tolerability and acceptance criteria and will be thoroughly examined in Chapter 8. These reference levels provide a basis for determining what are significant (high, low, acceptable, unacceptable) risks and what are not. As thoroughly discussed in Chapter 3, proper risk descriptions extend beyond probability-based risk metrics, which means that simple evaluations on the basis of comparisons with defined reference levels need to be used with care. A risk evaluation should not conclude what risk is acceptable or unacceptable. The risk assessment just provides insights based on a certain perspective – the decision-makers are informed by this perspective but also need to take into account other aspects, including the limitations of the risk analysis approach. The NASA example illustrates this. It would clearly be meaningless to let the Apollo project decision be based on whether the PRA showed that a specific probability number was met. The political support for the project needed a much broader and stronger foundation, covering both confidence in the technical feasibility of the project and an extreme desire to make a moment in history.

Similarly, national and international committees, such as those related to food safety and health issues, aim at providing risk evaluations on the basis of evidence and risk analysis without becoming political or including non-scientific issues. They provide professional judgments of the significance of the risks. However, in practice, this is an ideal which cannot be fully met, as these committees make conclusions about what is safe. That cannot be done without value judgments. The evaluation can build on values of the decision-makers (for example, how safe they require the food to be), but the processes are often challenging, as they depend on non-trivial judgments of uncertainties and interpretation of knowledge. The result is that the committees often operate in a 'no man's land' between science and policy and quickly become exposed to critique.

The risk evaluation also addresses ranking of alternatives and measures with respect to risk. The risk characterization provides the basis for this ranking, together with some criteria that have been established for how to conduct the ranking. An example would be a use of a type of cost-benefit analysis to compare options. We will discuss such criteria in Chapter 9. Again, the process is on the borderline between professional assessment and decision-making.

4.2.3 Use of risk assessment

As discussed in the previous section, the risk assessment informs decision-makers; it does not provide a clear answer on what to do. There is a leap from the risk assessment to the decision. We refer to this leap as management review and judgment, and it is defined as the process of summarizing, interpreting and deliberating over the results of risk and other assessments, as well as other relevant issues (not covered by the assessments), in order to make a decision. The management review and judgment is needed, as the risk assessment has limitations in capturing all aspects of risk, and there are other

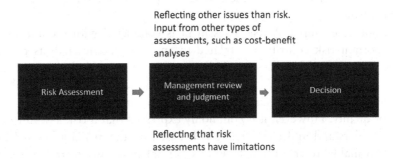

FIGURE 4.6 Illustration of the role of management review and judgment in risk decision-making

aspects than risk that are important for decision-making, as illustrated in Figure 4.6. Referring to the NASA example, the risk assessments (PRA) estimating a success probability for the moon project were not the only input to the decision to support the program. The limitations of the risk assessment in predicting the success rate and reflecting reality were acknowledged, as was the importance of other issues, like the prestige and importance of the United States being the first nation on the moon. Different methods exist for comparing options and evaluating the overall 'goodness' of an alternative or measure, including cost-benefit analysis. We will discuss these in Chapter 9.

Decision-makers do not in general have specific risk assessment competence. However, they are used to dealing with uncertainties and risk. In the management review and judgment, the knowledge supporting the findings is examined, in particular key assumptions made. Provided the risk assessment has been conducted in a professional way, giving sufficient weight to reporting and communicating the uncertainties and the knowledge basis, decision-makers have a proper basis for making the appropriate decision using relevant information and values.

4.3 QUALITY OF RISK ASSESSMENTS

In this section, we discuss what 'high-quality risk assessment' is. Is it sufficient to say that the quality of the assessment is high if the decision-maker and user of the risk assessment are pleased with the results and the analysis conducted? Is the quality of the assessment mainly determined by its ability to satisfy the decision-makers' expectations? Clearly this is not sufficient. Think about a situation where the risk assessment shows that the risk is very low compared to other similar type of activities, which are broadly accepted on the basis of calculations of expected losses. The decision-makers, who are not experts on risk science, may find the assessment and results trustworthy and solid. Yet the quality of the risk assessment can be poor, as seen from a risk science perspective. The decision-maker could be seriously misled by a risk characterization based on expected values only, as thoroughly discussed in Section 3.3.1. The quality always has to been seen in relation to what the current risk science knowledge is. The risk assessment analysts may be confident that they are in fact applying suitable risk assessment and management concepts, approaches, principles and methods, but this does not mean that this is actually the case, as the point of reference is the current risk science knowledge.

Take another example. A risk assessment is conducted with the aim of providing the important risk contributors related to an activity. The analysts perform the study, and it turns out that they overlooked a type of event A. This event has been shown to represent a serious risk in other comparable activities, but the analysis team lacked knowledge about this event. We would quickly conclude that the risk assessment was of poor quality. However, it is commonly expressed that a risk assessment should adequately understand and characterize risk based on the available knowledge. But what does 'available' refer to? Maybe in this example, it would be rather easy to gain the necessary knowledge about event A, but in other cases, it could be very resource demanding to obtain all relevant knowledge concerning a phenomenon or process studied. So where do we draw the line? And we can problematize this even further. Can the quality of a risk assessment be good if the knowledge available is very weak? Should the knowledge not be strengthened when it is poor? Yes: often this is done. Assessments cover activities to enhance knowledge, for example, through research projects, but in general, there will be limitations on what type of knowledge generation can be made within a specific risk assessment. It follows that the discussion of what constitutes a high-quality risk assessment needs to be seen closely in relation to what the knowledge supporting the assessment is.

Models play an important role in risk assessment, and their 'goodness' is critical. The Apollo PRA illustrates this clearly. Using comprehensive PRA models of the system in the design phase, a frequentist probability p^* for mission success (failure) was estimated. Comparing with the underlying 'true' unknown success (failure) rate p, we can speak about a model error $p^* - p$. We can refer to the uncertainty about the true value of this error as *model uncertainty*; see also Problem 3.4 and Section 10.3.1. An important topic of the quality of risk assessment discussion is model uncertainties. In this case, it was clearly very large. A model is by definition a simplified representation of the system; hence, there will always be a model error. The challenge is to ensure that it is sufficiently low, to nevertheless provide some new knowledge and be useful. The best guarantee for that is strong sciences understanding the phenomena considered. However, risk assessments also have a role to play to support decision-making when such understanding is not present; the uncertainties are large. Then it is particularly important that the limitations of the models be properly reflected. The use of knowledge strength judgments is a key instrument for doing this.

Some risk science guidance is provided by this book, but to be a highly skilled risk assessor, studies of fundamental risk science work are also needed. Examples of such work are included in the list of references. Practical issues for ensuring that risk assessments are appropriately planned, conducted and used are discussed in Chapters 10 and 11.

We refer to Problem 4.15 for a discussion of the terms 'validity' and 'reliability', which are closely linked to this discussions about the quality of risk assessment.

4.4 PROBLEMS

Problem 4.1

Accident data for an industry are reported. Would you see this report as a risk assessment?

Problem 4.2

The culture in a company is that risk assessments are to be conducted to satisfy regulatory requirements. Do you find this culture prudent?

Problem 4.3

Sketch a possible coarse or preliminary risk assessment for a road tunnel and related incidents (such as car accidents).

Problem 4.4

Figure 4.7 shows a tank that operates as a buffer storage for transportation of fluid from the source to the consumer. The fluid consumption level varies. An automatic system to control over-filling of the buffer storage works as follows: As soon as the fluid reaches a level 'normal high', then the level switch high (LSH) is activated and sends a closure signal to valve V1. The result is that the fluid to the storage then stops. If this barrier does not work and the fluid level continues to increase to an 'abnormally high level', the level switch high high (LSHH) is activated and sends a closure signal to valve V2. The result is that fluid to the buffer storage then stops. Simultaneously, the LSHH sends a signal to valve V3 to open so that the fluid is drained. The draining pipe capacity is higher than the capacity of the supply pipe.

  Illustrate the use of a failure mode and effect analysis (FMEA) by analyzing the components V1 and LSHH, addressing function, failure mode, effect on other units, effect on the system, failure probability, strength of knowledge and failure effect ranking.

Problem 4.5

Discuss the strengths and weaknesses of an FMEA.

Problem 4.6

The fault tree analysis (FTA) technique was developed by Bell Telephone Laboratories in 1962 when it performed a safety evaluation of the Minuteman Launch Control

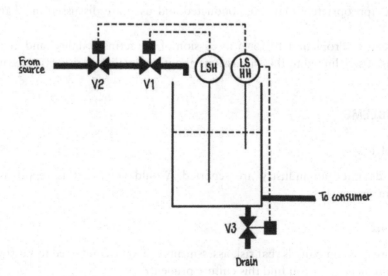

FIGURE 4.7 Storage tank example

System. The Boeing company further developed the method, and since the 1970s, it has become very widespread and is currently one of the most-used risk assessment methods for studying the reliability and safety of technical systems.

A fault tree is a logical diagram which shows the relation between system failure (also referred to as the top event) and failures of the components of the system, for example, as a result of technical failures or human error, using logical symbols.

AND gate: The output event (above) occurs if all input events (below) occur:

OR gate: The output event (above) occurs if at least one of the input events (below) occurs:

Construct a fault tree based on Figure 4.5.

For the tank example presented in Problem 4.4 and Figure 4.7, construct a fault tree for the top event 'overfilling of tank', based on the components V1,V2,V3, LSH and LSHH.

Problem 4.7

In a fault tree (and reliability block diagram), focus is on the minimal cut sets. A cut set is defined as a set of basic events which ensures the occurrence of the top event. A cut set is minimal if it cannot be reduced and still be a cut set. For the fault tree of the tank example in Problem 4.6, find the minimal cut sets. Why do you think we identify these sets?

Problem 4.8

Suppose the failure probabilities for the components V1, V2, V3, LSH and LSHH are 2%, 2%, 2%, 1% and 1%, respectively. What is then the probability that the top event will occur? Do you need to make an assumption to perform this calculation?

Compute an approximate probability of overfilling in one year, given that the fluid increases 25 times. Which of the components do you find to be most important? Why?

Problem 4.9

Draw an event tree starting from an initiating event A, with branches B1 and B2, that results in Y number of fatalities.

A: leakage, B1: ignition, B2: explosion, X: number of leakages.

If A, B1 and B2 occur, Y is either 2, 1 or 0, with probabilities 0.5, 0.3 and 0.2, respectively.

If A, B1 and Not B2 occur, Y is either 1 or 0, with probabilities 0.5 and 0.5, respectively.

If A and Not B1 occur, Y is equal to 0.

Assume P(B1|A) = 0.001, P(B2|A,B1) = 0.10.

Compute P(Y ≥ 1|A) and explain what this probability expresses in this case.

Problem 4.10

Event trees for the tank example can be drawn based on the initiating event 'fluid increases'. Specify possible branches for such trees.

Problem 4.11

With reference to the discussion in Section 4.1.1, how would you explain causality in relation to fault tree models (refer, for example, to the tank example discussed in Problem 4.6). Discuss to what extent the statistical analysis related to smoking proves that smoking causes lung cancer. Also discuss the concept of 'root cause' often referred to in the literature, expressing some type of basic cause that is the root or origin of the problem.

Problem 4.12 Use of Bayes' formula

Suppose a patient is tested when there is indication that the patient has a specific disease D. From general health statistics, we know that 2% of the relevant population is seriously ill from the disease, 10% is moderately ill and 88% is not at all ill from this disease. Suppose that the test gives a positive response in 90% of the cases if it is applied to a patient that is seriously ill. If the patient is moderately ill, the test will give positive response in 60% of the cases. In the case that the patient is not ill, the test will give a false response in 10% of the cases.

Now suppose the test gives a positive response. What is the probability that the patient is seriously ill?

Problem 4.13

Section 4.1.1 referred to the common misconception that a mission is thought of as a chain of links, and success is believed to be ensured by giving priority to the weakest links, and improving others is considered wasted effort. Explain why this is a misconception.

Problem 4.14

It is common to distinguish between a forward and a backward approach to risk analysis. The former starts with the events A' and analyzes the consequences C' following A'. The backwards approach, on the other hand, starts with some specified values or categories of C' – for example, only consequences with at least x number of fatalities – and studies how these specific values or categories can happen. Discuss the pros and cons of the two approaches.

Problem 4.15

When discussing the quality of risk assessments, the two terms *reliability* and *validity* are often used. The concept of reliability is concerned with the consistency of the 'measuring instrument' (analysts, methods, procedures), whereas validity is concerned with the success at 'measuring' what one sets out to 'measure' in the assessment. Discuss the meaning and applicability of these concepts in the context of risk assessment.

Problem 4.16

Refer to the local news in your area or university and make a list of five things that have 'gone wrong'. Refer to these as risk events that have actually occurred. Could each of these events have been avoided? Explain.

Problem 4.17

Suppose your university is debating a controversial decision: whether to close the university in response to a threat (pandemic, threat of violence, etc.). The university administration is asking for your help in organizing its risk assessment. To help, you have been asked to do the following: 1) Create a spreadsheet template for inputting information for a coarse risk assessment and 2) create a tutorial (document, slides or video) for the administration to use for inputting information and also making decisions using the template. Create this template and tutorial and share with your class.

KEY CHAPTER TAKEAWAYS

- A risk assessment is about improving the understanding of the risk studied in order to support relevant decision-making on how to handle the risk.
- The assessments can help us identify what might go wrong (or what can give positive outcomes), why and how; what the consequences are and how bad (good) they are.
- A risk assessment covers risk analysis plus risk evaluation.
- The risk analysis comprises the following main stages: identification of events (risk sources/threats/hazards/opportunities A'), cause analysis to investigate how these events can occur, consequence analysis to study the effects (C') of these events,

characterization of the risk (A',C',Q,K) and study of alternatives and measures that can be used to modify risk.

- The risk evaluation part makes judgments about the significance of the risk and ranks alternatives and measures based on comparisons with other risks and established criteria.
- Risk assessment informs decision-makers. There is a leap between risk assessment and decision-making. It is referred to as the management review and judgment.
- The management review and judgment takes into account the limitations of the risk assessments as well as concerns other than risk.
- A risk assessment is high quality if it is conducted in line with risk science concepts, principles, approaches, methods and models. It is not enough that relevant stakeholders be satisfied.
- Model error relates to the difference between model output and the underlying true value of the quantity considered. Model uncertainty is uncertainty about this value.

CHAPTER 5

More on risk assessment methods

CHAPTER SUMMARY

This chapter is a continuation of Chapter 4, presenting and discussing methods that can be used to assess and understand risk. It focuses on some of the most commonly used methods belonging to several categories of analysis, including sensitivity analysis, networks, simulation, probabilistic risk assessments (PRAs) and methods for assessing vulnerability and resilience.

A key element of any risk assessment is sensitivity analysis, showing how the risk results depend on its input, in particular assumptions made. The chapter reviews basic ideas and theory concerning such analysis and explains the difference between sensitivity analysis and scenario analysis. It also reviews related theory on importance analysis, which aims at identifying what elements of the system or activity studied are most important, for example, which contribute the most to risk.

The chapter also provides a brief introduction to Bayesian networks and Monte Carlo simulation, which are two well-established and commonly used methods to assess and compute risk. Both methods are based on models representing the activity or system to be analyzed.

Another topic of the chapter is probabilistic (quantitative) risk assessment. The discussion in the previous chapter concerning this type of assessment in relation to NASA is followed up on by providing more in-depth reflections of what the basic ideas of this method are, as well as challenges related to their use.

Vulnerability and resilience are fundamental concepts and aspects of risk, and the chapter also provides some further guidance on methods that can be used for studying these concepts. The COVID-19 case is used to illustrate some of the discussions. Applying the general (C',Q,K|A') vulnerability characterization, a classification system is introduced showing the multidimensionality of the vulnerability concept and the importance of approaching vulnerability from different perspectives and using different metrics. Indicators for assessing resilience are also discussed.

Finally, we reflect on methods for event identification, in general and in particular related to opportunities.

DOI: 10.4324/9781003156864-5

LEARNING OBJECTIVES

After studying this chapter, you will be able to explain

- the basic ideas of sensitivity and importance analysis, scenario analysis, Bayesian networks, Monte Carlo simulation and probabilistic (quantitative) risk assessment
- how vulnerability and resilience assessments can be conducted to improve the understanding of vulnerabilities, resilience and risk
- how hazard and threat identification can be used to identify opportunities

The chapter partly builds on Aven (2010, 2015b, 2017a, 2017b) and Thekdi and Aven (2021a).

5.1 SENSITIVITY AND IMPORTANCE ANALYSES

A sensitivity analysis in a risk assessment context is a study of how sensitive the risk characterization or risk metric is with respect to changes in conditions and assumptions made. Figure 5.1 shows an example of a sensitivity analysis of the calculated expected profit of a company as a function of the price of the key commodity produced by the company. In the risk assessment, the price was initially set to $100 per unit, but the price is subject to large uncertainties, and it provides insights to show how the results depend on this price. The actual price could deviate strongly from the initial assumption of $100.

A scenario analysis is not the same as a sensitivity analysis, but they are related. The sensitivity analysis shows changes in the risk characterization or risk metric due

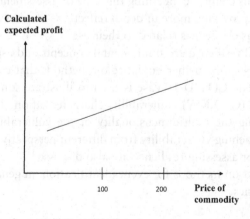

FIGURE 5.1 Example of a sensitivity analysis

to changes in input quantities, for example, an assumption. If risk, (C,U), is described by (C',Q,K), for example, using a metric P(C' > c) or E[C'], the sensitivity analysis may show how this characterization and these metrics depend on varying some assumptions in K, as illustrated in Figure 5.1.

A scenario analysis is a study of possible future states and development paths of the activity considered (Chermack 2011). For example, in the previous business case, three scenarios could be considered: the base case ($100), a 'worst case' of $10 and a 'best case' of $200. The analyst chooses which scenarios to study, acknowledging that some scenarios may be more likely than others and also acknowledging that cases may represent speculations about future conditions. The best case is not really the best case but a chosen value which is better than the base case and still considered somewhat reasonable – the term *plausible* is sometimes used. The precise meaning of this term is, however, not clear; refer to discussion in Problem 5.14.

Other types of scenarios could also be defined, for example

- A particular political party wins some critical amount of power in an election and radically changes the operating conditions for the company
- Another company releases a new technology
- Currency x strengthens strongly

In a risk assessment, these scenarios can be viewed as risk-influencing factors or events. If C and C' relate to the profit of the company, the uncertainties concerning these events are important to consider in order to assess the risk, although a probabilistic quantification is not used. The assessment may be founded on assumptions formulated on the basis of such scenarios. Possible deviations from these assumptions should then be addressed; refer discussion in Section 10.3.1. Sensitivity analysis could be used to study effects of changes in these assumptions on the company profit.

In scenario analysis, there is no search for completeness and characterizations of the uncertainties and risks, as in risk assessments. In the case of large uncertainties and a lack of accurate prediction models, the generation of such scenarios may provide useful insights about what could happen and potential surprises. Deductive (backwards) scenarios are of particular importance in this respect, where we start from a future imagined event/state of the total system and question what is needed for this to occur. System thinking, which is characterized by seeing wholes and interconnections, is critical if we are to identify potential surprises, as highlighted by many scholars of accident analysis, organizational theory and the quality discourse (Turner and Pidgeon 1997; Deming 2000).

A sensitivity analysis is not an uncertainty analysis, as a sensitivity analysis does not include any assessment of the uncertainties about the input (price in the previous example). The sensitivity analysis provides a basis for an uncertainty analysis with respect to the commodity price. If a probability distribution is assigned for the commodity price, we can use the sensitivity analysis to produce an unconditional expected profit. Alternatively, the uncertainty assessment could be a qualitative assessment of the commodity price, and the results are reported together with Figure 5.1.

To illustrate this discussion, we consider a model for analyzing the reliability of a system which comprises two units (labeled 1 and 2) in parallel (see Figure 5.2). The system may represent the reliability of a safety barrier in a process plant or the success or failure of an investment in two projects, where one project success is needed to ensure a successful investment. The reliability of the system h is given by

$$h(p) = 1 - (1 - p_1)(1 - p_2),$$

where $p = (p_1, p_2)$ and p_i is the reliability of unit i, i = 1,2; that is, p_i is the probability that unit i is in the success state 1. To simplify the analysis, it is assumed that the units are independent. Suppose that the reliability assessment produces values $p_1 = 0.60$ and $p_2 = 0.90$, which gives a system reliability h = 0.96. To study how sensitive the system reliability is with respect to variations in the unit reliabilities, a sensitivity analysis is conducted. Figure 5.3 shows the results. The upper line shows the system reliability h when the unit reliability of component 1 varies from 0 to 1 and unit 2 has reliability 0.90. The lower line shows the system reliability h when the reliability of unit 2 varies from 0 to 1 and unit 1 has reliability 0.60. We see that that relative improvement in system reliability is largest for unit 2. Hence, a small improvement of unit 2 gives the largest effect on system reliability. The relative increase is expressed by the slopes of the lines, which equal 0.1 and 0.4, respectively.

This information can be used in different ways, for example:

1   It shows that the input '$p_2 = 0.90$' is more critical than '$p_1 = 0.60$' in the sense that deviations from these values would give the strongest effect for unit 2.
2   It shows that a small improvement of unit 2 would give the strongest effect on the system reliability.

FIGURE 5.2 A parallel system of two units

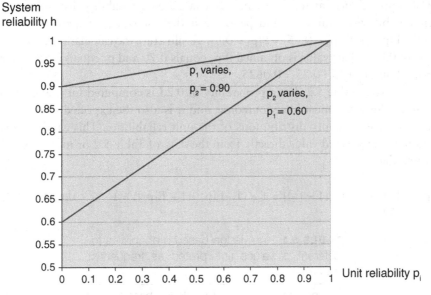

FIGURE 5.3 Sensitivity analysis with respect to the unit reliabilities for the reliability function $h(p) = 1 - (1 - p_1)(1 - p_2)$ (based on Aven 2010)

In this manner, the sensitivity analysis is being used as a tool for ranking the units with respect to criticality and importance. A number of such criticality and importance measures exist. In the previous analysis, the focus is the partial derivative $\delta h/\delta p_i$ of the object (here reliability) function, that is, the relative change in $h(p)$ relative to a change in p_i. In a reliability context, the measure is referred to as the Birnbaum importance measure (Birnbaum 1969).

Another common importance measure generated by the sensitivity analysis is the *improvement potential* (Aven and Jensen 2013), which is defined by the change in the object function when p_i is set to its maximum value, 1. In the previous case, this leads to the same ranking of the units, as both have the potential to make the system perfectly reliable. However, in most situations, this would not be the case, and the measure provides useful information about where the largest improvements can be obtained. See also Problem 5.2.

5.1.1 Sensitivity analysis in the context of uncertainty analysis

In the previous analysis, we did not clarify the interpretation of the p_is. They could either be knowledge-based probabilities or frequentist probabilities. In the latter case, they are assumed known; however, in practice, they would be unknown and subject to uncertainties. Suppose that is the case and a probability distribution is assigned for the p_is as shown in Tables 5.1 and 5.2. Thus we assume that p_1 can only take one of the values 0.5, 0.6 or 0.7 and p_2 only one of the values 0.85, 0.90 and 0.95.

The analyst's knowledge-based probabilities for these p values are 0.25, 0.50 and 0.25 respectively.

Using probability calculus, we can then calculate a probability distribution for the system reliability; see Table 5.2. Independence in the assessment of p_1 and p_2 is used, for example, $P(p_1 = 0.6 \mid p_2 = 0.95) = P(p_1 = 0.6)$. To illustrate the calculations, consider the h value of 0.925. This value is obtained if p_1 equals 0.70 and p_2 equals 0.85, and hence the probability is $0.25 \cdot 0.25 = 0.0625$.

We see from Table 5.2 that a probability of 0.25 is computed for h equal to 0.96. The value 0.96 is the most likely outcome, and it is also the expected value of h with respect to the uncertainty distributions of the unit reliabilities. This can be shown by computing the expected value directly from the data of Table 5.2 or more generally by observing that

$$E[h(p)] = E[1 - (1 - p_1)(1 - p_2)] = 1 - (1 - E[p_1])(1 - E[p_2]) = 1 - (1 - 0.6)(1 - 0.90)$$
$$= 0.96.$$

TABLE 5.1 Knowledge-based probabilities for different p_i values (interpreted as frequentist probabilities)

p_1 values	Knowledge-based probabilities
0.5	0.25
0.6	0.50
0.7	0.25
p_2 values	Knowledge-based probabilities
0.85	0.25
0.90	0.50
0.95	0.25

TABLE 5.2 Knowledge-based probabilities for the system reliability based on the input from Table 5.1

h values	Knowledge-based probabilities
0.925	0.0625
0.940	0.125
0.950	0.125
0.955	0.0625
0.960	0.250
0.970	0.125
0.975	0.0625
0.980	0.125
0.985	0.0625

The underlying system reliability could, however, deviate from the expected value. Based on the probability distributions of p_1 and p_2, we observe that the probability distribution of h has all its mass in the interval [0.925, 0.985]. The variance and standard deviation are measures of the spread of the distribution of h and take the following values (see Problem 5.1):

Var[h] = 0.000256 and SD[h] = 0.0160.

We have conducted an uncertainty analysis of the system reliability. Formally, *an uncertainty analysis refers to the determination of the uncertainty in analysis results* (here h) *that derives from uncertainty in analysis inputs* (here uncertainty about p_1 and p_2) (Helton et al. 2006). Using the general risk notation introduced in this book, we can formulate the uncertainty analysis problem as follows:

To analyze a quantity C', we introduce a model g(X), which depends on the input quantities X ($X_1, X_2, \ldots$), and set C' = g(X). An uncertainty analysis of C' means an assessment of the uncertainties about X and an uncertainty propagation through the model g to produce an assessment of the uncertainties about C'. Knowledge-based probabilities are used to express or characterize these uncertainties, accompanied with strength of knowledge judgments. In such a context, it is also common to perform sensitivity analyses, as will be discussed in the following.

A sensitivity analysis in this context refers to the determination of the contributions of individual uncertainty analysis inputs to the analysis results (Helton et al. 2006), or as formulated by Saltelli (2002): sensitivity analysis is the study of how the uncertainty in the output of a model can be apportioned to different sources of uncertainty in the model input.

Let us return to the previous system reliability example. As a measure of sensitivity of the input distribution for unit i, we may use the variance, here denoted $v_i(p_i) = $ Var[h| p_i]. The problem is, however, that the frequentist probability p_i is unknown. It will hence seem sensible to use the expected value, $E[v_i(p_i)]$, instead. Calculations show (see Problem 5.4) that using this measure, the uncertainty distribution of unit 1 has a stronger influence on the variation of the output result than the uncertainty distribution of unit 2. Thus, if we are able to reduce the uncertainties of p_1, this would have a larger effect on the uncertainties of h than reducing the uncertainties of p_2.

An alternative sensitivity measure (importance measures) is the correlation coefficient ρ between p_i and h, defined by $\rho(p_i, h) = $ Cov(p_i, h)/(SD[p_i] SD[h]); see Problem 5.5.

A common approach for showing the sensitivities of the component input to the output is to use scattterplots, as in Figures 5.4 and 5.5. To produce these plots, we used revised distributions for p_1 and p_2, with the probability values of Tables 5.1 and 5.2 replaced by intervals (for example, $p_1 = 0.5$ replaced by $0.45 < p_1 \leq 0.55$) and using a uniform distribution for this interval. Hence the knowledge-based probability that p_1 is less than 0.47 equals 2/10 of 25%, that is, 5%. By Monte Carlo simulation (see Section 5.3), drawing n random numbers on an Excel sheet, we obtain n values for p_1, p_2 and h, as shown in Figures 5.4 and 5.5. We see from these plots that p_2 and h are much more dependent than p_1 and h, as was to be expected from the previous study.

System reliability h

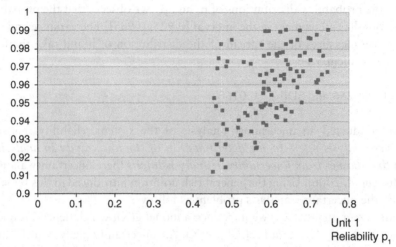

FIGURE 5.4 Scatterplot for p_1 and h (based on Aven 2010)

System reliability h

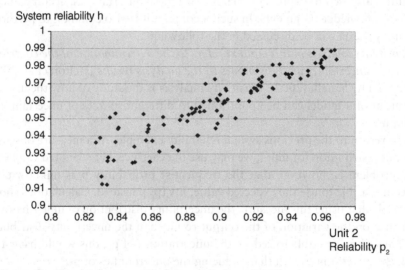

FIGURE 5.5 Scatterplot for p_2 and h (based on Aven 2010)

Many other techniques exist for sensitivity analysis and importance measures; see Problems 5.2 and 5.5 and Helton et al. (2006), Vose (2008), Saltelli et al. (2008), Aven and Nøkland (2010), Nøkland and Aven (2013), Aven and Jensen (2013) and Borgonovo and Plischke (2016).

5.2 BAYESIAN NETWORKS

Suppose a machine operating in a harsh environment is observed not to be working. This breakdown is either because of an external factor or some inherent engine defect.

The network in Figure 5.6 models the causal links. The network consists of three nodes: engine defect, external factor and machine not working. As a simplification, each node has two states only: the engine is either defective or not, it is impacted by some external factors or not and it is either working or not. We see from the arrows in the Bayesian network that engine defect and external factor are possible causes of the machine not working. It is common to use the terms 'parent' and 'child' for the two different levels of nodes in the network. For the three nodes in the example, we will thus say that 'engine defect' and 'external factor' are the parent nodes for 'machine not working'.

Bayesian networks are commonly used for quantitative analysis. They require that we specify a set of conditional probabilities. This is often done in tables called conditional probability tables (CPTs). The probabilities that the machine is not working given various combinations of engine defect and external factor are given in Table 5.3.

The probabilities in Table 5.3 could be based on available experience data or determined by expert judgments. All probabilities in the table are conditioned on the state of the parent nodes. In addition, we need to specify the unconditional probabilities for external factor and engine defect. Let us assume a probability of 0.10 is assigned for both of these events.

The task is now to calculate the probability that the engine is defective, given that we have observed that it is not working or, in other words,

P(engine defect | machine not working).

To find this probability, we make use of Bayes' formula. To simplify, we introduce the events A, B and C, expressing engine defect (A), external factor (B) and machine not

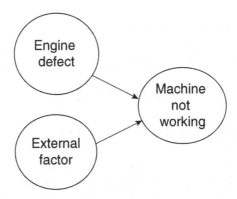

FIGURE 5.6 Example of a Bayesian network

TABLE 5.3 Conditional probability table for example, shown in Figure 5.6

	External factor		No external factor	
	Engine defect	Engine not defect	Engine defect	Engine not defect
Machine not working	0.95	0.85	0.90	0.02
Machine working	0.05	0.15	0.10	0.98

working (C). The complementary events are denoted not A, not B and not C, respectively. The task is to compute P(A|C).

Bayes' formula gives

$$P(A|C) = P(C|A) \, P(A)/P(C), \tag{5.2}$$

where P(A) has been assigned the value 0.10. Thus, it remains to determine P(C|A) and P(C). Let us first consider P(C|A).

The arrows in the Bayesian network show that in addition to being dependent on A, the event C is also dependent on B. Using the law of total probability, it follows that

$$P(C|A) = P(C|A,B) \, P(B|A) + P(C|A, \text{not } B) \, P(\text{not } B|A).$$

Assuming independence between the events A and B (in line with the network model shown in Figure 5.6), we obtain

$$P(C|A) = P(C|A,B) \, P(B) + P(C|A, \text{not } B) \, P(\text{not } B) = 0.95 \cdot 0.10 + 0.90 \cdot 0.90 = 0.905.$$

Hence, according to Bayes' formula (5.2), we have P(A|C) P(C) = P(C|A) P(A) = 0.905 · 0.10. Similarly, we obtain P(not A|C) P(C) = P(C| not A) P(not A)= 0.103 · 0.90, as

$$P(C| \text{not } A) = P(C| \text{not } A,B) \, P(B) + P(C| \text{not } A, \text{not } B) \, P(\text{not } B)$$
$$= 0.85 \cdot 0.10 + 0.02 \cdot 0.90 = 0.103.$$

By summing P(A|C) P(C) and P(not A|C) P(C), we obtain

$$P(C) = 0.905 \cdot 0.10 + 0.103 \cdot 0.9 = 0.1832.$$

We can then compute the desired probability:

$$P(A|C) = P(C|A) \, P(A)/P(C) = 0.905 \cdot 0.10/0.183 = 0.494.$$

Thus, there is a 49% probability that the engine is defective when it is observed that the machine is not working.

Bayesian networks can be used for many types of situations and applications, for example, in medicine, assisting in making diagnoses. A Bayesian network model for the relationship between various symptoms and analysis results is established by experts within the profession. Then, other physicians may submit analysis results and symptoms for individual patients into the model (thus locking some of the nodes) and calculate the probability of the patient having a disease or being healthy.

Another example is financial modeling to support credit assessments of customers. Factors that are deemed to influence capacities to pay, such as age and income, are reflected by the model. In dialogues with customers, individual nodes are locked, the model is updated and a probability of the customer not being able to pay within a given period is computed.

Strength of knowledge judgments can and should be added to the probabilistic analysis, as discussed in Chapter 3. For further discussions of Bayesian networks; see, for example, Weber et al. (2012) and Fenton and Neil (2018).

5.3 MONTE CARLO SIMULATION

Monte Carlo simulation represents an alternative to analytical calculation methods. The idea is to establish a computer model g of the activity or system to be analyzed and then to simulate the activity or the operation of the system for a specific period of time. The model g links a set of unknown input quantities $X = (X_1, X_2, \ldots X_n)$ to an output performance measure C', such that $C' = g(X)$. Probability distributions are specified for the X_i values. By sampling values from these distributions, observations are produced for C' using the model g. By repeating the sampling a large number of times, an accurate estimate of the probability distribution of C' is derived.

Consider the energy flow system in Figure 5.7. The system has three components, 1, 2 and 3, having as default design capacities 50, 50 and 100, respectively. In the case of component failure, the capacity is zero. Hence, the flow is 100 under conditions when all three components are working. If, say, component 1 fails, the system capacity is reduced to 50. We assume the flow is at the maximum level at any time. Let $X_i(t)$ denote the capacity of component i (i = 1,2,3) at time t, $t \geq 0$. Then the system capacity and flow at time t, denoted C'(t), can be expressed as (by the so-called max flow min cut theorem, Ford and Fulkerson 1956):

$$C'(t) = \text{minimum}\{X_1(t) + X_2(t), X_3(t)\}. \tag{5.3}$$

We assume that the capacity states of each component follow a process, as indicated in Figure 5.8, of consecutive 'uptimes' $T_1, T_2, \ldots$ and 'downtimes' (due to failure and repair) $D_1, D_2, \ldots$. The uptimes have a probability distribution F, for example, an exponential with mean μ, and the downtimes a distribution H, for example, a triangular distribution with mean τ. The Monte Carlo simulation is based on randomly drawing numbers from these distributions for each component and then calculating the flow C'(t) using formula (5.3). To simplify, we assume that the components are independent

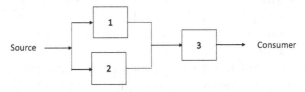

FIGURE 5.7 Energy flow network from a source to consumer

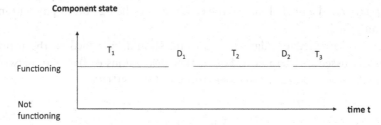

FIGURE 5.8 Simulated state process for a component, with consecutive uptimes T_1, T_2, . . . and downtimes D_1, D_2, . . .

of each other. Repeating the life of the system over a specific period of time, we can derive an accurate estimate of the probability distribution of $C'(t)$ for a fixed point in time t by computing the fraction of times $C'(t)$ takes the value 100, 50 or 0. Other metrics can also be derived, such as the total flow in the period considered [which equals the integral of $C'(t)$ in this period] or the fraction of time the flow is 100.

The operator of the system may use this model to study the effect of changes, for example, increasing the capacities of components 1 and 2. By comparing the results for different capacities and relating these to costs (refer to Chapters 8 and 9), a basis is made for choosing the best design alternative.

With a Monte Carlo simulation model, the time aspect and component dependencies are more easily handled than with an analytical method. A Monte Carlo simulation model may be a fairly good representation of the real world. The model of Figure 5.7 is simple, but it can be extended to include more components and different types of dependencies between the components. Think about an example where components may fail due to a common source or only one component can be repaired at the time.

Monte Carlo simulation generally requires detailed input data. For example, the uptime and downtime distributions must be specified. Mean (expected) values, as used in many analytical models, are not sufficient. On the other hand, the output from a Monte Carlo simulation model is very extensive and informative.

Developing a Monte Carlo simulation model could be resource demanding, and for some systems, the simulation times could be extensive and costly. To obtain accurate results using simulation, a large number of trials is usually required, especially when the system is functioning most of the time. The time and expense aspects are important if the model is to be used to study effects of changes in system configurations or if sensitivity analyses are to be performed. Fortunately, computers are constantly improving, and long simulation times have become a considerably less serious problem today than they were some years ago.

For comprehensive Monte Carlo simulation models, it is always a challenge to verify that the program has been written correctly and, therefore, if the result can be relied upon. Extensive testing is required to validate the model.

For further details on Monte Carlo simulation, see Zio (2013).

5.4 PROBABILISTIC RISK ASSESSMENT (QUANTITATIVE RISK ASSESSMENT)

For about 50 years, probabilistic analysis has been used as the basis for the analytic process of risk assessment of engineering systems, such as nuclear plants, spacecraft and chemical plants (see Rechard 1999, 2000; Apostolakis 2004, 2006; Meyer and Reniers 2013). The common term used is probabilistic risk assessment, also referred to as quantitative risk assessment. Its first application to large technological systems (specifically nuclear power plants) dates back to the early 1970s (NRC 1975).

A PRA/QRA answers the three basic questions (Kaplan and Garrick 1981; refer to Section 4.1.1)

- What can happen (i.e., what can go wrong)?
- If it does happen, what are the consequences?
- How likely is it that these events/scenarios will occur?

Following Apostolakis (2004), it is a top-down approach that proceeds as follows:

1   A set of undesirable *end states* (adverse consequences) is defined, for example, in terms of system failures, economic loss or fatalities.
2   For each end state, a set of disturbances or deviations from normal operation is developed that, if uncontained or unmitigated, can lead to the end state. These are called *initiating events* (IEs). For a process plant, such events may cover different sizes of leakages in different areas. It is critical that the IE be disjunct (non-overlapping) and not too numerous to obtain an overview and structure of the analysis.
3   *Event* and *fault trees* or other logic diagrams are employed to identify sequences of events that start with an IE and end at an end state. In this way, *accident scenarios* are generated. These scenarios include hardware failures, human errors and natural phenomena. The dependencies among failures of systems and components (common-cause failures) receive particular attention using inter alia various types of reliability methods. The number of steps in the event sequence and scenario depends on the number of barriers in the system. These barriers aim at preventing the events from resulting in serious consequences. Examples in a process plant include shutdown of processes to reduce the size of release and deluge activation to extinguish fire. The performance of these barriers is studied by addressing features like reliability, availability and robustness. Also, influencing factors and scenarios leading to the initiating events are commonly studied.

    Scenario development relies on understanding physical phenomena and processes. For example, in a process plant, the analysis needs to clarify the behavior of a gas dispersion. For this purpose, gas dispersion models that simulate the gas under various conditions, leak rates and so on will be used. Next, questions will be raised concerning potential combustible mixtures, then how a possible fire will develop and if the ignition produces an explosion.

4   The probabilities of these scenarios are evaluated using all available evidence, primarily past experience and expert judgment.

5   The accident scenarios are ranked according to their probability or expected frequency of occurrence. Common metrics include the probability that a specific person shall be killed due to an accident (individual risk), the expected number of fatalities expressed by indices such as potential loss of lives (PLL) and fatal accident rate (FAR), and f-n curves showing the expected number of accidents (frequency f) with at least n fatalities.

Apostolakis (2004) points to several benefits of using PRA/QRA: It considers a large number of scenarios that involve multiple failures, which allows for an in-depth understanding of system failure modes. The approach increases the likelihood of identifying interactions between events, systems and operators. It provides a common understanding of the risk issues faced, hence facilitating communication among various stakeholder groups. A PRA/QRA is an integrated approach of different disciplines (engineering, social and behavioral sciences, natural scientists, risk analysts). It creates a picture of what the experts know and do not know about the relevant risk issues. In this way, the analysis may provide valuable input to decisions regarding needed developments and research in diverse disciplines, for example, related to physical phenomena and human errors. The analysis also facilitates risk management by identifying the dominant accident scenarios, events and factors so that resources are not wasted on items that are insignificant contributors to risk. Sensitivity and importance analysis as discussed in Section 5.1 are basic tools for this purpose. Monte Carlo simulation is commonly used in PRAs/QRAs to simulate the performance of various systems and activities.

According to Apostolakis (2004), a peer review by independent experts is considered an essential part of a PRA/QRA process. Such a review is, however, not always conducted. The need depends on the importance of the assessment as well as the competence of the team performing the assessment.

The basic approaches and methods used in PRAs/QRAs have not changed much over the years. However, developments have been made on a number of more detailed analysis features, to reflect new types of systems and modeling insights (Zio 2009; NASA 2011). Many of the methods introduced allow for increased levels of detail and precision in the modeling of phenomena and processes within an integrated framework of analysis covering physical phenomena and human and organizational factors as well as software dynamics (Mohaghegh and Mosleh 2009; Zhen et al. 2020). Other methods are devoted to improved representation and analysis of the risk and related uncertainties in view of the decision-making tasks that the outcomes of the analysis are intended to support (see, e.g., Aven and Zio 2011). The present book, in particular Chapter 3, presents some of these methods.

PRAs/QRAs have commonly been used in relation to risk acceptance criteria or tolerability limits: If the calculated risk is above a predefined probabilistic limit, risk is judged unacceptable (or intolerable) and risk-reducing measures are required, whereas if the calculated risk is below the limit, it is concluded that no measure is required or, alternatively, that measures should be subject to a broader cost-benefit type of consideration, in line with the so-called as low as reasonably practicable (ALARP) principle

(Aven and Vinnem 2007; see Chapters 8 and 9, in particular Sections 8.3 and 9.5.1). At the same time, it is broadly acknowledged that the risk assessment is a tool to inform decision-makers about risk, not to prescribe decisions (Apostolakis 2004), as also highlighted in Section 4.2.3.

The probabilistic analysis underpinning PRAs/QRAs builds on frequentist probabilities and subjective probabilities, as explained in Kaplan and Garrick (1981). Subjective probabilities are used to express uncertainties about parameters of the probability models and the unknown frequentist probabilities.

As discussed in Section 4.1.1 for the NASA example, PRAs/QRAs have commonly been used to estimate the inherent, true risk (interpreted as frequentist probabilities) associated with the system studied. Using models, observational data and expert judgments, the aim is to accurately determine this risk. Specific uncertainty analyses are then used to express the uncertainties about the 'true' values of the risk (Paté-Cornell 1996).

PRAs/QRAs have been strongly criticized by many researchers (e.g., Renn 1998; O'Brien 2000; Aven 2011b; Rae et al. 2014; Goerlandt et al. 2017; Pasman et al. 2017). Many of the issues raised relate to the perspective that risk is to be accurately estimated by the assessments. In case of large uncertainties, the PRAs/QRAs would fail if this were the purpose of the study, as discussed in the NASA example of Section 4.1.1. In line with current understanding of a risk assessment, the main aim of the risk assessment should not be to accurately estimate the 'true' risk but to understand the risk. Apostolakis (2004) highlights this, but practice has been in line with the 'true risk estimation' perspective. Since the 1970s and 80s, contemporary risk science has provided new insights, and today it is acknowledged that probability alone is not enough to address the uncertainties. We need to incorporate the knowledge aspects of risk to a stronger degree than was done in early stages of PRA/QRA uses, as discussed in Chapters 3 and 4.

A main challenge of PRAs/QRAs relates to the use of causal chains and event analysis. This approach has strong limitations in analyzing complex systems, as it treats the system as being composed of components with linear interactions, using methods like fault trees and event trees. Complexity here refers to the difficulty of accurately predicting the performance of the system considered, based on knowing the specific functions and states of the system's individual components; see also Appendix A. It is argued that some of the key methods used in risk assessments are not able to capture 'systemic accidents'. Hollnagel (2004), for example, argues that to model systemic accidents, it is necessary to go beyond the causal chains – we must describe system performance as a whole, where the steps and stages on the way to an accident are seen as parts of a whole rather than as distinct events. It is not only interesting to model the events that lead to the occurrence of an accident, which is done in, for example, event and fault trees, but also to capture the array of factors at different system levels that contribute to the occurrence of these events. Alternative methods have been developed, of which the functional resonance analysis method (FRAM) and system-theoretic accident modeling and processes (STAMP) are among the most well known (Hollnagel 2004; Leveson 2004). Leveson (2007) makes her points very clear:

Traditional methods and tools for risk analysis and management have not been terribly successful in the new types of high-tech systems with distributed human and automated decision-making we are attempting to build today. The traditional approaches, mostly based on viewing causality in terms of chains of events with relatively simple cause-effect links, are based on assumptions that do not fit these new types of systems: These approaches to safety engineering were created in the world of primarily mechanical systems and then adapted for electro-mechanical systems, none of which begin to approach the level of complexity, non-linear dynamic interactions, and technological innovation in today's socio-technical systems. At the same time, today's complex engineered systems have become increasingly essential to our lives. In addition to traditional infrastructures (such as water, electrical, and ground transportation systems), there are increasingly complex communication systems, information systems, air transportation systems, new product/process development systems, production systems, distribution systems, and others.

Nonetheless, the causal chains and event modeling approach has been shown to work for a number of industries and settings. It is not difficult to point to limitations of this approach, but the suitability of a model and method always has to be judged with reference to not only its ability to represent the real world but also its ability to simplify the world. All models are wrong, but they can still be useful, to use a well-known phrase. It can also be discussed to what degree systems are in fact complex. Within many industries (e.g., nuclear, process and oil and gas), we can list at an overall level what type of undesirable events can occur, but how these will occur is not always straightforward. If we look at the accidents and near-accidents that have occurred, most systems and activities will be described as complicated – there are many components, but we understand and have good knowledge about how they relate to each other. This does not mean that surprises do not occur, but it is often because the knowledge in the case in question is weak or wrong and not because of fundamental deficiencies in our knowledge of the relevant phenomena and processes. To meet the potential surprises, resilience must be emphasized. Disruptions and errors of various kinds will occur, and then one must be able to withstand these and correct the problems quickly and efficiently. An important strategy here is to identify and follow up on relevant signals.

Furthermore, the causal chains and event modeling approach is continuously improved, incorporating human, operational and organizational factors, as mentioned previously. Mohaghegh and Mosleh (2009), for example, present a 'hybrid' approach for analyzing dynamic effects of organizational factors on risk for complex socio-technical systems. The approach links system dynamics, Bayesian belief networks, event sequence diagrams and fault trees.

5.5 VULNERABILITY ANALYSIS

We remember from previous chapters that vulnerability is an aspect of risk, defined by the consequences C – with associated uncertainties U – conditional on an initiating event A; that is, vulnerability is (C,U|A). To characterize or measure the vulnerability, we specify the consequences C and describe the uncertainties U – this is performed in risk and vulnerability assessments. In general terms, we are led to a vulnerability characterization of the form (C',Q,K|A'), where A' is one or more specified events, C' is some specified consequences, Q is a measure (interpreted in a wide sense) of uncertainty associated with A' and C', and K is the knowledge that Q and C' are based on. In this framework, Q is a measure of epistemic uncertainty. Variation in populations (sometimes referred to as stochastic or aleatory uncertainties), commonly represented by probability models, can be used to support the uncertainty assessment and characterizations, thereby allowing statistical analysis to update knowledge and describe vulnerability and risk.

Different metrics can be derived that are founded on this generic vulnerability characterization. An example is the expected value of C' given a threat A', E[C'|A'], or a probability distribution of C' given A', P(C' ≤ c|A'). In addition, strength of knowledge judgments (SoK) should be added.

The vulnerability assessments are conducted using methods as described previously for risk assessment, for example, event trees with barrier analysis, Bayesian networks and Monte Carlo simulation. In specific applications, like critical infrastructure, climate change, business and finance, specific models are developed to represent the systems and activities considered. These models can represent different types of phenomena, for example, physical, socio-technical or financial. The types of systems considered and the level of knowledge vary considerably in these vulnerability assessments. Consider the following three scenarios, as described in Thekdi and Aven (2021a):

Scenario #1: A transportation network with a single collision, causing congestion on the connected roadways.

Scenario #2: A region, such as Puerto Rico, that is impacted by a major hurricane. The hurricane causes thousands of deaths, widespread damage to the infrastructure and struggles to recover due to the terrain, infrastructure conditions, resources and political factors.

Scenario #3: An illness, such as the recent COVID-19, that is allegedly sourced from a single location and rapidly spreads across the globe, causing deaths, economic fluctuations, supply chain disruptions and other widespread repercussions.

Scenario #1 illustrates a widely studied and carefully engineered system, such that a disruption of this caliber can be modeled and managed using standard tools and management protocols. Scenario #2 illustrates a system that is somewhat documented, with a disruption that is unprecedented but known to be possible for the region. Scenario #3 involves a somewhat documented system, a disruption that is known to be possible, with rapid widespread consequences but subject to considerable uncertainties. The scenarios are characterized by different risk assessment and management features. Also, the vulnerabilities are different, and this is the issue addressed in this section. In all three cases, the vulnerabilities could be considered large, yet they are very different when it comes to the uncertainty and knowledge supporting this conclusion. In the following, we will present a vulnerability assessment for the COVID-19 pandemic, conducted early March 2020, illustrating the issues raised previously concerning modeling and analysis and in particular how these relate to the elements of the vulnerability characterization (C',Q,K|A'). The following presentation is based on Thekdi and Aven (2021a).

The COVID-19 pandemic

In mid-January 2020, the World Health Organization provided public guidance on a novel coronavirus, causing a disease later referred to as COVID-19. It was speculated that the virus was related to a seafood market in Wuhan, China (WHO 2020a). At that time, there were reportedly 41 cases and one death, with the death of a patient with underlying medical conditions. By late February, the same organization reported over 78,000 cases and over 2,700 deaths in China. Outside of China, there were reportedly over 3,000 cases and 54 deaths (WHO 2020b). By mid-March, there were over 191,000 cases and almost 8,000 deaths globally (WHO 2020c). By early August, there were over 18 million confirmed cases and over 700,000 deaths (WHO 2020d). The swift spread of the virus led to widespread temporary school and business closures, national border closures and major global economic turbulence (Otani et al. 2020).

The previous details were speculative and required further investigation over time, as would be the case for any newly identified and unstudied event or condition. While viruses in general had been well studied and treated in history, there was very limited information about how this particular virus incubates, spreads and impacts humans.

The vulnerability is assessed with respect to the consequences C', the Q-based metrics, and the strength of knowledge. Table 5.4 presents key assessment findings for this case.

TABLE 5.4 Vulnerability description for the COVID-19 case (March 2020) (based on Thekdi and Aven 2021a)

Vulnerability description given event A1	Relevant element	Example related to COVID-19 pandemic
$C' = f(x)$: consequences C' based on model $f(x)$ Vulnerability issues based on models of the consequences C'		
	C.1. Time dependency	Spread of the virus varies over time; effectiveness of interventions (behavior, drugs, healthcare, etc.) changes over time.
	C.2. One or a few components' failure causes failure of the system	Movements of a few people contribute to a rapid spread; behavior by a single person, organization or local policy causes cascading repercussions; single virus mutation can cause another pandemic.

Vulnerability description given event A1	Relevant element	Example related to COVID-19 pandemic
Q: Characterization of uncertainty g(q) Vulnerability issues based on characterizations using Q-based metrics		
	Q.1. Sensitivity	Scenario analysis to assess the sensitivity of transmission rates, hospitalization rates, fatalities and resource needs as a result of various social distancing strategies and other population-level intervention policies.
	Q.2. Importance/criticality	Geographic areas and populations can have varying levels of criticality for the spread of the virus.
K (knowledge) and SoK (strength of knowledge) Vulnerability issues based on judgments of the strength of the knowledge K		
	K.1. The degree to which the phenomena involved are understood	Minimal understanding about how the virus spreads, how long it lives on surfaces, if immunity is possible and how it impacts various populations.
	K.2. Accuracy of models	Model validation in progress but not yet available during the peak virus spread.
	K.3. Assumptions	Assumptions borrowed from similar viruses but not yet validated for COVID-19 during peak virus spread.
	K.4. Data and information	High visibility of recent data and information, with even non-peer reviewed study findings being published in online formats.

A model is developed for the consequences C'; hence, we can write C' = f(x), where f(x) represents an epidemiological function (model) that measures a virus spread or fatalities under a strict set of assumptions. Additionally, we assume a metric g = g(y|A') that is derived from the vulnerability characterization (C',Q,K|A'). The components of y may include aspects such as time, probabilistic representations of events and consequences, intervention strategies and knowledge strength for the components. In this case, we assume this function is given by the expected value, g = E[C'|A',K], or the probability distribution of C', g = P(C'<c|A',K). In the former case, using the function C' = f(x), we have g = E[f |A',K]. Thus, if f(x) and C' represent the number of deaths, g expresses the expected number of deaths. If f(x) is deterministic (all parameters are assumed known), then g = f for this particular choice of g. These metrics can, for example, be used to predict the number of deaths under different assumptions and study effects of various interventions.

5.5.1 Vulnerability issues based on models of the consequences C'

A model C' = f(x) for COVID-19 can be developed using a variety of features of the health conditions across regions and globally, such as in Ferguson et al. (2020) and

Coburn et al. (2009), where measuring cases per day and resulting fatalities per day per 100,000 population. The components of the model f can include factors such as those included in Ferguson et al. (2020): Population demographics with factors such as population density, contacts with individuals, age, household distribution size, class sizes and workplace sizes; infectiousness based on the timing related to onset of symptoms for patients who are symptomatic and based on timing of infection for patients who are asymptomatic; transmission events such as contacts made within a household, at the workplace or in the broader community while adjusting for social distancing; and incubation periods based on fixed assumptions found in related research.

This function $f(x)$ allows the analyst to model performance metrics that may include daily new cases, daily fatalities, daily % of positive test results, cumulative cases or cumulative fatalities. These metrics allow for reflecting specified geographic boundaries, for example, a region or country.

However, $f(x)$ can also be used to inform other aspects of C', which highlight the much broader contextual impacts of the virus. For example, it can be used to project or forecast impacts, including economic impact due to closures and workforce shortages; educational impact due to school closures and loss of productivity; community impact due to disruptions in social services; societal impact due to changes in crime and law enforcement; sector-specific impact, such as to healthcare services; broader health impacts due to potential shortage of healthcare services, coinfection issues or access to medical care; and indirect impacts, such that when seemingly minor or larger risk-related events coincide with pandemic, the resulting C' can be exacerbated.

There are several important elements that are essential for understanding the consequences C'.

Element C.1. Time dependency is a major issue due to the fact that the existence and spread of the virus implies that the number of cases varies over time, the impact of the virus depends on the time horizon considered and uncertainties imply that that long-term outlooks can be wildly inaccurate and untrustworthy. Additionally, the effectiveness of interventions, including human behavior, pharmaceutical advances, healthcare availability and others may rapidly change over time.

Element C.2, suggesting that failure of one or a few components can cause failure of the system, is important to note for this example. As the virus was allegedly sourced from a single source and quickly grew into a major pandemic and potential economic catastrophe, it can be questioned whether this type of system response is typical. Presumably, as the virus was highly contagious, the movements of a few people could contribute to a rapid spread, which is thought to be exponential (Wu et al. 2020). As risk mitigation directives involve requesting populations to abide by social distancing and other protective measures, breakdowns in this behavior by a single person, organization or local policy can cause massive and cascading repercussion to the affected populations. In the case of the pandemic being politicized, or as fatigue with mitigation measures develops, social pressures can cause breakdowns in this behavior to grow within populations. Additionally, a single virus mutation has the capability to invalidate any developed immunity or pharmaceutical advances, thereby causing another pandemic.

In the general approach presented by Thekdi and Aven (2021a), several other elements are referred to, including 'reorganization' (system is to varying degrees able to reorganize following loss of functionality for sub-components) and 'other resilience aspects' (the system is to a varying degree able to restore functioning following a disturbance).

5.5.2 Vulnerability issues based on characterizations using Q-based metrics

The study of Q-based metrics in relation to COVID-19 first involves a closer investigation of the metric $g = g(y|A') = E[C'|A',K]$, where C' expresses number of deaths. The metric can be used to predict fatality rates across a variety of intervention scenarios. Intervention scenarios may include introducing policies on business operating procedures, mask-wearing, social distancing, quarantines and school closures. Additionally, interventions can include policies and standards related to the introduction of vaccines and medication to treat or avoid the illness.

Reference is made to models of the type presented by IHME (2020), which makes projections of virus spread over time. We interpret the models' study of virus spread projections, combined with aspects of intensive care unit (ICU) bed and ventilator availability, as characterizations of g using the notion of $g = E[C'|A',K]$. The analysis includes a projection value for fatalities, ICU bed usage and ventilator usage. These projection values are supplemented with a band of uncertainty, defined as

> Uncertainty is the range of values that is likely to include the correct projected estimate for a given data category. Larger uncertainty intervals can result from limited data availability, small studies, and conflicting data, while smaller uncertainty intervals can result from extensive data availability, large studies, and data that are consistent across sources.
>
> (IHME 2020)

This band can be interpreted as a prediction interval [a,b] for the number of deaths, C', such that the probability for C' to be included in the interval is high, for example, 90%; thus, $P(a \leq C' \leq b \,|A',K) = 0.90$.

In both methods (Ferguson et al. 2020; IHME 2020), there is minimal discussion of knowledge strength. While citation of peer-reviewed literature related to other viruses and also COVID-19 work in various stages of peer review can be used to inform model formulations, this supporting knowledge is weak (March 2020).

When discussing vulnerability based on uncertainty-based metrics, both elements Q1 and Q2 are important; see Table 5.4. Element 'Q.1. Sensitivity' remains a major issue involving f and g, recognizing that these functions serve as only a projection based on a set of assumptions. The model f will be inaccurate or incorrect in both foreseeable and unforeseeable ways, and g may be based on weak or wrong knowledge, leading to poor predictions. However, sensitivity analysis enables the analyst to understand the impact of these inaccuracies and limitations on results and decisions. For example, expressions of g could include scenario analysis to assess the sensitivity of transmission rates as a result of various social distancing strategies (CDC 2020).

Similarly, the COVID-19 epidemic calculator demonstrates how fatality rates and hospitalization rates change as assumptions related to populations, transmission rates, incubation times and so on change (Goh 2020). These studies demonstrate that small amounts of variation within the scenarios of populations, transmissions and other factors can have drastically large impacts on the overall global impact of the virus.

There is also potential for 'Q.1: Sensitivity' to be further studied and leveraged for risk-informed policy decisions. There is opportunity to look across published approaches and models to see how these differ, potentially finding variability in the prediction of virus cases and fatalities. Additional sensitivity studies can explore the prediction interval of virus spread projections (IHME 2020) to understand what changes in assumptions are associated with the largest changes in projection values, such as one scenario of 100,000–200,000 fatalities versus 1 million–2 million fatalities.

The element 'Q.2. Importance/criticality' is also relevant, as, for example, particular virus carriers, geographic areas or professions can have varying levels of influence over the virus spread. Basic reproduction numbers, R_o, are used to understand the number of secondary cases that result from a single infectious case (Dietz 1993) as a component or output of both f and g computations.

These R_o values can differ based on diseases, interventions (such as social distancing, school closures, etc.), geography and populations. While they can be computed differently based on the modeler and application, it is generally assumed that $R_o > 1$ implies that an outbreak will continue, while $R_o < 1$ implies that an outbreak will end. For example, R_o can have relatively low values associated with social isolation, while it can be relatively high when associated with high physical contact within populations. Some populations, behaviors, geographies and so on can be particularly critical or important within the modeling of overall R_o values. As another example, a high R_o resulting from activity near major international airports implies a higher level of criticality or importance of large-scale population movements among geographically separated nations. This is particularly an issue because screening interventions are largely ineffective (Normile 2020) and may in fact contribute to the virus spread by slowing airport activities, thereby interfering with social distancing measures.

5.5.3 Vulnerability issues based on judgments of the strength of the knowledge K

A major issue when assessing the strength of knowledge is element 'K.1: The degree to which the phenomena involved are understood'; refer to Table 5.4. As the virus was new, there was little to no understanding about how the virus spreads, how long it lives in surfaces, if immunity is possible and how it impacts various populations. For example, there was wide disagreement on whether 'herd immunity' (Cohen and Kupferschmidt 2020; Fine 2011) was possible or feasible for COVID-19. As a result, there are also issues with elements 'K.2: Accuracy of models' and 'K.3: Assumptions', as there was limited to no information that could be used to inform the models and also assumptions existing within those models. However, as was seen with the virus studies of Ferguson et al. (2020) and Goh (2020), information from past related conditions, such as influenza, combined with adapting to a variety of model assumptions could provide insight for decision-makers. The issue of accuracy and assumptions also related to the use of directing reopening of businesses and schools by using metrics such as "# of cases" or "% of positive cases". These types of metrics assume some uniformity or randomness of testing the population. These tests may be performed on population groups who are already exhibiting symptoms, those who have not yet been tested during the timeframe of the pandemic, those who can afford the test and those with a particular interest in receiving the test. Additionally, there may be potential for these types of metrics to be manipulated by making tests more or less accessible in particular regions or demographic groups. Element 'K.4: Data and information' may also be relevant, as the high visibility of the virus allows for the most recent data and information to be publicized, while preliminary non-peer reviewed study findings can be published in online formats.

5.5.4 Additional issues

Thekdi and Aven (2021a) also point to an alternative approach for assessing and characterizing the vulnerabilities, which will be briefly outlined in the following. It is based on similar ideas for risk ranking as described in Section 3.2.2, using the concepts of general knowledge (GK) and specific knowledge (SK). Here, the GK involves available knowledge associated with previously known viruses – in particular coronaviruses – in general. GK associated with viruses includes physical properties, how they generally spread, whether or how long they live on various surfaces, how they can be removed or destroyed and so on. GK associated with the transmission of viruses includes population demographics (age, pre-existing conditions, population density, etc.) and population behaviors (attendance in schools/workplaces, hygiene, etc.). GK associated with the treatment of viruses includes the healthcare system (capacity of hospitals/ICUs, health insurance availability, available prescription medicines that can treat the virus, etc.) and home treatment (effectiveness of over-the-counter drugs). GK associated with intervention strategies includes human psychology and behavior to understand how populations will comply with any suggested or mandatory social distancing, feasibility of social distancing based on known family structures and population densities and feasibility/legality of school and workplace closures. GK associated with broader global impacts includes economic issues to understand the monetary implications of interventions, social issues to understand how to care for populations with food insecurity or lack of availability of healthcare and political issues related to how the voting public perceives the actions and decisions of those in political power. Finally, historical issues include learning from past incidents, such as the 1918 influenza pandemic. GK associated with the influenza pandemic includes hypothesized factors that contributed to the spread, such as distancing within populations and medications to treat the illness; social or political issues influencing how populations comply with mitigation directives; best practices for protecting populations; and timings associated with transmission of the virus.

While there was limited to no SK when the virus was initially discovered, accelerated scientific research can provide some initial insight. Unfortunately, some types of scientific research were decelerated due to university closures, quarantines, illness, paper retractions and competing needs for scientists in healthcare settings. Additionally, the scientific field can be inherently slow in developing credible insights due to the critical need for thorough experimentation and peer review. However, as research and testing continue, the SK becomes increasingly strong. Also, SK is developed by observing and analyzing the virus transmission and response in other regions and countries.

The overall vulnerability assessment depends on several factors, including probability, consequences and knowledge. While the probability of this virus release may have been judged negligible, there is literature that implies otherwise. For example, a study of the history of coronaviruses in 2005 (Kahn et al. 2005) states: "Additional human coronavirus strains will very likely be discovered, which stresses the need for further investigation into the virology and etiology of these infectious organisms". Viruses that have been widely circulated and known may not have severe consequences due to population immunities, but they do have potential to be catastrophic. For example, consider the 1918 flu pandemic that is estimated to have resulted in 50 million fatalities (National Archives 2020) and related estimates relating the H1N1 pandemic to massive economic disruptions (Santos et al. 2013). Given the variation in impacts of various viruses and the very low SK about the particular COVID-19 virus, there is very weak knowledge regarding consequences and probability.

Table 5.5 shows each of the vulnerability criteria as defined by Thekdi and Aven (2021a), implying that the classification is deemed *High* if one or more of the listed criteria are met. Compelling evidence of this *High* vulnerability classification is primarily based on the fact that there was weak knowledge about this particular virus, due to the fact that it was previously unknown. Also, it is compelling to note that while

TABLE 5.5 Applicability of vulnerability criteria; vulnerability is deemed *High* if one or more criteria are met (Thekdi and Aven 2021a)

Criterion	Applicability of criterion	Evidence
a. The vulnerability is judged to be high when considering consequences and probability	Yes	Consequences of known coronaviruses have historically been minor, but comparable incidents shown to have significant consequences, such as the 1918 influenza pandemic; probability of this new virus is thought to be high by some researchers
b. The vulnerability is judged as high, considering the potential for severe consequences and significant uncertainty (relatively weak knowledge)	Yes	Consequences of known coronaviruses have historically been minor, but comparable incidents shown to have significant consequences, such as the 1918 influenza pandemic; relatively weak SK due to no historical information about this new virus

Criterion	Applicability of criterion	Evidence
c. Lack of resilience	Yes	It was speculated that the global ability to react to this type of virus was poor (Gates 2015) and resilience has been questioned after previous virus incidents (Lurie 2009)
d. Weak general knowledge about (B, C), where B refers to barriers and C the consequences given A'	No	Relatively strong GK about barriers to and consequences of virus infections
e. Weak specific knowledge about (B, C)	Yes	Very weak SK about consequences of COVID-19, as the novel coronavirus was not previously known
f. Strong general knowledge about undesirable features of (B, C)	'Weak' Yes	Some early warnings about this type of virus (WHO 2015)
g. Strong specific knowledge about undesirable features of (B, C)	'Weak' Yes	Indications that COVID-19 is very contagious

forewarnings existed using information from historical epidemics and pandemics, the system resilience was relatively weak.

5.5.5 Discussion

Traditionally, vulnerability is used as a rather static concept expressing an inherent property of the system analyzed. If the vulnerability is high, the system, if exposed, would fail, and this potentially leads to severe consequences. The present assessment approach and classification system builds on a more general perspective on vulnerability in which vulnerability is seen as conditional risk given the occurrence of an event (or risk source) A'. Hence, uncertainty is a main component of the vulnerability concept, in the same way as uncertainty is considered a main component of risk. This conceptualization leads to a clear distinction between the concept of vulnerability and how the vulnerabilities are measured or described. For the latter part, the knowledge supporting the measurements also needs to be reflected.

By considering vulnerability as $(C',Q,K|A')$, we are led to a holistic approach with input from all three components C', Q and K. Earlier work has focused on C' (through models f) and/or Q (through metrics g), but an integrated perspective which also highlights the knowledge dimension K (with GK and SK) represents an alternative perspective.

The focus on the knowledge aspects is important to properly understand the dynamics in relation to time, as illustrated by the COVID-19 case. As SK about the virus evolved over time, understanding of the consequences increased. Additionally, as is the case with any time series projections or forecasts of a system, the models become increasingly inaccurate or uncertain for time periods that are further into the future. In the case of COVID-19, initial discussions used the term 'flatten the curve', provoking nations to implement strict and severe interventions to ensure virus cases did not overwhelm the healthcare system. However, future projections implied that flattening the curve at the early onset of the virus results in an even greater spike of cases at a later time. This type of behavior is not a new phenomenon, as it is speculated that the 'second wave' of the 1918 flu pandemic was the most deadly (Taubenberger and Morens 2006). Of course, there is increased uncertainty about appropriate assumptions that can be used to model future time period, as healthcare resources, pharmaceutical interventions, and virus spread are increasingly difficult to predict. Also, intervention methods and compliance with intervention methods may become increasingly effective or ineffective over time.

Overall, there is value in incorporating dimensions of time within the $(C',Q,K|A')$ vulnerability definition. In particular, we see that knowledge and uncertainty can vary over time, particularly for risk events that have a relatively long duration, such as pandemics or slow-changing climate conditions. Over expanses of time, often, knowledge related to both GK and SK will likely increase, allowing for more informed decision-making and more effective risk mitigation strategies. However, the characterization of uncertainty through sensitivity or importance/criticality assessments may not change in any predicable direction throughout the duration of a risk event. For example, the sensitivity of the system may in fact increase as a result of a particular risk event coinciding with other risk events. For example, a pandemic can make the system more sensitive to disruptions related to natural disasters, socio-political events and supply chain inefficiencies. The suggested approach is able to separate the Q and K issues which do not in reality behave in similar or predictable ways. Studying Q and K separately allows for time-related issues to be addressed in ways that are specific to characteristics of the system and threats.

The approach presented is conditional on an event (risk source) A'. At first glance, this may seem to limit the analysis scope to study vulnerabilities only for specific, known events. However, this is not the case. Aspects of C', Q and K in Table 5.4 and GK/SK in Table 5.5 involve characteristics of the system and characteristics of the underlying uncertainty and knowledge, not necessarily linked to a specific threat. In other words, a novel virus with limited information or history inherently involves reasonably similar C', Q and K (in particular GK) as other potentially different but similar viruses. Thus, a single classification can be applied to a collection of events $A' = (A'_1, A'_2, \ldots, A'_n)$, with some of these events potentially not yet known or identified. With this recognition, we reinterpret the event-specific descriptions of vulnerability $(C',Q,K|A')$ as being more nuanced, suggesting that vulnerability assessment and classification do not necessarily need to be specific to a single event but rather a collection of events with similar features. Overall, this transition from an event-specific vulnerability assessment and classification system to a collection-of-events–specific system better allows decision-makers to gauge risk based on features of consequences, uncertainty and knowledge instead of relying on information about unknown or surprising events that cannot be adequately foreseen. This is a similar and meaningful direction of risk research that aligns with non-event–specific resilience strategies.

We refer to the classification system (Tables 5.4 and 5.5) as a risk-vulnerability classification system. Per the definition, there is a difference between risk and vulnerability, as vulnerability is conditional on the occurrence of an event. However, this kind of dichotomy can be challenged, as we also talk about risk related to the coronavirus consequences in later stages of the pandemic, hence in situations where the virus has already spread to some level. This indicates an overlap in the interpretations of the two concepts. One way of maintaining a difference is to limit the consequences C' in the vulnerability case to more system performance-related quantities, like the failure or not of the system considered, and the consequences C' in the risk case to captures all effects, short term and long term, of the events. The frameworks presented are general and allow the user to define what are the appropriate consequences to consider in a particular situation.

The 'Q.1 Sensitivity' aspect of the risk-vulnerability approach particularly highlights the importance of independently investigating the sensitivity of the system and how system sensitivity influences overall vulnerability. The COVID-19 case study showed that the 'Q1 Sensitivity' was relatively high, here referred to as a *highly sensitive system*. It was shown that fatality rates and case rates for the disease were particularly sensitive to the behavior of a few virus cases, the behavior associated with a few social distancing policies and other interventions. While a *highly sensitive system* can be detrimental, as seemingly minor behaviors can cause disastrous consequences, the identification of this system property can aid in risk-management decisions. For example, this sensitivity in the COVID-19 case study can direct mitigation strategies toward population-level policies, such as closing national borders, closing schools and issuing stay-at-home orders for particularly sensitive regions. More generally, if risk-based investments are able to reinforce small, yet critical, components of the system or functions within the system, this can have a major improvement in the system's ability to avoid destruction and also support an efficient system recovery. Similarly, knowledge of the *highly sensitive system* can also guide decision-making and policies for reopening businesses, schools and other important elements of societies and economies.

5.5.6 Conclusions

We have presented two approaches for assessing and classifying vulnerabilities, reflected by Tables 5.4 and 5.5, respectively. The former approach in particular allows for aspects of vulnerability to be decoupled into distinct characteristics, such as time, sensitivity and underlying knowledge of the system. The classification approach can be applied to events such that the specific events may not yet be known or identified, while also not being limited to a single event, but rather be applied to a collection of events with similar characteristics. The use of distinct characteristics also highlights the importance of acknowledging anomalies in individual characteristics and how those anomalies can contribute to system vulnerability. For example, the COVID-19 case study demonstrated that a *highly sensitive system*, defined by excessively high levels of 'Q.1 Sensitivity', is an important factor in assessing the overall system vulnerability.

Uncovering this *highly sensitive system* property allows for more informed risk management and resulting decision-making.

The latter approach allows for the distinction between general knowledge and specific knowledge to guide overall vulnerability assessment. The COVID-19 case study demonstrated that both GK and SK are dynamic, such that these knowledge types can change over time. Most notably, we recognize that it is possible for both GK and SK to decrease over time or behave in unpredictable ways. There may also be a cyclical relationship between GK and SK, such that SK can question or provide additional guidance for the study of GK and so forth. Thus, active assessment of this knowledge is important, while recognizing that expert disagreement on GK and SK can be important factors in informing classification of overall system vulnerability.

The two approaches presented are designed to be complementary. Approach 1 is intended to be used for detailed analysis of overall system characteristics. Each distinct element is intended to be understood separately from others, with no intent to summarize into a general vulnerability classification level. Approach 2, which is informed by approach 1, is intended for a single assessment of vulnerability, such as *high, medium* and *low*, which can provide a more generalized description that is more easily understood and utilized by high-level decision-makers.

The approaches presented can be used to understand and characterize vulnerability. While the goal of the present chapter is not to discuss how to reduce vulnerability, there is potential to use the characterization to guide decisions for achieving improved (lower) vulnerability. The classification system can be used to guide efforts for improving the knowledge and strength of knowledge (K.1–K.4) in the four categories (in approach 1) and GK/SK (in approach 2). Knowledge can be obtained from many sources, including the purchase/collection of data, interviews with experts, surveillance technologies and research. When considering near-term decisions, there is value in understanding the current state of vulnerability in the current risk landscape. Decision-makers can benefit from studying *why* a particular system exhibits some level of vulnerability, which can influence near-term decisions. Recall that the approach characterizes vulnerabilities given the risk source. Suppose a given system is exposed to two risk sources and the vulnerability is characterized as *High* for both risk sources A_1 and A_2. Suppose vulnerability related to A_1 is judged to be *High* when considering consequences and probability, based on strong general and specific knowledge. The vulnerability for A_2 is also judged to be *High* but has weak general knowledge, without a vulnerability issue related to consequence and probability. While ranking the relative importance of the risk sources is not the issue here, one can see value in using this information to direct knowledge procurement and resource allocation. In the case related to A_1, there may be a need to prioritize further study for risk mitigation investments because there is already a strong knowledge base. In the case related to A_2, there may be need to invest resources into increasing general knowledge, using methods such as data collection or research.

More generally, the approaches presented will enable decision-makers to carefully assess the dynamic properties of system vulnerability while focusing on critical elements of uncertainty and knowledge. The approaches enable the risk process to focus studies on characteristics of systems and threats. This equips analyses to assess unforeseen or potential surprising events and also recognize that differing events may in fact have very similar vulnerability characteristics, as shown in approach 1. Thus, a single vulnerability assessment and classification process can evaluate vulnerabilities for a collection of similar behaving risk events, thereby promoting overall efficiencies in the risk management process.

5.6 METHODS FOR ASSESSING RESILIENCE

We remember from Section 2.3.2 that resilience refers to the ability to return to the normal state given a risk source, or alternatively the ability of a system to sustain or restore its basic functions following a risk source or an event (even unknown). What method can we use to assess this ability?

As an example, consider the human body (refer to Problem 2.3). A healthy individual is considered resilient in their ability to persevere through infections or trauma. Even for individuals with severe illnesses, critical life functions (such as breathing) can be sustained and the body can potentially recover, often adapting by developing immunity to further attacks of the same type (Linkov et al. 2014). However, the human body can also be considered vulnerable. History has shown that if medical advances like penicillin had not been made, the consequences of some bacterial infections would have been devastating.

This may suggest a characterization of the human body as *quite resilient* but not *highly resilient*, but how should we distinguish between *quite resilient* and *highly resilient*? Or, in more general terms, how should we measure the degree of resilience?

A number of methods and metrics have been suggested for this purpose; see, for example, Henry and Ramirez-Marquez (2012), Francis and Bekera (2014) and Hosseini et al. (2016). Some of these methods and metrics are simple, others very complicated, as these papers show. As an illustration of a simple metric, consider the definition by Hashimito et al. (1982), who define the resilience of a system as the conditional probability of a satisfactory (i.e., non-failure) state in time period t + 1 given an unsatisfactory state in time t.

Many authors have argued that resilience cannot simply be measured in a single unit metric, for example, Haimes (2009). According to Haimes, the question "What is the resilience of system x?" cannot be answered, as it would require knowing whether system x would recover following any attack (stressor) y (also unknown types of attacks). What can be done, however, is to study how the system functions – what the outputs (the consequences) of the system are – for any specific inputs (threats, stressor).

The point made by Haimes is important. We cannot see resilience independently of the stressors (threats). Say that a system can be subject to two types of stressors, A_1 and A_2. The system is resilient in relation to stressor A_1 but not to A_2. Now suppose A_2 will occur with a probability of 0.000001 and A_1 with a probability 0.999999. Is the system resilient? Yes, with high probability. If we allow for unknown and surprising types of stressors, we cannot conclude in the same way, as there is no basis for making the probability judgments. Haimes' statement on the difficulty of measuring resilience is thus reasonable. We are faced with a basic problem, which stressors to include in the resilience judgments. With respect to some stressors, the system considered may have shown itself to be resilient in the sense that it was able to recover and sustain its basic functionality. However, for other stressors, the system may not have shown itself to be resilient, and when facing the future, it may turn out that the system will also experience problems when faced with other stressors of known or unknown type.

Despite these issues, we can perform resilience management, as we will discuss in Sections 8.6 and 8.7. We know that resilience can be improved in many ways, for example, with a strengthened immune system, safety barriers, different layers of protection, redundancy and diversification. We do not need to identify all possible threats and assess their likelihoods to understand that a specific resilience arrangement or measure will be useful in many situations. However, there is a need for ways of describing and comparing the resilience for alternative arrangements and measures. Resource limitations mean that we have to prioritize – where should we improve the resilience? There could be many areas in which the resilience can be improved, but which should be selected and given weight? Many resilience metrics exist, as mentioned previously, but a basic feature of risk is not captured by these: What stressors will in fact occur? For the previous example with the two events A_1 and A_2, we can think of a specific arrangement that could significantly improve the resilience with respect to stressor A_2, but its effect on risk (interpreted in a wide sense) could be marginal. The arrangement could still be justified, but some type of considerations of risk seems useful. Also in the case where we have difficulties in identifying the relevant events and in assessing likelihoods, we still need to make prioritizations. The question is, rather, how can we make these considerations of risk informative?

To be more concrete, let us return to the example presented in Section 2.3.2, where a system has four states, functioning as normal (3), intermediate state (2), intermediate state (1) and failure state (0). We think about a specific case and interpret these states in the following way:

- State 3: normal functioning state
- State 2: event of a 'known type' has occurred (e.g., known type of failure of a component or known type of threat/attack)
- State 1: an event of an 'unknown type' or having surprising features has occurred (e.g., unknown type of failure, a new type of threat/attack)
- State 0: loss of function (e.g., accident, attack with many fatalities)

Resilience is about the degree to which the system is able to quickly return to state 3 if a stressor causes it to go to state 2 or 1. There could be different types of scenarios and uncertainties related to which scenarios lead to a return to state 3 and which lead to loss of function (state 0). If the system is in state 2, measures can be developed to ensure that the system returns quickly to state 3, and the performance of these can be analyzed using standard assessments as the events are of known types. The system can be made resilient in state 2, and the level of vulnerability is low.

The challenge is state 1. Many of the measures considered in state 2 would also be useful for this state. For example, redundancy and diversification are fundamental principles in design of systems and used for meeting both known and unknown types of stressors. The question is what amount of additional resources should be used for meeting potential events in category 1. As discussed, it is relevant to discuss the risk and probability related to this state, recognizing that resources are limited. Clearly it is not straightforward to specify probabilities in situations for this type of event to occur. One approach is to introduce two categories of judgments of uncertainty: 'requires due attention' and 'normal attention', depending on the outcome of some judgments addressing the strength of the knowledge supporting the risk assessment. The criteria suggested in Section 3.2 can be used as a checklist for the type of assessment. The category 'requires due attention' may for example, be assigned if one or more of the following criteria apply:

1   Key assumption(s) made lack justification
2   Limited amount of relevant data/information
3   Different views among experts
4   Weak understanding of the phenomena and processes being studied (for example, on natural phenomena or on the capacity and intention of potential attackers)
5   Existing knowledge not being scrutinized
6   Signals and warnings

It is stressed that the point here is not to accurately determine the probabilities but to 'dig into' the knowledge that is available and unavailable and reveal risk aspects therein.

The conclusion of such an analysis could be that state 1 requires due attention, events of this type should not be excluded and additional measures should be considered. As these events are of the unknown type, we need to rely on resilience knowledge – identify measures that can improve the resilience of the system. We could, for example, introduce more redundancy, flexibility and diversification; however, such measures could threaten system performance and efficiency. Hence, a balance has to be found. For different types of applications, considerable work has been and is currently being conducted to identify the key criteria or indicators to use to be able to assess the level of resilience. It is common to consider different aspects of resilience, which include *responding* to regular and irregular threats in a robust yet flexible (adaptive) manner, *monitoring* what is going on, *anticipating* risk events and *learning* from experience (Hollnagel et al. 2006). The resilience criteria and indicators would relate to all four aspects. Consider the following three examples.

5.6.1 Personal life

Resilience is here interpreted as the ability of the individual to returning to one's original state of mental health or wellbeing when exposed to some adversity or stress.

There exist many key resilience criteria (e.g., Windle et al. 2011), including (Connor et al. 2003)

- Personal competence
- Acceptance of change and secure relationships
- Trust/tolerance/strengthening effects of stress
- Control
- Spiritual influences

and (Friborg et al. 2003):

- Personal competence
- Social competence
- Social support
- Family coherence
- Personal structure

The development of these criteria is based on psychological theories and empirical statistical testing. The level of individual resilience can be assessed by evaluating an individual with respect to these criteria. An assessment may show that a person scores low on some of these criteria, which may point to the need for a resilience strengthening program on these criteria.

5.6.2 Community resilience assessments

Communities involve activities and systems subject to a variety of hazards and threats that can result in significant disruption to critical functions. Community resilience assessment is a tool used to help communities better prepare for, respond to and recover from disruption (Gillespie-Marthaler et al. 2019). A number of criteria and indicators have been suggested and are used for this purpose. Gillespie-Marthaler et al. (2019) summarize and structure these. The authors refer to indicators reflecting different aspects:

Social (community composition, governance, policy and planning, services), economic (micro-/mesoeconomic efficiency; macroeconomic stability), environmental (built, natural, general) and resilience domains (survival, well-being, preparedness) and resources and drivers.

Examples are shown in the following for *community* and *built* (Gillespie-Marthaler et al. 2019):

Community: Relative health of community-led organizations; levels of trust, inclusion, awareness and cohesion; demographic characteristics (population stability, race, household make-up, etc.); spatial distribution of populations (access to key services, mobility, etc.)

Built: Quality of and access to critical infrastructure and networks (communication, transportation, power, etc.), physical safety and security (protective structures, housing, building codes, emergency shelter, land use planning and zoning, community blight/renewal, etc.)

An assessment of a specific community can then be carried out by providing scores on each indicator and using this to evaluate status and consider where to look for improvements.

The degree to which such indicators do in fact measure resilience can, however, be discussed. The validity of the indicators is a critical point, in particular for activities of this type consisting of so many types of systems and elements. For example, it is not obvious how to provide scores on trust; refer to discussion in Chapter 6, where we argue that some type of critical trust should be the ideal, highlighting some level of skepticism. In fact, a number of the issues captured by such indicators can be debated, and they may depend strongly on values and political stands. Yet the list of indicators provided can be useful as a point of departure for a discussion of what issues to address in the resilience assessment. It may be concluded to adjust some indicators and focus on just some selected ones.

Finally, in this section, we outline examples of indicators for transportation infrastructure (Hughes and Healy 2014), which is more detailed than the community resilience study presented previously. The assessment considers the following resilience aspects:

Technical resilience

Robustness (structural, procedural, interdependencies)
Redundancy (structural, procedural, interdependencies)
Safe-to-fail (structural, procedural)

Organizational resilience

Change readiness: Communication and warning, information and technology, insurance, internal resources, planning strategies, clear recovery priorities, proactive posture, drills and response exercises, funding, situation awareness (sensing), learning
Networks: Breaking silos, leveraging knowledge, effective partnerships (external)
Leadership and culture: Leadership, staff engagement and involvement, decision-making authority, innovation and creativity

The assessment is based on deriving scores for each aspect and summing up an overall resilience metric, which is used for identifying weaknesses and improvement areas. The method can also be used to study the effect of specific measures. The issues raised previously on validity also apply to this case.

5.7 OTHER METHODS

The risk science literature covers a number of risk assessment methods. Only some of them have been discussed in this book. There exist generic methods and more specific methods meeting the needs of different applications. This book highlights the former category of methods but has also looked into some examples of more domain-oriented techniques, for example, related to engineering safety studies. For further studies of risk assessment methods, the reader is referred to Zio (2007, 2016, 2018), Vose (2008), Yoe (2012), Meyer and Reniers (2013), Aven (2015b) and Rausand and Haugen (2020).

In the following, we will discuss techniques for event (hazard/threat/opportunity) identification, which is a special and important category of risk assessment methods. We have addressed some of them in previous chapters (including SWIFT, red teaming and FMEA), but would like to provide some additional comments in the following. See also Problem 5.15.

The first method we would like to draw attention to is anticipatory failure determination (AFD) techniques introduced by Kaplan et al. (1999); see also Aven (2014) and Jensen and Aven (2015, 2017). These methods apply different theories for creative problem solving, such as I-TRIZ (Sternberg 1999). The relevance of this methodology to risk analysis is rooted in the fact that revealing failure scenarios is fundamentally a creative act, yet it must be carried out systematically, exhaustively and with diligence (Kaplan et al. 1999). Traditional failure analysis addresses the question, "How did this failure happen?" or "How can this failure happen?" The AFD goes one step further and poses the question, "If I wanted to create this particular failure, how could I do it?" The power of the technique comes from the process of deliberately "inventing" failure events and scenarios (Masys 2012).

"…and by tomorrow, I'll need a list of specific unknown risk events that we'll encounter with this project"

These methods can be supported by different types of analysis approaches, for example, actor network theory (ANT) (Latour 2005; Masys 2012). ANT allows an understanding of the dynamics of the system by following the actors – it asks how the world looks through the eyes of the actor undertaking the work. As underlined by Dekker and Nyce (2004, p. 1630) and Masys (2012), using this approach, issues emerge pertaining to the roles that tools and other artifacts (actors) play in the actor-network in the accomplishments of their tasks.

Brainstorming is an established methodology for creating ideas and is often used for finding solutions to problems. It is also effective in relation to identifying events (hazards, threats, opportunities). The method is about generating ideas, not evaluating them. The more the better. Unusual ideas are of particular interest. Building on others' ideas is encouraged – take someone else's idea and tweak it. Combine suggestions. There should be few rules and procedures, yet techniques have been developed to spur brainstorming creativity. See the box subsequently presenting the Osborn (1963) questions used for this purpose.

There are several pitfalls to using brainstorming. Two of the most critical ones are group domination and group thinking. The former is about one of the participants of the group dominating the process and suppressing the creative energy of the group, whereas the latter is about the group members seeking consensus and in this way hampering creativity. See Yoe (2012) for further discussion on brainstorming in a risk context and related methods that can be used to stimulate creativity.

Alex Osborn (1963) posed questions to help stimulate brainstorming creativity that were adapted by Bob Eberle into an easy-to-remember acronym, SCAMPER (Yoe 2012):

- *Substitute*: What else instead? Who else instead? What other materials, ingredients, processes, power, sounds, approaches or forces might I substitute? Which other place?
- *Combine*: What mix, assortment, alloy or ensemble might I blend? What ideas, purposes, units or appeals might I combine?
- *Adjust/Adapt*: Does the past offer a parallel? What else is like this? What other idea does this suggest? What might I adapt for use as a solution? What might I copy? Who might I emulate?
- *Modify, magnify, minify*: Could I add a new twist? What other meaning, color, motion, sound, smell, form or shape might I adopt? What might I add? Could I duplicate, multiply or exaggerate it? Could I subtract something? Could I make it lower, smaller, shorter or lighter?
- *Put to other uses*: What new ways are there to use this? Might this be used in other places? Which other people might I reach? To what other uses might this be put if it is modified?

> - *Eliminate*: What might I remove or eliminate? Could I disaggregate or decompose it?
> - *Reverse, rearrange*: What might be rearranged? What other pattern, layout or sequence might I adopt? Can components be interchanged? Should I change pace or schedule? Can positives and negatives be swapped? Could roles be reversed?

Finally, this section provides reflections on methods for opportunity identification and related analysis. Our focus in this book is on risk – where there is a potential for undesirable consequences of the activity considered. However, the framework adopted for conceptualizing, assessing and handling risk is general in relation to the consequences and also allows for and encourages considerations with respect to positive, desirable consequences. Using, for example, the SWIFT technique or brainstorming, opportunities could be highlighted along with problems and failures.

Think about a sport activity, like a soccer match. A main strategy is to produce many chances, providing opportunities for success, meaning that there is a relatively high probability of scoring goals. Plans are made for how to produce such chances. In the same way, an enterprise has some main functions and objectives (the main being to be profitable), and strategies are developed to meet these objectives. One such strategy is to produce many 'chances' to be successful in this regard. That would involve identifying what type of chances could be possible and effective and ways to produce these chances. One example could be to start production in another country. Risk analysis and risk science are not the key instrument for identifying these chances but could be important when evaluating them, as the risks related to the strategies and plans could be critical for the actions to be taken.

Failures represent opportunities to think differently. If you are the coach of the soccer team and suddenly your team is down by two goals, you need to do something that can fundamentally change the game. Your team is likely to lose unless drastic measures are implemented. You are willing to take some risk. One idea is to completely change team formation, another to substitute several players. Analogously, for the enterprise, negative events represent opportunities to make changes that could have positive effects for the company in a longer perspective. A fire in a production system may, for example, lead to rethinking of the need for this system and implementation of other more efficient solutions. Hence, hazard and threat identification may also be seen as processes for opportunity identifications. Risk analysis and risk science provide support for how to assess and handle the risks related to strategies and plans.

5.8 PROBLEMS

Problem 5.1

Verify the calculations of Table 5.2 and that $Var[h] = 0.000256$ and $SD[h] = 0.0160$.

Problem 5.2

Is there a best importance measure for all situations? Can you establish a guideline for when to use different measures.

Problem 5.3

Consider the reliability example of Section 5.1. Let us go one step further and model the unit reliability p_i. Suppose an exponential probability distribution $F_i(t) = 1 - \exp\{-\lambda_i t\}$ is used, where λ_i is the failure rate of unit i, i = 1,2. The interpretation of the model is as follows: $F_i(t)$ represents the fraction of lifetimes less than or equal to t when considering an infinite (very large) population of similar units of type i, and $1/\lambda_i$ is the mean lifetime in this population. In this model $p_i = \exp\{-\lambda_i t\}$, where t is a fixed point in time. If T_i denotes the lifetime of unit i, we have $P(T_i > t) = 1 - F_i(t) = \exp\{-\lambda_i t\}$. On the system level, we introduce the lifetime T, and we can write $P(T > t) = 1 - (1 - \exp\{-\lambda_1 t\})(1 - \exp\{-\lambda_2 t\})$.

Sensitivity analyses can now be conducted with respect to the failure rates λ_i, as discussed in Section 5.1, and with respect to the choice of the exponential distribution. Discuss how this can be done in relation to distribution choice.

Problem 5.4

Refer to the uncertainty measure $E[v_i(p_i)]$ discussed in Section 5.1.1. We know from probability theory (Ross 2009, p. 119) that $Var[h] = E[v_i(p_i)] + Var[u_i(p_i)]$, where $u_i(p_i) = E[h| p_i]$. Use this formula to show that the uncertainty distribution of unit 1 has a stronger influence on the variation of the output result than the uncertainty distribution of unit 2.

Problem 5.5

We consider again the reliability example of Section 5.1. A common sensitivity measure (importance measures) is the correlation ρ between two unknown quantities, here the coefficient between p_i and h. Hence $\rho = \rho(p_i,h) = Cov(p_i,h)/(SD[p_i] SD[h])$. Explain the meaning of the measure and compute the index for the example in Section 5.1.

Problem 5.6

Tornado charts for different methods of sensitivity analysis are commonly used. One is based on a correlation measure. Use this measure to present a tornado chart for the reliability example of Section 5.1.

Problem 5.7

In Bayesian networks, the probabilities specified are commonly subjective probabilities. Can they be frequentist probabilities?

Problem 5.8

Consider the Bayesian network example of Section 5.2. Suppose the probabilities in the network are frequentist probabilities. Suppose some are known, others unknown. Introduce uncertainty distributions for the unknown and use Monte Carlo simulation to produce relevant metrics.

Problem 5.9

Consider the example studied in Section 5.3. Explain how we can calculate an approximate value for the fraction of time the system performance level is below 100 in the case of highly available units.

Problem 5.10

Consider an industry with extensive experience of construction and operation of facilities over several decades. A new facility is to be built which is of the same type as earlier facilities operated. Do you see the need for conducting a comprehensive PRA/QRA? Why/why not?

Problem 5.11

Present some resilience indicators for individuals in relation to financial crises.

Problem 5.12

In relation to COVID-19, what do you consider the most important contributors to vulnerability?

Problem 5.13

Perform a brainstorming session in your class using the SCAMPER guidance on what could hamper a successful realization of your current study program.

Problem 5.14

What are the main elements of a typical human health risk assessment?

Problem 5.15

Explain the basic ideas of the HAZOP technique.

Problem 5.16

In the case of large uncertainties, the term *plausibility* is sometimes used. Does it make sense to say that it replaces likelihood and knowledge strength judgments?

Problem 5.17

Consider the vulnerability description given in Table 5.4. Use outside resources to study the evolution of some system over time, for example:

- Space flight between the 1970s and present
- Safety in automobiles (seat belts, air bags, building materials, etc.) between the 1920s and present
- Cybersecurity between the 1990s and present
- Safety in manufacturing between the mid-1800s and present

Take the perspective of a system stakeholder responsible for understanding and managing risk for the chosen system. Complete the following tasks:

1   Complete a table, formatted similarly to Table 5.4, to characterize the vulnerability of the system.
2   Present your findings to the class.
3   Compare your description to classmates who studied similar systems. What similarities and dissimilarities exist in your characterizations? Why?

Problem 5.18

Consider the system you studied in Problem 5.16. What are examples of general knowledge (GK) and specific knowledge (SK) for this system?

Problem 5.19

In recent years, we have seen an increasing use of machine learning and artificial intelligence (AI) types of methods in risk assessments. Consider the literature on this topic (e.g., Ale 2016; Choi and Lambert 2017; Guikema 2009, 2020; Nateghi and Aven 2021). Explain the difference between supervised and unsupervised learning. Why have these methods gained increased interest recently?

KEY TAKEAWAYS

- Sensitivity analysis should be conducted to show how risk results depend on key factors and assumptions.
- A sensitivity analysis is not an uncertainty analysis.
- Scenario analysis can in some cases be seen as a sensitivity analysis but not in general.
- Importance measures are useful tools for identifying what are important risk contributors, where improvements can be made and so on.
- Different situations call for different sensitivity and importance measures.

- Bayesian networks and Monte Carlo simulations are well-established methods for modeling and analyzing large systems and activities with dependent elements.
- Vulnerability assessments need to address issues related to all of the elements C', Q and K.
- Vulnerability and resilience assessments should not be seen in isolation from risk.

PART III

Risk perception and communication

CHAPTER **6**

Risk perception

CHAPTER SUMMARY

Some people enjoy the adventure of hang gliding or cliff diving, while others are hesitant to even travel by airplane. Some people perceive the risk related to vaccinations as high and do not allow their children to be vaccinated, despite strong scientific evidence of a negligible risk due to the vaccination compared to the risk posed by getting an illness like measles. More people die from heart attacks than terrorist attacks, yet some fear the latter risk more than the former. Risk perception science offers explanations for such observations. It is the topic of the present chapter.

We look closer into what risk perception actually means, in particular how it relates to the professional judgments discussed in Chapter 3–5. In short, we define risk perception as a person's subjective judgment or appraisal of risk, taking into account social, cultural and psychological factors. In contrast to risk perception, professional judgments about risk are not to include emotions or feelings nor conclusions about risk (un)acceptability. Risk perceptions can in some cases identify aspects of risk which are not properly reflected by professional risk assessments. People's concerns could be justified, although analysts have not yet identified the problem or they have judged it unimportant. The point being made is that risk perception is not only about emotions and feelings; it can also capture conscious judgments of uncertainties and hence risk. An historical example is the risk related to nuclear power plants, where the experts argued that the risks were relatively small and people's concerns were all about perceptional aspects. However, the risk perception was also capturing important aspects of risk which the professional judgments at that time basically ignored.

Risk perception research has shown many examples of what is referred to as misjudgment of risk. This research has shown that when faced with situations characterized by factors such as lack of control, catastrophic potential, delays in effects, new and unknown – often summarized by dread and newness – people tend to raise their risk figures about how likely the actual events are to happen to them, and there is a potential for overreaction.

DOI: 10.4324/9781003156864-6

The chapter reviews this important insight about risk perception but also warns against a naïve use of it: How can we conclude that we 'misjudge the risk'? What is the reference for comparison when faced with situations characterized by large uncertainties? This issue is interesting from a risk science point of view, and several examples will be presented in this chapter to illustrate the dilemmas faced.

Risk perception studies are important for identifying concerns but not so much for measuring the seriousness of the risk. People's risk perception suffers from a number of measurement problems, which will be discussed in this chapter. A main challenge relates to the fact that people's ability to draw conclusions from probabilistic formulations and information is in many respects subject to strong biases. People uses different types of heuristics (including the availability and representativeness heuristic; see box in Section 6.1.3), which can lead to poor judgments and understanding of risk.

To understand how people perceive risks and why there could be a difference between risk perception and professional risk judgments, it is important to understand how people think. The chapter will explain the difference between 'thinking fast and slow', as reflected by the dichotomy between the *System 1* mode of thought, which operates automatically and quickly, instinctively and emotionally, and the *System 2* mode of thought, which is slower, more logical and more deliberative. Risk perception is often strongly influenced by System 1, whereas professional risk judgments are based on System 2 thinking.

Risk perceptions may influence decision-makers, as the perceptions express what is important for people. However, care should be shown taking risk perceptions as clear guidance for how to handle the risks, as risk perceptions could lead to resources being used unwisely, for example, by financing costly risk reduction measures that are high on the public agenda but may only marginally improve human health and the environment or by not addressing serious risks because these are not perceived as serious in the public eye. As perceptions are based partially on biases or ignorance and are highly subjective, they should not be used as strict yardsticks for risk reduction. At the same time, however, these perceptions reflect the real concerns of people and may include important aspects of risk not fully captured by risk assessments. Risk perception needs to be addressed to properly understand, communicate and handle risk.

LEARNING OBJECTIVES

After studying this chapter, you will be able to explain

- the meaning of risk perception
- the difference between risk, risk characterizations and risk perceptions
- the difference between professional risk judgments and risk perceptions
- the meaning of the two modes of thought, System 1 and System 2, and their link to risk perception and risk assessments
- common biases and heuristics in relation to risk perceptions
- the basic ideas of the psychometric model explaining risk perception through factors like lack of control, catastrophic potential, delays in effects, new and unknown
- the relationship between risk perception and trust
- the basic ideas of 'social amplification of risk'

The chapter is organized as follows. First, we provide some examples to illustrate what risk perception knowledge is about and the difference between risk perception and professional risk assessments. Then we review basic risk perception theory, covering the different topics mentioned in the summary. We also provide an example illustrating how to conduct a risk perception study. Section 6.2 is partially based on Visschers and Siegrist (2018) (mainly (Sections 6.2.2 and 6.2.3), Chauvin (2018) (parts of 6.2.4) and Fjæran and Aven (2019, 2020a, 2020b) (Sections 6.2.1 and 6.2.5).

6.1 EXAMPLES ILLUSTRATING WHAT RISK PERCEPTION AND RISK PERCEPTION RESEARCH ARE ABOUT

6.1.1 Traveling by plane

Let us return to the example from Chapter 3 where Anne and Amy consider a plane trip from Washington, DC, to San Francisco. Amy is hesitant because she views the trip as being risky. In Chapter 3, we discussed this example from a professional risk characterization and assessment point of view. Now we take a risk perception perspective. Amy is afraid of flying because she perceives the risk as high. Her stance was formed years earlier on a trip with her brother who pilots small planes as a hobby. During a Piper Cherokee trip involving flying over mountains and fjords, Amy was startled by her brother's piloting with aerobatic maneuvers. While such maneuvers were not atypical for this type of flight, they led to Amy feeling very uncomfortable. She perceived the risk as very high then, and she does the same for all flights now. She knows that traveling on a commercial airliner is quite different than flying in a smaller plane; nonetheless, she remains uncomfortable entering any plane. She acknowledges that her risk perception is greatly influenced by her feelings and that flight with her brother.

Clearly, such reflections can be made without being an expert on risk science. As non-risk professionals, it is natural to talk about how we feel in relation to risk while at the same time having an idea about a 'real risk', which provides the proper (professional, scientific) reference for making judgments about whether the activity is dangerous. Intuitively, this real risk is linked to the history and observations, and some would also relate the risk to probabilities. However, few non-risk professionals would be able to explain precisely what the risk perception concept captures and how it differs from more professional risk judgments. Risk science and in particular risk perception knowledge are needed for this. This science provides concepts and knowledge that can be used to obtain a higher level of understanding on these matters: what risk judgments are really about, why a person perceives one risk as high and not others and what the difference is between risk perception and professional risk judgments.

As with any science, new knowledge is continuously gained through research and development. For risk perception, some major insights were obtained in the 1970s and 1980s. A main challenge for the research was to identify the key factors that influence the perception of hazards and threats. Factors included voluntariness of risk, immediacy of effect, knowledge of risk of those who are exposed, scientific knowledge, control over risk, newness, dread potential and high severity of an incident. Based on extensive studies using questionnaires and statistical analysis, researchers concluded that there are two main contributors: dread and newness (Fischhoff et al. 1978; Slovic 1987). From this result, a well-known map was developed, which shows a number of hazards in a two-dimensional space with these two factors; see Figure 6.1. It is common to refer to this research as the psychometric paradigm.

The psychometric paradigm helps us to better understand why lay people (non-risk professionals) are concerned about some hazards but not about others (Visschers and Siegrist 2018). Gene technology and nuclear power are perceived as dreadful and unknown hazards. Consequently, they are perceived to be high risk. However, hazards such as alcoholic beverages or swimming are considered familiar and non-dreadful. As a result, these hazards are perceived as not being very risky.

During Amy's experience with flying with her brother, she was influenced by dread and newness. This was her first flight in such a small plane. For her, lack of control over the risk was also very important. Sitting in the back seat, she just had to wait and see what would happen next. Her brother, on the other hand, had full control over the flight, including the aerobatic maneuvers. For him, the risk was perceived as small, with low scores on dread and newness, as well as lack of control over the risk.

Is there a 'real risk' related to Amy's flight? In case there is, what is it and how can it be determined? We thoroughly discussed this issue in Section 3.2.3 when looking into flight risks and professional risk characterizations. We argue that accident data provide a good reference point for the term 'real risk' when talking about flight risks. Data exist for general aviation and for commercial airliners. Amy admits that her perception of risk being high for commercial flights is because of her feelings. For Amy's trip with her brother, there is some additional relevant information. Her brother may have been a good pilot, but he did some aerobatic maneuvers. It is difficult to define a 'real risk' in this case. Amy's perception of the risk is also high here, but was this perception only

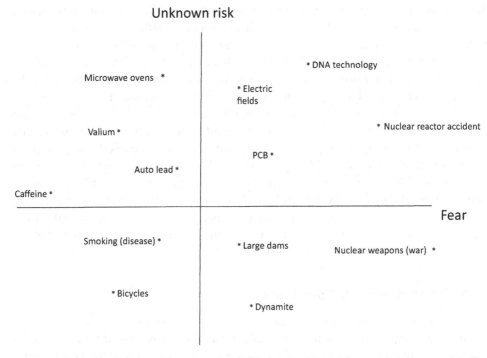

FIGURE 6.1 Risk perception judgments relative to the factors fear and unknown risk (the degree to which the risk is understood), specified for a set of activities and technologies, on the basis on a study of US students (based on Slovic 1987). It is observed that a nuclear reactor accident scores high with respect to fear, whereas smoking scores rather low for both dimensions

a result of feelings about the commercial flight? No, and this we will discuss in more detail in the coming examples. Risk perception may also capture conscious judgments of risk and uncertainties, which has nothing to do with feelings or affect.

6.1.2 The nuclear industry risk

In the 1970s and 1980s, there was a strong debate about the risk related to nuclear power plants. Some experts argued that the risk related to possible serious accidents was minor and there was no need to be concerned. However, many people remained concerned. Risk psychologists explained why: Nuclear activity 'hits all the hot buttons' for a high perceived risk and potential misjudgment of the risks. These buttons include lack of control, catastrophic potential, delays in effects, new and unknown (summarized by the two dimensions dread and newness); see discussion in the previous section. Because of such factors, people tend to raise their risk estimates – how likely it is to happen to them.

At the same time, it is acknowledged that assessing the risk posed by the nuclear industry is difficult. How can we then conclude that we 'misjudge the risk'? What is the reference for comparison when faced with situations like this characterized by rather large uncertainties? The nuclear industry was then still in a rather early stage. This issue is interesting from a risk science foundation point of view. Seeing uncertainty as a key component of risk (refer to Chapters 2 and 3), it is not sufficient to use historical data and probabilistic estimates as a reference for what is the 'real risk'. This was exactly what happened. The experts built their conclusions on statistics and probability numbers, but these did not convince the public about the plants being safe. Now we know that the public concerns made sense, as the experts' perspective on risk failed to properly capture important aspects of risk. Peoples' risk perception reflected perceptional factors like fear and dread, as mentioned, but also conscious judgments of uncertainties and risk that were not adequately reflected by the professional judgments. Hence, there is no basis in stating that the public concerns related to the nuclear risk represent a misjudgment of risk. The history with several serious incidents in the nuclear industry (e.g., Chernobyl, Fukushima) has shown that the concerns raised had also a rational basis.

6.1.3 The emergence of a pandemic

Inspired by the development of the coronavirus in 2020, we can think more generally about the emergence of a possible pandemic caused by a new type of virus, where the uncertainties about the seriousness of the virus are large, for example, on how the virus incubates, spreads and impacts humans. The situation is similar to the nuclear power plant case discussed previously – the epidemic again 'hits all the hot buttons' for a high perceived risk and potential misjudgment of the risks. The fact that people are repeatedly informed through the media about cases and deaths has the same effect. We know from the risk perception research that people use some rather primitive cognitive techniques when assigning probabilities and making judgments about risk, as in this situation; see text box subsequently. As we can easily retrieve from memory a number of 'alarms', the conclusion is that the danger is large. The representativeness is not questioned. Every year, hundreds of thousands are killed by influenza, but it is not given as much attention. The risk perception literature explains: we tend to either 'round down' the probability to 'basically zero' and we possibly underreact, or we focus on the worst-case outcome, which gives us a strong feeling, so we may overreact (Kahneman 2011; Slovic 1987; Fisher 2020).

As for nuclear risk, we can question how it is possible to conclude that we 'misjudge the risk' given the large uncertainties. There exists no truth reference at this stage of the development that we can compare with. There were some public concerns in relation to the coronavirus as early as February 2020 (Fisher 2020) – it turned out that these concerns were justified and that the indications about overreaction and misjudgment of the risk were unsubstantiated.

Of course, this coronavirus example and the nuclear example do not show that public concerns are always right and the experts wrong. The point made is that we should be careful concluding that people overreact or misjudge the risk when the situation considered is subject to uncertainties. Rather we should acknowledge that the risk is amplified, but the reason for this could be that the professional risk judgments have not adequately reflected all risk aspects.

BIASES AND HEURISTICS

When making judgments about probabilities, people tend to use some rather primitive cognitive techniques, often referred to as heuristics. The result of using such heuristics is typically that the assessor unconsciously puts 'too much' weight on some factors. Three of the most common heuristics are (Tversky and Kahneman 1974):

Representativeness: The assessor assigns a probability by comparing with the stereotypical member of the population considered. The closer the similarity between the two, the higher the judged probability of membership in the population. In the case of a new virus epidemic, the probability of severe impacts could be judged low if comparisons are made with a typical flu.

Availability: The assessor's probability assignment is to a large extent based on the ease with which similar events can be retrieved from memory. Events where the assessor can easily retrieve similar events from memory typically lead to higher perceived probabilities than events that are less vivid or unknown to the assessor. In the plane example previously, Amy's mind is focused on her unpleasant trip with her brother, amplifying her probability of thinking something will go wrong.

Adjustment and anchoring (see Problem 6.11): The assessor chooses an initial anchor and adjusts information and beliefs in relation to this anchor. A failure rate estimation could, for example, be strongly affected by questioning whether the rate is smaller or larger than a specific low or high number.

6.2 RISK PERCEPTION KNOWLEDGE

In risk perception research, one of the most important theories and approaches developed is the so-called *psychometric paradigm* (Fischhoff et al. 1978; Slovic 1987), which was mentioned in the chapter summary and discussed in Sections 6.1.2 and 6.1.3. The basic idea of the approach is to identify the main factors that influence the perception of hazards and risks. The following factors were considered: voluntariness of risk, immediacy of effect, knowledge of risk of those who are exposed, scientific knowledge, control over risk, newness, number of people killed in an incident, dread potential and high severity of an incident. Then statistical analysis was conducted, showing that there were two main contributors: dread and newness. From this result, the map in Figure 6.1 was established, which shows various hazards in a two-dimensional space with these two factors as axes.

This map is one of the most popular figures in risk science. It represents scientific knowledge about risk perception. However, knowledge is not static; it reflects the most warranted statements and justified beliefs of the field at a particular point in time. It is a basic feature of any science that the current beliefs are scrutinized in order to obtain new knowledge and improved models. The dread–newness map is a model, and, like all models, it is a simplification of the 'world' – with limitations and weaknesses. Since the development of the dread–newness map late in the 1970s, considerable work has been conducted to gain further insights on the issue; see, for example, Sjöberg (2000, 2003), Siegrist et al. (2005) and Visschers and Siegrist (2018). Two main directions of research can be identified. The first focuses on revealing variations in perceptions as a result of differences concerning specific features of the hazards or risks, whereas the other focuses on identifying why individuals perceive risks differently, addressing factors like the perceived benefits, trust, knowledge, personal values and fairness.

Research shows that for many types of hazards, the higher the perceived benefit, the lower the perceived risk, and vice versa. Smoking, for example, tends to be judged as high in risk but rather low on benefit, whereas antibiotics are commonly judged as high in benefits but relatively low when it comes to risk. This inverse relationship between perceived benefit and perceived risk is understandable in view of the fact that we are more prone to accept risks if the benefits are high than if the benefits are small. If an industry leads to more jobs and a better standard of living, the accident risks are perceived as less bothersome than if there are minor benefits from the industry.

For some hazards, the perceived benefits seem to be more important than the perceived risk in explaining people's acceptance of this hazard. However, for other types of hazards, for example, in relation to artificial sweeteners and food colors (potentially causing tumors and cancer), the opposite conclusion is made: The perceived risk is more important than the perceived benefits in explaining people's acceptance of the hazards (Visschers and Siegrist 2018).

These results are not surprising. Our attitudes to an activity, with its risks, depend on our judgments and perceptions of the risks and benefits. There is no simple formula explaining these attitudes, as these judgments and perceptions relate to uncertainties and values, which are highly subjective. In the last example of artificial sweeteners and colors, people may agree about the benefits of the hazard but not about their risks. There is a rather strong element of uncertainty, and hence the risk is high, which makes people concerned.

Visschers et al. (2011) show that for the nuclear power hazard, the perceived benefits could be more important than the perceived risk for explaining the acceptance of the industry. The example illustrates the importance of considering these types of results as findings depending on the context. The study by Visschers et al. (2011) was based on interviews from Switzerland (both the German- and French-speaking parts of Switzerland) before the Fukushima disaster in March 2011. Would this event change the risk perception? As will be discussed in Section 6.2.1, not necessarily. But we know that following Fukushima, Germany decided to phase out its nuclear power plants by the end of 2022 (Ethik-Kommission 2011; Aven and Renn 2018). The risks and uncertainties became an important issue in Germany.

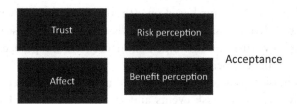

FIGURE 6.2 Model linking acceptance of an activity with perception and underlying key factors (trust and affect) (based on Visschers et al. 2011)

More knowledge about a hazard may influence the risk perception strongly. Obviously, it could lead to either an increased or decreased perceived risk depending on the type and strength of the new knowledge. If strong evidence is provided for a drug causing (not causing) cancer, it could obviously increase (decrease) the perceived risk. In some cases, more knowledge would have a minor influence on the risk perception, as it is mainly determined by values. For the nuclear industry, more knowledge about the risks would probably not change people's risk perception very much, as it is to a large extent determined by how much we dislike this risk and weigh it relative to the industry's benefit.

For many situations and cases where risk perception is an issue, knowledge about the risks is weak. Then the factors trust and affect are of special importance to explain the risk perception (see Figure 6.2, which shows a common model for understanding lay people's acceptance of an activity). Trust and affect influence risk and benefit perceptions, which in their turn influence the conclusion on acceptance or not. Trust and affect also interact. The term *affect* is often used as an umbrella word to describe anything related to emotions – a person's basic sense of feeling. More specifically, affect commonly refers to a subtle feeling of goodness or badness, which underlies all emotional experience, whereas emotions (e.g. joy, fear, sadness) are more intense, typically shorter in duration and of a more bodily character (Russell and Barrett 1999). Most of the time we do not experience emotions but find ourselves in a calmer state, guided by more subtle and less intense feelings – namely affect.

6.2.1 Trust and risk perception

Research shows that trust is formed on the belief that the relevant actors share the same values (value similarity; Earle and Cvetkovich 1995). The stronger belief, the stronger trust that the risks would be appropriately handled. People particularly rely on trust when making judgments about a hazard or risk when they have little knowledge about that hazard or risk (Siegrist and Cvetkovich 2000). As trust is related to values, which are fundamental to people and do not vary much over time, it is to be expected that the link between trust and perceived perception also remains relatively stable. Visschers and Siegrist (2013, 2018) provide an example related to nuclear power. The accident in Fukushima did not significantly affect the Swiss people's trust in, perception of or acceptance of the industry. However, as noted, the accident prompted Germany to phase out its nuclear power plants based on a thorough evaluation with a strong ethical foundation (Ethik-Kommission 2011). See Visschers and Siegrist (2018) for other examples showing how trust in stakeholders influences the acceptance of a hazard and risk through risk and benefit perception.

In risk research, there is a general understanding that trust affects how one understands and perceives risks and risk events and also risk response. There is, however, some disagreement about the strength of this relationship. Nonetheless, trust is commonly associated with acceptance of risk-related messages, compliance and effective functioning of democratic processes and societal functions. Distrust, on the other hand, is often related to heightened public concern, risk amplification, questioning of the work of risk regulators, enacting measures for risk reduction or avoidance and selective use of information sources (Walls et al. 2004).

Today, many risks are regulated in what has been characterized as a landscape of social distrust, a distrust to a large extent caused by the failure to take into account public concerns when assessing, managing and communicating about risk (e.g., Löfstedt 2013; Frewer 2017; Frewer and Salter 2010). There is, however, also evidence pointing to stable or increasing levels of trust (e.g., van de Walle et al. 2008). Yet the conception of the existence of public distrust in today's society strongly influences research and the political discourse and work of many governments, policy makers and risk managing institutions. Strategies like stakeholder involvement, public participation and communication of scientific uncertainties in risk governance processes are increasingly drawn upon in order to rebuild or increase levels of public trust. It can, however, be discussed to what degree these strategies have had the desired effects on trust. Maybe we are attempting to rebuild something that never existed (Wynne 2006). Or maybe the cases we are 'talking about' are not expressions of trust or distrust but of something more complex and multidimensional, and trust or distrust is not necessarily descriptive of how the public perceive and relate to risk managing institutions and information coming from these. This is in line with research of Poortinga and Pidgeon (2003) demonstrating the co-existence of trust and distrust in the public perception of government and its policies and of Walls et al. (2004) holding that "The binary opposition of trusting or not trusting is inadequate to understand the often ambiguous and contradictory ideas people possess" (p. 133).

There are good reasons to question the dominant understanding of trust and distrust as two mutually exclusive states, of 'full' trust as a complexity-reducing factor in society and an ideal 'state of affairs' and of distrust as the opposite: a negative and complicating factor and situation that should be prevented or counteracted.

An alternative perspective is to recognize distrust as a potential resource in risk assessment and risk managements contexts, as discussed by Fjæran and Aven (2020a). Understanding distrust in more positive terms corresponds with the ideas and statements of, for instance, Barber (1983) and Tuler et al. (2017). A certain amount of distrust is necessary for political accountability in a participatory democracy. Distrust serves important functions, for instance, ensuring social and political oversight, generating alternative control mechanisms and holding in check the power of elites and technical experts.

Such a way of relating to distrust requires an acknowledgment of the complexity of the trust concept and that distrust and trust are not 'either–or' states. Trust and distrust exist along a continuum, ranging from critical emotional acceptance at one end of the extreme to downright rejection at the other (Walls et al. 2004). Between these two extremes on the continuum of trust lies what is defined as a healthy type of distrust, reflecting that the public can rely on institutions and at the same time possess a critical attitude towards them. To illustrate the multidimensional and complex character of the trust concept, reference is made to the typology of trust put forward by Poortinga and Pidgeon (2003). In a study of public trust in governmental risk regulation, Poortinga and Pidgeon found that different degrees of trust coexisted with different degrees of skepticism. The typology combines varying degrees of the two independent dimensions, general trust and skepticism, into different categorizes of trust; see Figure 6.3. The dimension of general trust covers aspects of competence, care, fairness and openness, while skepticism, the second dimension, involves a skeptical view of the process by which policies are brought on and put into practice and concerns the credibility and reliability of the enactor. Skepticism also includes the 'vested interest' factor, put forward by Frewer et al. (1996) as a measure of integrity, and has an affective character. The typology ranges from full trust (acceptance/trust) to deep distrust (rejection/cynicism). The category of trust called critical trust in the typology is similar to what Walls et al. (2004) describe as a healthy form of distrust. Critical trust is defined in Pidgeon et al. (2010) as a practical form of reliance on a person or institution combined with a degree of skepticism.

Increased knowledge of an issue makes trust less influential in affecting risk perception and subsequent responses. When one is well informed and knowledgeable about a topic, one can use this available knowledge when making decisions, and trust becomes superfluous (Earle et al. 2012). Hence, by facilitating consumers and the general public to actively and skeptically relate to information and by increasing their knowledge and awareness of risks, the role of trust when it comes to how these groups interpret and respond to risk-related information can be diminished.

At the same time, exposing and emphasizing uncertainties and knowledge gaps when discussing and communicating risk-related information can generate risk amplification

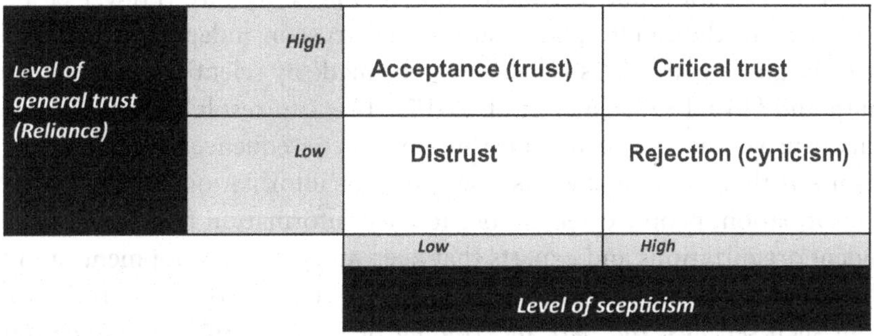

Level of general trust (Reliance)	High	Acceptance (trust)	Critical trust
	Low	Distrust	Rejection (cynicism)
		Low	High
		Level of scepticism	

FIGURE 6.3 Typology of trust (based on Poortinga and Pidgeon 2003; Fjæran and Aven 2020a)

(higher judged or perceived risk) in different ways. Risk analysts and managers may appear less competent and less in control and may lead the public to interpret risk as higher than following a more 'narrow' probabilistic approach to risk. Studies have, for example, shown that when there is initially trust, reception of knowledge is associated with more concern and higher risk perception (Malka et al. 2009; Earle et al. 2012). Also, stimulating careful evaluation of the information about risks can serve to amplify signals or aspects previously ignored or overlooked and affect how this information is perceived and how the public reacts. It may, for example, lead to cynicism or rejection of information, increase risk perception (Frewer et al. 2003; Jansen et al. 2019) and negatively affect levels of trust.

It has been hypothesized by many (e.g., Frewer et al. 2003; Van Asselt et al. 2009) that the fear of increasing public distrust lies behind much of the unwillingness to disclose uncertainties. However, studies have also demonstrated that the general public is familiar with and capable of handling uncertainties (e.g., Wynne 1992, 2006; Frewer et al. 2003) and that uncertainty constitutes a central element in how the public understands and relates to risks (Fjæran and Aven 2019). Generating some distrust and amplification in the early life of risks may prove an important investment in the long run. Honestly and openly displaying uncertainties, stimulating skepticism of information and enabling public awareness at an early stage can act to avoid or reduce later amplification and increased risk perception later on; see discussion in Section 6.2.5. And, as indicated by, for example, Earle et al. (2012) and Malka et al. (2009), when the background is characterized by skepticism, low trust or distrust, the reception of more or new knowledge does not necessarily entail increased risk perception and concern. It is when there is initial trust that the impact on risk perception is greatest.

Presenting the public with an 'objective' picture of risks and stressing safety aspects where uncertainties, concern and low trust exist can have opposite effects of those expected. Referring to risk assessments and estimates in such settings may increase concern and lead to amplification – higher perceived risks. A consequence of not taking into account and incorporating public concerns into assessments and presenting risks in 'objective' terms is increased distrust in the motives of regulators, science and industry (Frewer and Salter 2010), as well as the belief that information has been distorted and that the source is protecting its own interests rather than providing good information out of concern for the public welfare (Frewer et al. 1996, 2003). In general, the public places substantial trust in independent scientists but gives little weight to statements it believes to be made by scientific guns for hire (Jenkins-Smith and Silva 1998; Tuler et al. 2017). This can result in distrust of sources traditionally providing risk-related information. A consequence of such distrust – and skepticism – is that the public looks elsewhere for information. When there is conflicting information, people often choose to trust information from the 'watchdogs': independent organizations and experts that keep an eye on developments and inform the public about potential consequences (Pidgeon et al. 2010). According to Slovic (1999), in such settings, the bare mentioning of possible links or associations and statements of potential risks outweigh any statement of lack of evidence of causal effects and low probabilities.

The response of the public and the degree of distrust or skepticism these reflect can often be tied to the failure of risk assessors and managers to recognize the role of uncertainties in the way the general public understands and perceives risks. The technical language used by those in charge of assessing and managing the risks and the understanding of risk does not match the public interpretation of the risks. Risk numbers and probabilities do not always reflect how the public fundamentally understands risk. The public considers risks in a broader value context than the technical probabilistic notion of risks. An uncertainty-based risk perspective, as presented in this book, we argue, resonates better with public understanding of risk through its broader understanding of risk and its more 'humble' attitude to knowing the truth about risk.

If risk assessors and managers proactively address uncertainties of the consequences and knowledge limitations and consider these aspects in their risk communication, there is potential to reduce or avoid amplification. An example of amplification could involve the public turning to risk protestors and 'watchdogs' for information. Such a perspective could also potentially 'block' or pre-empt some of the amplification (increased risk perception) generated by distrustful stakeholder groups and those opposing the risks. Following Van Asselt et al. (2009), an unintended consequence of avoiding addressing and not recognizing the importance of uncertainties is the increased distrust among risk protestors themselves. These actors may exaggerate uncertainties and/or misuse information in ways that may produce unnecessary amplification. If these groups 'reveal' camouflaged or downplayed information, presenting risks as mismanaged and attenuated, this may seriously harm public trust. This point illustrates how amplification and the extent of such amplification can be tied to the degree of prior attenuation (Poumadere and Mays 2003); see also discussion in Section 6.2.5.

6.2.2 Affect and risk perception

Affect and emotions influence how people perceive risk and the degree to which they accept a hazard/risk or an activity, for example, a new technology. Instantaneous affective responses may be as important as cognitive assessments and deliberations on the risks (L'Orange Seigo et al. 2014; Slovic et al. 2004; Strack and Deutsch 2004).

The risk perception literature refers to an affect heuristic, which can be seen as the mental shortcut in which people rely on the positive or negative valence associated with a hazard to judge its benefits and risks (Finucane et al. 2000). If negative feelings overrule, then people will associate the hazard with low benefits and high risks, and vice versa. The heuristic also explains why perceived benefits and risks are inversely related for most hazards, as discussed previously. Affect has a type of indirect influence on acceptance of a hazard or an activity (for example, a technology) as trust; see Figure 6.2.

Studies indicate that affect can explain the observation that people seem to be more concerned about man-made (i.e., anthropogenic) catastrophes compared with natural disasters (Siegrist and Sütterlin 2014). There can be very different reactions to an event consisting of birds being killed because of an oil spill compared to the event of birds being killed due to weather activities. The fact that we have some control over the risk in the former case and not in the latter may to a large extent explain this difference.

A common method to study people's affective reaction to a hazard or risk is to explore their affective associations with this hazard or risk. In a study, individuals were asked several times which association came spontaneously to their mind when thinking of a hazard or risk – and how negative or positive this association was to them (Szalay and Deese 1978). In the case of nuclear power, male proponents of nuclear power plants strongly associated nuclear power with its positive impacts and with the need for this energy, whereas male opponents of nuclear power plants thought about health risk and nuclear waste. Female opponents showed associations with accidents and military use of nuclear technologies (Keller et al. 2012).

The negative emotions of fear and anger have been strongly related to risk perception. Fear is experienced when a person encounters an unpleasant situation that is accompanied by uncertainty (Lerner and Keltner 2001). The result is a pessimistic risk perception and wanting to escape or avoid the hazard or risk. Anger is evoked when a negative situation can be blamed on someone else and can, for example, result in risk-seeking behaviour. Studies have shown (e.g., Böhm 2003) that people perceived global human-caused hazards (e.g., chemical dumps) to be high on the basis of moral emotions, such as anger and contempt, whereas natural hazards (e.g., earthquakes) were perceived low on moral emotions and higher on emotions such as fear and hopelessness, related to future impacts (so-called 'prospective consequential emotions'). Fear is commonly associated with higher perceived risks, whereas anger often reduces the perceived risks, benefits and acceptance of a hazard (Dohle et al. 2012; L'Orange Seigo et al. 2014).

Human-caused creeping hazards (i.e., slowly evolving, unobserved hazards, such as species extinction and nuclear power) are strongly associated with 'prospective consequential emotions'. These types of hazards are also associated with emotions related to 'retrospective consequences', such as regret and sadness. Ethical emotions induced aggressive and retaliation responses, whereas both types of consequential emotions were related to help and prevention responses (Böhm and Pfister 2000).

The affect heuristic can obviously lead to biased or 'poor' decision-making. Siegrist and Sütterlin (2014) provide an example where identical negative outcomes are evaluated differently depending on the type of activity. We write 'poor' in quotes, as what is poor depends on the perspective taken. Care has to be shown when discussing rationality in this context, as illustrated by the examples of Section 6.1.2 and 6.1.3; see also Section 6.2.4. In case of large uncertainties and weak knowledge, there is no reference for deciding objectively what the best decision is.

Different modes of thinking: System 1 and System 2

Affect and other heuristics are examples of what is referred to in risk perception research as System 1 thinking. It operates automatically and quickly, instinctively and emotionally, in contrast to System 2, which is slower, more logical and more deliberative. A considerable body of literature exists linking this dichotomy to risk analysis, communication and management. It is argued that, to properly respond to and handle risk, both types of thinking are needed; they are complementary. Daniel Kahneman (2011) explains the concepts in detail, building on works going back to the 80s.

The real self is both System 1 and System 2. As Epstein (1994) noted:

> There is no dearth of evidence in everyday life that people apprehend reality in two fundamentally different ways, one variously labelled intuitive, automatic natural, non-verbal, narrative and experiential, and the other analytical, deliberative, verbal and rational.
>
> (p. 710)

For System 1, think, for example, about detecting that one object is more distant than another, detecting hostility in a voice or driving a car on an empty road (Kahneman 2011). Some mental activities become fast and automatic through prolonged practice. System 1 has learned associations between ideas; for example, most people will think of Paris when the capital of France is mentioned. Other activities, such as chewing, are susceptible to voluntary control but normally run on automatic pilot. For System 2, think, for example, about focusing on the voice of a particular person in a crowded and noisy room, monitoring the appropriateness of your behavior in a social situation or checking the validity of a complex logical argument (Kahneman 2011).

Table 6.1 compares these two modes of thinking, based on Slovic et al. (2004) and Epstein (1994), in which the authors refer to these modes as "experiential system" and "analytical system". These two modes correspond to Kahneman's System 1 and System 2, respectively.

In a risk context, we can distinguish between many types of thinking, including

1   perception of risk as high, fearful and unacceptable
2   judgments of the risk as high

TABLE 6.1 Comparisons of the two modes of thinking (Slovic et al. 2004, based on Epstein 1994)

Experiential system (System 1)	Analytic system (System 2)
1 Holistic	1 Analytic
2 Affective: pleasure-pain oriented	2 Logical: reason oriented (what is sensible)
3 Associationistic connections	3 Logical connections
4 Behavior mediated by "vibes" from past experiences	4 Behavior mediated by conscious appraisal of events
5 Encodes reality in concrete images, metaphors and narratives	5 Encodes reality in abstract symbols, words and numbers
6 More rapid processing: oriented toward immediate action	6 Slower processing: oriented toward delayed action
7 Self-evidently valid: "experiencing is believing"	7 Requires justification via logic and evidence

3 judgments of the risk as too high
4 judgments of the uncertainties as high
5 judgments of the uncertainties as too high
6 professional judgments of how large the risk is, based on historical data and probabilistic analysis
7 professional judgments of how large the risk is by comparing with other activities

Here 1–5 could be examples of System 1 thinking or System 2, depending on the setting, whereas 6–7 are associated with System 2. Think, for example, about 3 and the examination case presented in Section 3.4. When faced with the complacency signal, the automatic thinking of System 1 could have triggered an instinctive and automatic reaction: re-check the exam carefully, as complacency is an indicator that something could go wrong. However, statement 3 can also be a result of System 2, when concluding that the risk is too high given the signal, using professional risk assessment methods with associated deliberations. If, however, the 'hard data' are the focus of the analysis, such softer information is easily ignored. In general, risk assessment is based on System 2 thinking, whereas risk perception is strongly influenced by System 1.

6.2.3 Other factors

Values relate to principles or standards expressing what is important in life. Values influence our attitudes, thinking, beliefs and behavior. Hence, values would also affect people's risk perception. This is thoroughly discussed in the literature. Theories have been developed, including the cultural theory of risk perception (Douglas and Wildavsky 1982) and cultural cognition theory (Kahan et al. 2007). In these theories, individual and societal differences in risk perception are explained by differences in underlying cultural values. It is commonly referred to as four characteristics or worldviews: hierarchy vs. communitarianism (i.e., the importance of the group) and individualism vs. egalitarianism, but several other categorizations have been suggested; see Problem 6.4. Research indicates, for example, that people with strong communitarian and egalitarian values are typically more worried – perceive the risk as higher – than people with strong hierarchical and individualistic values about issues like gun possession, environmental pollution and climate change (Kahan et al. 2007, 2012; Shi et al. 2015), the reason being that these hazards and risks can have severe negative consequences on society as a whole and consequently oppose egalitarian or altruistic values. Typical political stands can be explained by these worldviews; for example, individualists correspond to libertarians, egalitarianists to progressives/socialists and hierarchists to traditional conservatives.

Not surprisingly, values are more important for the perception of some hazards than others. For example, values strongly influence the risk perception linked to climate change, but not so much when it comes to nuclear power (e.g., Peters and Slovic 1996; Visschers and Siegrist 2014; Kahan et al. 2012; Shi et al. 2015). The reasons for this could be many, as discussed in the literature. One important aspect is obviously the fact that there are strong aspects of egalitarianism related to seeing climate change as a serious risk requiring strong mitigation measures, whereas nuclear power energy seems acceptable for people with different value orientations.

Another factor that can explain risk perception and the acceptance of a hazard or a related activity is perceived fairness. The fairness factor can, for example, relate to the degree to which the potential consequences of a hazard are equally distributed over different people and regions (i.e., outcome or distributional fairness; Tyler 2000). It can also relate to the fairness of the decision process, for example, to what extent people are allowed to have a voice and contribute during this process (i.e., procedural fairness; Tyler 2000). Risk perception research indicates that people who experience a fair process are in general more positive in accepting the hazard ('the fair process effect'; Van den Bos 2005). In addition, there is interpersonal fairness, which refers to the perception that the stakeholders are trustworthy and respect the public's views (Besley 2010). It is also common to talk about informational fairness or justice, which is people's belief that they are well informed.

6.2.4 The difference between professional risk judgments and risk perception

The early research on risk perception, and in particular the studies from the psychometric paradigm, showed that experts' judgments of risk differed systematically and markedly from those of nonexperts (Slovic 1987). For instance, in Slovic et al. (1979), the authors asked a group of experts and lay people (non-experts) to rate the riskiness of 30 activities and technologies. The expert group viewed, for example, electric power, surgery, swimming and X-rays as riskier than did the lay people, but, for example, nuclear power, police work and mountain climbing as much less risky than did lay people. On average, experts rated risk as lower than lay people. Many later studies have shown similar results (see Chauvin 2018).

A common explanation provided for this difference was that when experts judge risks, they tend to relate the risks to probability of harm or expected loss (for example, mortality). The expert responses correlated strongly with estimates of annual fatalities. In contrast, the lay people's judgments were found to have a broader conception of risk – a risk perception – also capturing considerations of uncertainty and factors such as dread, catastrophic potential, equity and risk to future generations (Slovic 2000). From this, it is not surprising that

- considerable differences between experts' and lay people's risk judgment were identified
- experts' recitations of historical data and risk numbers commonly did little to change people's perceptions and conflicts over risk (Slovic 2016)

Seemingly, the experts' perspective – mainly based on System 2 thinking – is more rational than that of the lay people, which is to a large extent rooted in System 1 thinking. But, as discussed in Sections 6.1.2 and 6.1.3, the issue is not that straightforward. The nuclear power example of Section 6.1.2 is illustrative. Experts seemed to believe that they possessed the truth about risk and that lay people were influenced strongly by fear and other perceptional factors. Following this view, the public should be better educated and informed to rely less on emotions, to be aware of the bias in their judgments and to be open to new knowledge and evidence that could alter and improve their perceptions. The aim would be to align lay people's misjudgments with the experts' 'objective' risk characterizations to reduce or illuminate the differences among groups and the conflicts that arise from these differences. Some actors, mainly industrialists and scientists, held this stance for a while (see Kraus et al. 1992; Slovic 1987), but the opposition was strong, and it increased over time. The ideas were rejected, the key point made that disagreements about risk would not disappear in the presence of more knowledge or evidence. For rare events, the knowledge is weak, and there is no truth expressing what the most severe risks are.

The foundations of the technical scientific risk assessment also came under scrutiny. It was recognized that there is also subjectivity, biases and value-laden issues in professional risk judgments (e.g., Merkelsen 2011; Savadori et al. 2004). Social scientists argued strongly that the measurement of risk is inherently subjective (Slovic 1999). They were right, as thoroughly explained and discussed in Chapter 3. However, the professional judgments are not as subjective as risk perceptions. They do not include perceptional factors, nor judgments about how one likes or dislikes the various aspects of risk (for example, the events, consequences and uncertainties). Think about the traveling by plane example of Section 6.1.1. The historical numbers for commercial airplanes provide a rather objective reference for the risk levels. The levels of subjectivity, biases and value-laden issues are not comparable with those of lay people's risk perception. For simple risk problem where accurate predictions can be made, such as commercial flying, the difference between professional risk judgments and lay people's risk perception is sharp and well defined.

Risk perception takes into account all types of uncertainties related to the threat or activity considered and captures all types of judgments and feelings related to how the person likes or dislikes the risk, including questions about acceptance or unacceptance. Hence, we cannot expect risk perception results to be comparable to professional judgments of risk, which do not reflect feelings and judgment about how we like or dislike the risk, nor judgments about risk acceptability and risk handling in general.

Figure 6.4 illustrates the difference between risk, professional risk descriptions and risk perceptions using the terminology from Chapters 2 and 3. Risk perception may be influenced by professional risk judgments; the person's own professional risk description (if the person is able to do this); how the person judges, likes or dislikes aspects of risk; affect; trust and acceptability issues.

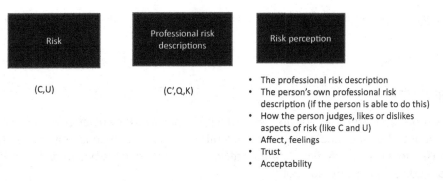

FIGURE 6.4 Illustration of the difference between risk, professional risk descriptions and risk perceptions

Challenges arise when facing situations and activities that are subject to uncertainties. Then, accurate predictions cannot be made. Models and probabilistic analysis depend on a knowledge base which could be rather weak and also include erroneous beliefs and assumptions. Risk assessments are based on judgments which to varying degrees are subjective and value laden. Even the choice of risk metrics to be used is based on judgments. However, following risk science knowledge, guidance is provided for how risk should be professionally characterized. There is always a possibility of misuse of risk assessments, as for most instruments, but if the science works as intended, there should be clarity about the judgments and the basis for these judgments. As argued for in this book, risk assessment in cases of uncertainties is not about presenting an objective truth. Instead, characterizations of risk are a prudent way of reflecting what we know and what we do not know. In this way, risk assessment is more of a tool for dialogue and communication about risk issues than accurate risk measurements.

Considerable work has been conducted to reveal differences in values between experts and lay people; see Chauvin (2018). For example, the Mertz et al. (1998) study revealed that senior managers of a major chemical company perceived the risk to be similar or lower than those of British toxicologists working in industry. Both groups showed similar individualistic cultural worldviews. In contrast, toxicologists affiliated with academia tended to have a higher risk perception and dissimilar cultural worldviews. While professional risk analysts would strive for fair characterizations of the risks independently of other stakeholders' opinions, experience shows that, in many cases, there could be a strong affiliation bias; see, for example, O'Brian (2000). These problems can be rooted in the idea that risk assessment is a tool for accurate risk estimation. If, however, the perspective is that risk assessment is just a way of helping us improve our knowledge about relevant issues, risk assessments can still contribute with useful decision support despite their subjectivity and limitations.

In the examples of Section 6.1, we pointed to the fact that peoples' risk perception is not only influenced by System 1 thinking but can also capture conscious judgments of uncertainties and risk which are not reflected by the professional risk assessment

judgments. The more narrow the risk assessment perspective adopted, the more important these lay people's judgments.

REFLECTION

You are concerned about an issue which is sensitive for society, but it is not
discussed by the media. You are a scientist, but your science is only indirectly
linked to this issue. A friend states that your risk perception is based on fear and
prejudices. Is this person right?

*The person could be right but also wrong. Your concern could be based on con-
scious judgments about risk and other relevant aspects. Risk perception does
not need to be founded on affect and perceptional factors. Your friend's way of
speaking could be inappropriate and in fact a ruling technique.*

The risk perception research has many implications for risk communication and
risk handling, as will be discussed in coming chapters.

6.2.5 The social amplification and attenuation of risk

In the last decade, there have been a number of events (for example, *Bovine spongiform
encephalopathy* [BSE], acrylamide, aspartame, melamine milk, COVID-19) that have led
to risk amplification and public distrust in regulatory bodies and industries. The social
amplification of risk framework (SARF), introduced by Kasperson et al. (1988), provides
a description of how such responses come about and explains in detail how risks or risk
events, assessed by experts as low risks, still produce significant public concern that often
has large societal impacts. This tendency reflects the complexity in risk judgment and the
fact that risk is more than just quantitative expressions in the form of probabilities or risk
estimates. The framework brings together technical analysis and social experience of risk
and provides a description of how risk amplification occurs at two levels. First, it happens in the processing of risk-related information and, second, in societal responses. It is
explained how certain characteristics of risks or risk events are formed into a message.
This message is communicated to someone, and, as part of the communication of this
message, risk signals (images, signs, symbols) in the message interact with a wide range of
processes in ways that either intensify or weaken perception of the risk and its manageability (Kasperson et al. 2003). Risk amplification involves intensifying or increasing the
importance or 'volume' of certain risk signals and symbols. It is generally associated with
heightened perception of risk and tends to trigger risk-reductive measures. The concept
of risk attenuation refers to opposite tendencies to those associated with amplification
and involves weakening or decreasing the importance or 'volume' of certain risk signals
and symbols. Risk attenuation commonly contributes to lowered perception of risk and
works to compromise risk reduction and risk regulation.

The process in which risks and risk events are interpreted and responded to is comprehensively described in the SARF. It is also explained how this process is repeated, as the responses go through different rounds of interpretation, spurring secondary and even tertiary effects. These waves of effects spread the impacts of the original risk event far away from where it initially took place and are referred to in the framework as ripple effects. This process is depicted in Figure 6.5.

While amplification processes and effects have been thoroughly described in the SARF and widely demonstrated empirically, the concept of attenuation has received less attention in the framework and related research. The mechanisms, processes and effects of attenuation are less apparent and often hard to identify and describe. Risks and risk events associated with attenuation are of a different character than the ones generating amplification. Where, for instance, an accident or a report showing increased numbers of injuries may result in more restrictive regulations and/or distrust of risk managing institutions and regulators, attenuation often produces the opposite effects. The interpretation of a piece of scientific research demonstrating no harmful effects of exposure to a certain chemical or a report on a decline in injuries may, for example, manifest itself in relaxed regulations and increased credibility of risk assessors and regulators.

At the stage where risks become amplified, risks and risk events are usually well defined. Here, risk problems are often already exacerbated, and the positions of involved actors are polarized (Poumadere and Mays 2003). However, risks develop and can change over time and are less noticeable and difficult to define at the earlier stages, refer, for example, to the coronavirus in early 2020: risk was not a major issue. In a study of the responses of local villagers to the Moirans-en-Montange fires, Poumadere and Mays (2003) found that the wide amplification of these fires depended upon the prior repression and denial of an earlier risk event: the burial of an electric cable. This finding "introduces another level of dynamic into the SARF: the degree of amplification may sometimes be a function of the degree of prior attenuation in the given social context" (Poumadere and Mays 2003, p. 226). The example points to the need to look into the dynamics and phases that precede and shape risk and risk events.

Essential in this regard is the way the risk assessors conceptualize and describe risk. Adopting a narrow probabilistic approach to risk could easily lead to attenuation; refer to the examples of Sections 6.1.2 and 6.1.3 on nuclear power plants and epidemics. Adopting an uncertainty-based perspective to risk, as presented in this book, we argue that risk attenuation can be reduced or even avoided. Rosa (2003) highlights the importance of clarifying the perspective on risk when applying the SARF to different contexts. In the SARF, risk is seen partly as an objective threat or harm to people and

FIGURE 6.5 Illustration of the social amplification of risk framework (SARF) (based on Kasperson et al. 1988; Fjæran and Aven 2019)

partly as a product of culture and social experience (Kasperson 1992). Rosa (2003) points to the fact that SARF, on the one hand, refers to a 'true' or 'objective' baseline risk and, on the other, to a subjective risk that deviates from the true risk due to amplification or attenuation distortion. This is misleading, according to Rosa, as it may be taken to imply that a 'true' risk exists and "opens up the SARF to the challenging questions: what is being amplified: something real or only claims about something real?" (2003, pp. 49–50).

The following remark illustrates the double-sided nature of attenuation, "Whereas attenuation of risk is indispensable in that it allows individuals to cope with the multitude of risks and risk events encountered daily, it also may lead to potentially serious adverse consequences from underestimation and underresponse" (Kasperson et al. 1988, p. 179). For situations where there is high uncertainty and potentially serious consequences, such as health risks associated with the use of chemicals in food and feed, risk attenuation may downplay important risk signals and create an impression of effective risk management that can have severe consequences for safety and crisis management (O'Neill et al. 2016).

Thus, in order to understand processes of risk amplification and its mechanisms, consequences and effects, it is important to address the period preceding the risks or risk events generating this amplification (Fjæran and Aven 2019). Analyzing the phase before the risk amplification stage to a greater extent facilitates 'turning the lens' towards those who are involved in defining the risks from the very start. The importance of such a focus of analysis has also been recognized by Hilgartner (1992), as he argues that, in order to understand processes of displacing and emplacing risk objects (processes similar to attenuation and amplification), one should focus on the 'system builders' and move analysis 'upstream' into the arenas of specialized professionals and technical experts. He stresses that it is in these arenas that risks are constructed, created and controlled and that the public generally has little power, resources or effect on constructing or defining what come to be considered risks worth or not worth 'attention'. In line with Hilgartner's arguments, focus should be on the 'system builders', which represent risk producers and assessors, as well as risk managers and regulators, in risk attenuation and amplification processes. In relation to the coronavirus, this points to the importance of studying the risk attenuation and amplification processes in late 2019 and early 2020.

In addition, we need to highlight the role and importance of the media and of public in the societal agenda-setting of risks and in the shaping, redefining and amplification of risks. For such processes, the SARF proves an invaluable tool, and its applicability has been proven in a range of empirical studies.

In order to make 'visible' and illustrate the significance of the more hidden phase or period prior to amplification, it has been suggested to add a sequence to Figure 6.4; see Figure 6.5. What happens or does not happen in this period, represented by the first line in Figure 6.5, often goes more or less unnoticed and is more easily identified in retrospect: after a risk event has generated amplification. The attenuation phase covers less defined and less clear events and responses. The SARF commonly proceeds from clearer and more defined risks or risk events; attenuation often springs from what to a large extent more resembles a 'non-event', which in Figure 6.6 is referred to as a 'risk and uncertainty source' (box 1). This term covers many of the same type of factors, events, elements, situations and so on as those captured by the concepts of risk sources (see Section 2.3.1) and risk-influencing factors (see Section 4.2.2). In the first line in the figure, we see how the processing and interpretation of these 'risk and uncertainty sources' (box 2) are often manifested in some 'non-responses' that maintain and substantiate current risk-management practices (box 3). Although less visible, these responses or effects, where 'business continues as usual', are absorbed and interpreted in ways that serve to preserve the regulatory and societal status quo and the perception of risks as low (box 4). These secondary responses or effects, in turn, contribute to a third wave of unclear or invisible effects (box 5), represented by a societal drift away from focusing on these sources of uncertainty as potential risks or issues meriting attention. This backdrop of societal attenuation represents an important part of the context in which new risks and risk events (box 1, line 2) are interpreted and responded to and may contribute to explaining why these sometimes can generate considerable amplification and large ripple effects. The coronavirus case can be used to illustrate the attenuation phase; see Problem 6.6. In early 2020, the risks were broadly framed as minor.

The issues raised by SARF and also the extended version presented in Figure 6.5 fit the traditional 'narrow' probabilistic perspective on risk. Adopting an uncertainty-based risk understanding can strongly influence the development process. Focusing on the uncertainty and knowledge aspects of risk can work as a filter to prevent attenuation from occurring and/or from spreading from one level to another by 'breaking the chain of attenuation' (from risk producer to risk assessment and from risk assessment to risk management and regulation). Following SARF terminology, the adoption of such a perspective as the starting point for assessing and managing risks can prevent attenuation at both levels of the framework: in the interpretation of messages of risks or risk events and in the responses to these. Still, the most important contribution of an uncertainty-based perspective on risk lies in how it changes the portrayal of the risk or risk event and in how it affects the content of the message itself. This way, an uncertainty-based risk perspective also has implications for the underlying principle

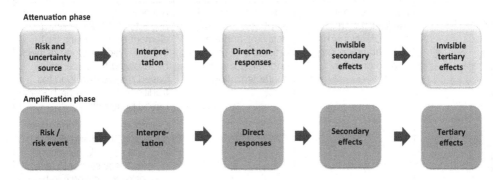

FIGURE 6.6 An extension of the SARF highlighting the early stages of a risk development (based on Fjæran and Aven 2019)

and foundations of the SARF: namely the discrepancy between the risk understanding and judgment of the scientific assessors and the larger public, where the risks assessed as low by experts are judged to be high by the public. As discussed, such an uncertainty-based understanding of risk involves injecting a level of amplification into the risk assessment and management processes in ways that may entail a higher risk judgment. This way, it may serve to narrow the gap between the risk understanding of technical experts and risk assessors, on one hand, and the public, on the other.

Some risk amplification at an early stage of these processes may reduce the degree of later amplification and the associated consequences. As for many risks, they develop, grow and/or change character over time. Risks may be attenuated for years; they may go through a brief or a long period of focus and attention, where some actors intensify signals downplayed by others and vice versa, before they again are put to rest, forgotten, ignored or attenuated until a new risk event puts them on the agenda again and so on. The dynamics of these continuously ongoing processes and cycles challenge the chronological structure of the SARF.

To properly reflect the dynamics and complexities of risk development processes, the time dimension needs to be properly reflected. The issue is discussed in detail by Fjæran and Aven (2020b), who refer to risks going through waves of risk amplification or attenuation in which different actors and stakeholder groups dominate the risk communication. The downplaying and intensification of risks and risk signals occur concurrently in these phases, both within the communication of a specific risk message or risk event by a specific actor and, on a larger scale, in the competition played out in the media between stakeholders over who gets to frame and control the risks. Table 6.2 summarizes key aspects of the previous discussion, pointing to differences between a traditional 'narrow' probabilistic perspective on risk and an uncertainty-based perspective on risk.

TABLE 6.2 A summary of differences related to risk perception and judgments for the traditional probability-based and uncertainty-based risk perspectives

	Traditional probability-based	Uncertainty-based perspective	Comments
General	Large 'gap' between the 'real risk' and the professional risk understanding and characterization	Considerably smaller 'gap' between the 'real risk' and the professional risk understanding and characterization	Acknowledging the uncertainty aspects of risk as a central element of risk means that the 'real risk' is judged higher by following the uncertainty-based perspective than the traditional technical probability-based perspective
			Communicating uncertainties and using broad and diverse knowledge to a larger extent enables the assessor to take on a risk-moderating role than the risk-attenuating role typically associated with the traditional probability-based perspective

	Traditional probability-based	Uncertainty-based perspective	Comments
Early stages, framing	This gap is not broadly visible; it is to a large extent suppressed The uncertainty dimension of risk is downplayed in the professional risk understanding and characterization	This gap is to a larger extent acknowledged and communicated	The suppression in the traditional perspective case 'builds energy', and there is a potential for a strong outburst
Middle stage, responses along the way	Gap is indicated on several occasions. Indications are repeatedly downplayed, ignored or explained away Information and risk signals are disregarded, and risk characterizations and evaluations remain unchanged	Risk assessment can be used proactively and absorb and adapt to changes as risk develops. This way, the gap can be reduced as new knowledge emerges	Failure to recognize less visible risks or risk events and to make use of knowledge other than 'objective' historical data continue to substantiate the gap and to accumulate 'energy' Traditional perspective can produce ignorance and periods in which risks incubate
Later stages	New knowledge, risk signals and risk events call attention to the gap, and strong risk amplification may occur For example, independent scientists argue that the actual risk is not negligible, that we should be concerned and that risk assessments are founded on incomplete data	Risk events may occur and new knowledge may emerge, but effects can be reduced, as the gap, to a large extent, is already acknowledged and communicated In addition, if a gap exists, it can be reduced. The risk concept is broad and flexible and can capture changes and new knowledge	For the traditional technical perspective, the amplification to some degree is 'justified', as key uncertainty aspects of risk have been suppressed Also, at this stage, by remaining in the role as a risk attenuator, the risk assessor contributes to increased amplification

The traditional, probability-based risk perspective is characterized by a 'gap' between the 'real risk' and the professional risk understanding and characterization. This assertion is based on the argument that there exist considerable aspects of risk that are not captured by a pure probability-based approach. When adopting the uncertainty-based approach, these aspects are acknowledged as risk or risk contributors. The implication is that this perspective adds a component to the risk concept which is not included by probability-based thinking. Ignoring this component means that real aspects of risk are camouflaged.

In the early stages of the development of a risk or risk event, this gap is not broadly visible following a traditional perspective; it is to a large extent suppressed. The professional risk understanding and characterization downplay the uncertainties, as these uncertainties are not considered an integrated element of the risk. This downplaying creates energy and increases the potential for a strong uprising wave of risk-related worries and concerns. However, for uncertainty-based thinking, this gap is to a large extent acknowledged and communicated. It is also considerably smaller than what we would have when using a probability-based perspective.

Over time, as new knowledge is revealed, or emerges, this gap may be indicated in different ways and on different occasions. Although this may act to leave 'the ocean' quiet for periods of time, this way of relating to risk-related information contributes to maintain and uphold the gap. This way, these periods or troughs of attenuation can also be understood as periods of incubation in which the force and energy of the next wave(s) are gradually accumulated by the systematic non-responses to uncertainties and less observable risks and risk events.

In later stages of the risk-development process, we again may experience clearer indications and stronger signals calling attention to the gap. This often is the case when different stakeholder groups enter the arena of the media. Typical examples of such indications and signals include independent scientists arguing that the risks discussed are not negligible and should not be ignored. When the perspective held by those in charge of assessing the risks is a traditional one, the result is often that strong risk amplification occurs. The amplification here observed can to some degree be seen as justified, as key uncertainty aspects of risk have been systemically suppressed along the way. For the uncertainty-based perspective, effects would potentially be of a more limited character, as the gap is already largely absorbed and reflected by the risk framework used from the very start of the risk assessment process.

6.3 HOW TO CONDUCT A RISK PERCEPTION STUDY

Risk perception research studies how people perceive risk. Typically, a theory or model is developed to describe how people in real life perceive and make decisions in relation to risk. Next, data are collected using various methods, including surveys, which allows the gathering of information from a sample of individuals, for instance by questioning how these individuals perceive the risk associated with some defined activities (Sjöberg 2003). Statistical analysis is then conducted to make inference on the basis of the data and the underlying theory (model).

The development of the psychometric paradigm is based on this type of analysis, as well as most other studies of risk perception. As for all statistical analysis of this type, there are pitfalls and limitations, as was briefly addressed in Section 6.2. Yet the models and theories can be useful for our understanding of how people perceive risk.

For demonstration, we show a multi-step risk perception study that explores relevant risks and how those risks are perceived and performs statistical analysis on results.

We begin by creating a survey that collects several pieces of information from respondents. We first seek demographic information, such as age and level of education. Then, we ask questions about the perceived level of risk for some hazards. Finally, we seek information on how the respondent responds to the various hazards. A sample survey is as follows:

Step 1: Gather demographic factors, such as age, level of education, occupation and so on. Sample questions are:

- 1.1 What is your age [enter number]
- 1.2 What is your level of education [some high school, high school, undergraduate, graduate]

Step 2: Rate perceived level of risk for some specific hazards. Sample questions are:

- 2.1 Consider a potential wildfire impacting your hometown in the next five years. How would you judge:
  - 2.1.1 The risk of this wildfire [1 = Low, 2 = Medium, 3 = High]
  - 2.1.2 The probability of this wildfire [enter number from 0 to 100]
  - 2.1.3 The strength of knowledge for your previous response [1 = Low, 2 = Medium, 3 = High]
- 2.2 Rate your level of concern/worry about this risk [1 = None, 2 = Somewhat, 3 = A lot]
- 2.3 Consider a potential cyber attack that disrupts power for your region in the coming five years. How would you judge:
  - 2.3.1 The risk of this cyber attack [1 = Low, 2 = Medium, 3 = High]
  - 2.3.2 The probability of this cyber attack [enter number from 0 to 100]
  - 2.3.3 The strength of knowledge for your previous response is [1 = Low, 2 = Medium, 3 = High]
- 2.4 Rate your level of concern/worry about this risk Are you worried about this risk [1 = None, 2 = Somewhat, 3 = A lot]

Step 3: Gather statements about behavior in response to hazard

- 3.1 How likely are you to do the following because of the risk related to wildfires?
  - 3.1.1 Purchase insurance [1 = Difficult to say, 2 = Unlikely (less than 10%), 3 = Somewhat likely (in the range 10% to 90%), 4 = Very likely (more than 90%)]
  - 3.1.2 Adopt more sustainable practices, such as lowering your carbon footprint or reducing waste [1 = Difficult to say, 2 = Unlikely, 3 = Somewhat likely, 4 = Very likely]
- 3.2 How likely are you to do the following because of the risk related to cyber attacks impacting your region's power?
  - 3.2.1 Invest in preparedness supplies [1 = Difficult to say, 2 = Unlikely, 3 = Somewhat likely, 4 = Very likely]
  - 3.2.2 Lobby politicians for proactive efforts to avoid an attack [1 = Difficult to say, 2 = Unlikely, 3 = Somewhat likely, 4 = Very likely]

Step 4: Perform statistical analysis on the results from previous steps. Computations include:

- Average responses for questions

  - Across all responses
  - Specific to groups (age ranges, levels of education, etc.)

- Variability in responses for questions
- Comparisons between groups (age ranges, levels of education, etc.) using hypothesis testing for difference in means and proportions

Table 6.3 summarizes hypothetical results, including a comparison across age groups and educational groups. The table shows a total of 150 respondents. The majority of respondents are in the 40+ and 22–39 age groups. Most respondents have attained a high school or undergraduate-level education.

Figure 6.7 shows a scatterplot of the probability of a cyber attack (question 2.3.2) and the strength of knowledge (question 2.3.3). The plot shows that the respondents' probability judgments are distributed on the whole interval [0,1] for all three SoK categories but somewhat less spread for the high SoK judgments. For this highest SoK category, the majority of the probability judgments exceed 50%. It seems that if the assessors have strong knowledge about the issue, they think it is likely (more than 50%) that a cyber attack will occur.

Using this information, one can perform statistical testing that can be used to study whether differences in risk perception exist among the various age and educational groups. The foundation of these hypothesis tests involve the study of some population parameter, μ, which represents the population average when contemplating a very large – in theory, an infinite population of respondents. The sample average, $\bar{x}$, is the estimator for the population parameter, as shown in Table 6.3, based on the data from the study. For theoretical discussion of statistical sampling and hypothesis testing, we refer the reader to Anderson et al. (2020).

First, consider a hypothesis test to compare the mean between age groups. We can ask questions, such as: "Do respondents with a graduate level of education rate the risk of a cyber attack differently than those with a high school education?

We formulate this question using a two-tailed hypothesis test that asks the questions:

$$H_o : \mu_{Graduate} = \mu_{High\ School}$$

$$H_a : \mu_{Graduate} \neq \mu_{High\ School}$$

The population parameter, μ, represents the underlying theoretical population average score for question 2.3.1 related to the perceived risk level for this cyber attack [1 = Low, 2 = Medium, 3 = High]. In the null hypothesis, we assume the theoretical population average score for the two educational groups is the same. The test aims at potentially identifying a statistically significant difference between the two age groups.

TABLE 6.3 Summary statistics for risk perception study related to wildfire and cyber attack risk

Group	Sample size		Step 1		Step 2								Step 3			
			1.1	1.2	2.1.1	2.1.2	2.1.3	2.2	2.3.1	2.3.2	2.3.3	2.4	3.1.1	3.1.2	3.2.1	3.2.2
All respondents	150	Average	42.36	n/a	1.78	45%	1.93	1.99	1.97	51%	1.89	1.90	2.66	2.52	2.57	2.32
		St. deviation	14.34	n/a	0.76	29%	0.66	0.58	0.70	33%	0.73	0.76	1.07	0.86	0.92	0.84
		Maximum	67.00	n/a	3.00	100%	3.00	3.00	3.00	100%	3.00	3.00	4.00	4.00	4.00	4.00
		Minimum	1.10	n/a	1.00	0%	1.00	1.00	1.00	0%	1.00	1.00	1.00	1.00	1.00	1.00
				n/a												
Age groups																
18–21	10	Average	19.30	n/a	1.90	32%	2.10	2.00	2.10	55%	2.50	1.80	3.40	2.50	2.50	2.60
		St. deviation	1.06	n/a	0.74	26%	0.74	0.67	0.88	24%	0.53	0.63	0.70	1.18	0.85	0.70
22–39	54	Average	31.02	n/a	1.69	49%	1.85	1.96	1.93	50%	1.72	1.81	2.87	2.67	2.50	2.44
		St. deviation	5.39	n/a	0.77	30%	0.66	0.58	0.70	32%	0.76	0.83	0.95	0.95	0.93	0.88
40+	86	Average	52.65	n/a	1.83	45%	1.95	2.01	1.98	51%	1.93	1.95	2.44	2.52	2.53	2.21
		St. deviation	8.57	n/a	0.75	28%	0.65	0.58	0.69	35%	0.70	0.73	1.12	0.88	0.84	0.81
Education groups																
Graduate	22	Average	37.32	n/a	2.00	44%	1.86	1.95	1.95	55%	2.00	1.41	2.77	2.64	2.77	2.50
		St. deviation	11.93	n/a	0.82	23%	0.71	0.65	0.84	36%	0.82	0.59	1.07	0.90	0.81	0.91
High school	54	Average	42.52	n/a	1.78	44%	1.85	2.04	1.93	53%	1.94	2.04	2.43	2.44	2.33	2.37
		St. deviation	13.81	n/a	0.77	30%	0.68	0.55	0.70	34%	0.66	0.75	1.14	0.82	0.85	0.85
Some high school	15	Average	44.27	n/a	1.60	33%	2.20	1.87	1.93	31%	1.73	2.13	2.73	2.93	2.73	2.33
		St. deviation	15.99	n/a	0.63	32%	0.77	0.52	0.70	28%	0.70	0.74	0.96	0.80	0.88	0.62
Undergraduate	59	Average	44.32	n/a	1.75	50%	1.95	2.00	2.02	52%	1.85	1.88	2.81	2.58	2.54	2.20
		St. deviation	14.16	n/a	0.76	28%	0.57	0.62	0.66	32%	0.78	0.77	1.03	1.04	0.88	0.85

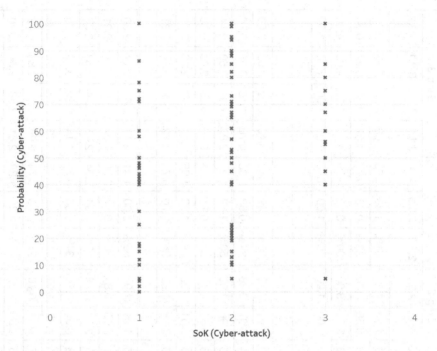

FIGURE 6.7 Probability (%) of a cyber attack vs. strength of knowledge

We assume the null hypothesis is true until we have enough evidence to reject that null hypothesis in favor of the alternative hypothesis.

For this demonstration, we assume the standard deviations are equal for both populations (graduate and high school). This assumption can be questioned. Therefore, students will have the opportunity to perform this analysis without this assumption as an exercise (see Problem 6.7). The pooled sample variance, s_p^2, can be computed as follows:

$$s_p^2 = \frac{\left(n_{Graduate}-1\right)s_{Graduate}^2 + \left(n_{High\ School}-1\right)s_{High\ School}^2}{n_{Graduate}+n_{High\ School}-2} = \frac{(22-1)*0.84^2+(54-1)*0.70^2}{22+54-2}$$

$$=0.55$$

Then, the t-statistic can be computed as follows:

$$t = \frac{\overline{x}_{Graduate}-\overline{x}_{High\ School}}{s_p\sqrt{\dfrac{1}{n_{Graduate}}+\dfrac{1}{n_{High\ School}}}} = \frac{1.95-1.93}{0.55\sqrt{\dfrac{1}{22}+\dfrac{1}{54}}} = 0.14$$

Because the standard deviations for both populations are assumed to be equal, the degrees of freedom, v, can be computed as:

$$v = n_{Graduate}+n_{High\ School}-2 = 22+54-2 = 74$$

The critical value is found using a t-distribution and assuming some level of significance, α. Here, we assume $\alpha = 0.05$. The critical value equals

$$t_{1-\alpha/2,v} = t_{1-0.05/2,74} = 1.993$$

If $|t| > t_{1-\alpha/2,v}$, then the null hypothesis can be rejected, thereby concluding that there is a difference in the theoretical population average score between the two populations. Because the test statistic 0.14 is not greater than the critical value of 1.993, the null hypothesis is not rejected. Therefore, we believe there is no statistical difference in risk score between the two groups. In other words, because we assumed the null hypothesis was true, the data do not allow us to conclude that there is a difference in risk score between the two education groups.

When comparing proportions, we can ask questions such as "Does the proportion of people who view the risk of wildfires as 'high' differ between the youngest and oldest studied age groups?" We study this question using a hypothesis test for the difference between proportions. We define the sample proportion using question 2.1.1 as:

$$\bar{p}_{Age\ group} = \frac{Number\ of\ respondents\ in\ this\ age\ group\ who\ percieve\ risk\ as\ 'high'}{Sample\ size\ in\ this\ age\ group}$$

Like with the previous hypothesis test, we assume that there is some true population proportion of individuals who view the risk of wildfires as 'high' for each studied age group. We infer about this true population proportion using the sample proportion, as shown in the previous computation.

Table 6.4 provides an analysis of proportions to demonstrate the testing, which combines the 18–21 and 22–39 age groups. We see that 53% of the sampled 40+ age group perceives this risk as 'high' compared to only 20% of the sampled 18–39 age group. While these results alone can be convincing, note that these results are based on only sample data. A hypothesis test is still necessary to make an inference about the true population proportions.

We then pose the following hypothesis test:

$$H_o : p_{40+} = p_{18-39}$$

$$H_a : p_{40+} \neq p_{18-39}$$

where p is the theoretical underlying true proportion for the different groups. We choose to adopt the normal distribution to approximate the binomial distribution here,

TABLE 6.4 Sample proportions for proportion hypothesis test

Age group	# answered '3' for question 2.1.1	Sample size (n)	Sample proportion
18–39	13	64	0.20
40+	46	86	0.53

as the product of the sample size and sample proportion is greater than 5. For example, for the 18–39 age group, $n = 64$ and $\bar{p}$ is 0.2, so $n\bar{p} = 64 * 0.2 = 12.8$.

We compute the pooled sample proportion and estimated standard error as follows:

$$Pooled\ Proportion\ \bar{p} = \frac{\left(\bar{p}_{40+} * n_{40+} + \bar{p}_{18-39} * n_{18-39}\right)}{n_{40+} + n_{18-39}}$$

$$= \frac{\left(0.53 * 86 + 0.20 * 64\right)}{86 + 64} = 0.39$$

$$Estimated\ Standard\ Error = \sqrt{\bar{p} * \left(1 - \bar{p}\right) * \left[\frac{1}{n_{40+}} + \frac{1}{n_{18-21}}\right]}$$

$$= \sqrt{0.39 * 0.61 * \left[\frac{1}{86} + \frac{1}{64}\right]} = 0.08$$

Finally, the test statistic can be computed as:

$$z = \frac{\bar{p}_{40+} - \bar{p}_{18-21}}{Estimated\ Standard\ Error} = \frac{0.53 - 0.20}{0.08} = 4.13$$

The critical value using a 0.05 level of significance is:

$$z_{\alpha/2} = z_{\frac{0.05}{2}} = 1.96$$

Therefore, the null hypothesis can be rejected. There is a significant difference between the 40+ and 18–39 age groups in perception of a 'high' risk due to wildfires. This also confirms the initial assessment made previously, seeing that the sample proportion of respondents in the 40+ age group was much higher than that of the 18–39 age group.

By reformulating the hypotheses, we can formally also show that the 40+ age group scores higher than the 18–39 age group. The hypotheses formulated are then

$$H_o : p_{40+} = p_{18-39}$$

$$H_a : p_{40+} > p_{18-39}$$

The significance level can then be reduced to 2.5%, since initially we could also consider the alternative hypothesis $p_{40+} < p_{18-39}$. The test statistic is unchanged, and the critical value is computed as:

$$z_\alpha = z_{0.025} = 1.96$$

We can conclude that the 40+ age group's perception of a 'high' risk due to wildfires is significantly above that of the 18–39 age group.

6.4 PROBLEMS

Problem 6.1

Perform a study in class making judgments about risk perception for different hazards/threats and activities, for example, ranking the following activities: sports, traffic, industry-related, violence, terrorism and natural hazards. The issue is the individual risk as perceived in relation to serious injuries or deaths.

Discuss in class to what extent the judgments made are based on affect or conscious judgments of risk and uncertainties.

Problem 6.2

It has been stated by well-known scientists that *risk is the same as risk perception* (Jasanoff 1999), *risk coincides with the perceptions of it* (Douglas and Wildavsky 1982) and *because risks are risks in knowledge, perceptions of risks and risk are not different things, but one and the same* (Beck 1992, p. 55). Discuss these statements.

Problem 6.3

Consider the bus risk example in Kahneman (2011), Chapter 30, and related commentaries and perspectives provided by Aven (2015a, 2018b). Do you think people overrated the risk when not taking the bus? When Kahneman drove away more quickly than usual when the light changed, would you see that as mainly based on System 1 thinking or System 2? Should he be chagrined by his behavior?

Problem 6.4

In Section 6.2.3, when discussing cultural risk theories and values, it is referred to four worldviews: hierarchy vs. communitarianism (i.e., the importance of the group) and individualism vs. egalitarianism.

Where would you place climate change risks?

Another common classification refers to individualistic, egalitarian, hierarchical, and fatalistic worldviews. Summarize what characterize these four worldviews by pointing to what types of risks are typically feared.

Problem 6.5

Explain the meaning of the terms *risk amplification* and *risk attenuation*.

Problem 6.6

Try to apply the SARF, with extensions as described previously, in relation to stages of the coronavirus case.

Problem 6.7

Consider the hypothesis test conducted in this chapter:

$$H_o : \mu_{Graduate} = \mu_{High\ School}$$

$$H_a : \mu_{Graduate} \neq \mu_{High\ School}$$

In the text, the hypothesis test assumed the standard deviations were equal for the two populations (graduate and high school). However, that assumption may not be supported in some cases. Perform this hypothesis test using the assumption that the standard deviations are *not* equal.

Problem 6.8

Suppose you are a risk perception consultant for a regional emergency management agency. You have been tasked with using the data from Table 6.3 to suggest how to prioritize investments in activities that promote emergency preparedness. Create a presentation that discusses the following topics:

- Which demographic groups are in highest need of emergency preparedness investments? Why?
- What types of investments are needed to prepare the highest need groups?
- What additional data should be collected and analyzed?

Problem 6.9

Part 1: Conduct a risk perception data collection by surveying your classmates. Design the survey to comment on topics including:

- Demographic information
- Perception of some risk (e.g., wildfires, pandemics, etc.). The data collection should include questions that evaluate:

  o   Risk level
  o   Probability
  o   Strength of knowledge
  o   Level of concern/worry about the risk

Part 2: Perform a statistical analysis on the results. In your analysis, include at least one hypothesis test.

Part 3: Present the findings to your class

Problem 6.10
Study the overview papers by Siegrist and Árvai (2020) and Siegrist (2020). Identify at least two topics highlighted in these paper on risk perception research not addressed in the present chapter. Prepare a talk on these topics for your class.

Problem 6.11
Study the definition of a heuristic presented in Kahneman (2003). Would 'adjustment and anchoring' be a heuristic according to this definition?

KEY CHAPTER TAKEAWAYS

- Risk perception is understood as a person's subjective judgment or appraisal of risk taking into account social, cultural and psychological factors (related to affect, trust, acceptability).
- People uses different types of heuristics (including availability and representativeness heuristics), which can lead to poor judgments and understanding of risk.
- Risk perception can also include conscious judgments of uncertainty, which are not properly reflected by the professional risk assessments.
- The psychometric paradigm relates risk perception to key features of the hazards or activities such as lack of control, catastrophic potential, delays in effects, new and unknown – summarized by the two dimensions dread and newness (referred to as 'all the hot buttons' for a high risk perception).
- People tend to overrate risk and overreact when faced with situations that hit 'all the hot buttons'.
- Care should be shown when using these results for situations with large uncertainties, as there is no objective reference for what constitutes a misjudgment of risk.
- Trust and affect are important factors to explain risk perception.
- System 1 thinking operates automatically and quickly, instinctively and emotionally, whereas System 2 is slower, more logical and more deliberative.
- Risk perception is often strongly influenced by System 1, whereas professional risk judgments are based on System 2 thinking.
- The SARF, with extensions as described previously, provides insights on the dynamics of risk perception over time, through different stages of risk attenuation and risk amplification.

CHAPTER 7

Risk communication

CHAPTER SUMMARY

While the previous chapters of this book have discussed how risk analysts and experts can assess risk related to an activity and also how to understand how non-experts perceive risk, this chapter discusses how to bridge the understanding of risk among these types of stakeholders. This task involves the concept of 'risk communication', which studies the exchange or sharing of risk-related data, information and knowledge. Stakeholders involved with risk communication efforts often include analysts, experts, decision-makers, regulators, consumers, media and the general public.

Effective risk communication requires several elements. First, there is a need for understanding of the target audience, a credible or trusted source and a properly formed message. Equally important is the quality of the risk analysis and risk characterizations that support the communication. All stakeholders could be pleased with the sharing of relevant data, information and knowledge, but the risk communication would still fail if the risk is seriously misrepresented or misjudged because of poor risk science. With few exceptions, such as proprietary information or that which may damage public security, an open, transparent and timely risk communication policy is needed. Such a policy demonstrates respect for the target audiences and is essential for building trustworthiness and legitimacy of the sources.

The chapter also reviews common methods for developing and testing risk communication messages, including surveys, focus groups, interviews and experiments.

LEARNING OBJECTIVES

After studying this chapter, you will be able to explain

- the meaning of risk communication
- the different functions and goals of risk communication

DOI: 10.4324/9781003156864-7

- what successful risk communication means
- the importance of building risk communication on risk science
- the key ideas and features of central models and theories of risk communication
- what characterizes good risk communication messages
- common methods for developing and testing risk communication messages

The chapter is organized as follows. First, in Section 7.1, we provide some examples to motivate the coming discussions. Then, in Section 7.2, we provide some basic theory of risk communication, partly based on Bostrom et al. (2018) (Section 7.2.2 in particular). Section 7.3 looks into methods for conducting risk communication studies and research. The final Section 7.4 provides some problems.

7.1 SOME ILLUSTRATING EXAMPLES

7.1.1 Poor risk assessment foundation hampers good risk communication

This is a story about Trevor, an experienced risk analyst who failed to communicate risk when presenting the result of a risk assessment to the leadership of the company where he worked (inspired by Aven and Reniers 2013).

Trevor was challenged by one of the top managers of his company after a large industrial accident in a competitor's facility, one with very similar infrastructure and business practices. While Trevor's company had a large focus on safety, the management was willing to spend a considerable amount of money on improving the safety for its facilities. It was decided to conduct a comprehensive risk assessment as a basis for making judgments about the risk level and identifying those areas where one could most effectively reduce risk. Trevor had a leading role in the analysis and presented the results of the assessment to the top management of the company. While showing a diagram with probability numbers for different loss of lives categories, one of the managers asked him: "What do these probability and risk numbers really mean?"

It was a good question. The company planned to spend large sums of money on the basis of this assessment, and to adequately use its results, the decision-makers needed to understand what the numbers really expressed. The problem was that Trevor could not give the manager a good answer. He was the risk analysis expert but did have not clear and convincing answers to this basic question, also addressing how to understand uncertainties related to these numbers. He was not able to communicate the results of the assessment in a way that was trustworthy.

The managers, that is, decision-makers, in this example were informed by the probability and risk numbers, but the lack of clarity on what these numbers expressed made them question how to use the results of the analyses. The interpretation discussion was not an academic one; it was considered a critical issue to be solved in order for the management to be able to properly understand what the risk analysis was actually communicating. Without clarity on the meaning of the probabilities and risk numbers reported, the mangers did not know how to use them in the decision-making process.

Risk science, as presented in this book, aims at providing clarity on the foundational issues of risk assessments and characterizations; see Chapters 3 and 4. The example points to the importance for key concepts to be properly defined and interpreted during risk communication between analysts and decision-makers. Without a solid risk scientific basis, the value of the information provided will be reduced, the results will be questioned and the trustworthiness hampered.

7.1.2 The nuclear industry

We refer to discussion in Section 6.1.2. In the 1970s and 1980s, risk analysts in the nuclear industry tried to convince populations that the industry was safe and the risks were small and acceptable. However, people were not convinced. Their perceived risk was high. The communication failed, as the nuclear industry experts claimed that they knew the 'truth' about the risk and that people would agree if they just were adequately informed. There is, however, no 'objective' reference for specifying a true risk here, and the analysts' approach failed to properly reflect important aspects of risk. The people were affected by perceptional factors, but their concerns were also rooted in conscious judgments of uncertainties and risk. As discussed in 6.2.5, the nuclear industry experts' perspective represented an initial attenuation of the risk, the result being amplification at a later stage, as explained by the social amplification of risk framework.

7.1.3 Smoking risk

The evidence when it comes to the negative health effects of smoking is today very strong. For decades, cigarette advertising on television and radio has been prohibited in many countries. Typically, cigarette packages contain warnings attesting to the dangers of smoking. There are many reasons smoking remains prevalent; social, economic, personal and political influences all play important roles in determining who starts smoking, who discontinues smoking and who continues. Risk communication science can help in developing governmental policies on how to reduce smoking in society, balancing the need for proper handling of a major societal problem and acknowledging the personal freedom to choose to smoke. History has shown that it is not a straightforward task.

7.1.4 Climate change risk

Few threats can compare with the scale and potential impact of global climate change. The related risk communication is, however, challenging. What is its objective? There should be no discussion about the need for 'objective' risk characterizations, but, as discussed in Chapter 3, such objectivity is not possible to establish for non-trivial situations. Yet there seems to be rather broad agreement among many scientists, bureaucrats and politicians worldwide that some type of objectivity exists, and it is summarized by the Intergovernmental Panel on Climate Change (IPCC). A main task of the IPCC is to inform governments and decision-makers at all levels about the state-of-the-art scientific knowledge on climate change risks. The risk communication can be considered successful in the sense that most governments now take serious action aiming at following up the IPCC findings and recommendations.

However, many people are still skeptical about the validity of some of the key insights provided by the IPCC. A key problem relates to the degree to which human activity is the main source for the observed climate change. Risk communication science and knowledge have no view on such issues. Their scope is related to how communication can be improved, subject to some defined objectives. One such objective could be to inform people about the issue, here what the risks are and how large. This leads us to discussions as conducted in Chapter 3 on how to characterize the risk (see Problem 3.24), as well as more specific communication challenges on how people should be educated about the risk given that the main ideas and principles of the risk characterization are established. Clearly the risk communication would fail from a risk science perspective if risk characterizations, for example, were based on metrics that ignore important aspects of risk; see discussion in relation to Problem 3.24. In the coming Section 7.2, we will discuss risk communication knowledge that provides insights concerning the latter issue (how to communicate risk when risk is described in a specific way). These insights are based on both theoretical analysis and empirical research showing how people react to different types of information sharing and tools used for this purpose.

7.1.5 Epidemic

We return to the example of Section 6.1.3, related to the early development of the coronavirus in 2020. We consider the emergence of a possible pandemic caused by a new type of virus, subject to considerable uncertainties about how the virus incubates, spreads and impacts humans.

Based on risk perception theory and the fact that the epidemic 'hits all the hot buttons' for a high perceived risk, risk communication faces a serious dilemma. By communicating this type of message, a potential misjudgment of the risk can be avoided, thereby avoiding an over-amplification of the risk. However, there is also a chance that a serious risk will be ignored or attenuated by key actors such that the emerging epidemic risk could develop without the necessary and timely implementation of mitigation measures. Care has to be shown in the design of the risk communication. We will discuss this issue in more detail in the coming section.

7.2 RISK COMMUNICATION KNOWLEDGE

7.2.1 The functions and aims of risk communication

Risk communication has different functions or goals, including (Renn and Levine 1991):

1. enlightenment function (to improve risk understanding among target groups);
2. right-to-know function (to disclose risk-related information to potential victims);
3. attitude change function (to legitimize risk-related decisions, to improve the acceptance of a specific risk source or to challenge such decisions and reject specific risk sources);
4. legitimization function (to explain and justify risk management routines and to enhance the trust in the competence and fairness of the management process);
5. risk reduction function (to enhance risk reduction through information about risk reduction measures);
6. behavioral change function (to encourage behavioral change);
7. emergency preparedness function (to provide guidelines for emergencies or behavioral advice during emergencies);
8. involvement function (to educate decision-makers about concerns and perceptions);
9. participation function (to assist in reconciling conflicts about risk-related disputes and controversies).

To be able to evaluate the success of risk communication, we need to see it in relation to what function it intends to meet. Risk communications range from telling a child to keep away from the warm oven, to informing neighbors of an industrial plant about potential risks from its operation, to international collaborations and debates for addressing global risks such as climate change. Risk communication messages and strategies that work for some situations will fail in others. A good message for one population group may be ineffective in another.

Are the anti-smoking campaigns we have observed over the last few decades in many countries good examples of successful risk communication?

Clearly, they have contributed to reduce smoking and saved many lives. As such, they can be seen as successful. The perspective is a health policy one (risk reduction function 5). However, from a risk communication science perspective, the answer to this question is not so straightforward. Is this science normative, saying what is right and wrong? No, but it can be used for such a purpose. We need to distinguish between the risk communication research and science in itself on the one hand and how it is being used on the other. The previous functions and goals relate to the applications. Risk communication research and science provide concepts, principles, approaches, methods and models for communicating risk. They have no view on smoking or any other current risk issue. These concepts, principles, approaches, methods and models can be used for all types of situations. From a risk communication science perspective, the actual risk communication can be considered successful if it is in line with the state-of-the-art risk communication knowledge. In practice, risk communication has been and is still commonly used to 'educate' people (refer to the nuclear example of Section 7.1.2), despite the fact that risk communication experts highlighted the role of *risk communication as an interactive process to improve the understanding of the risks, in order to make appropriate judgments and decisions* (Árvai 2014; Renn 2014).

Risk communication is partly prescribed by laws and regulations, partly required by public pressure and stakeholder demand (Aven and Renn 2010). In the light of engagement by consumer and environmental groups, people expect governmental agencies and industry to provide relevant information and guidelines for consumers, workers and bystanders. Risk communication is embedded in an industrial and political thinking of openness and "right to know" policy framework (Baram 1984). In addition, international trade and globalization make it mandatory that potentially hazardous products are identified, properly labeled and regulated. Exposed people should have sufficient information to cope with the risk.

Three levels of risk debate

Risk debates can be conducted at different levels. A common classification for such debates is shown in Table 7.1, with requirements (including elements for evaluation), based on work by Ortwin Renn and others (see, e.g., Renn 2008).

The **first level** is the technical and scientific one based on evidence, risk measurements and characterizations. The goal of the communication on this level is to provide an accurate and neutral picture of factual knowledge, including the treatment of uncertainties. The objective here is to transfer knowledge, yet there is a need for two-way communication to ensure that the message has been understood and that the technical concerns of the audience have all been addressed (Aven and Renn 2010).

TABLE 7.1 The three levels of risk debate and their communication needs and evaluation criteria (partly based on Renn 2008; Aven and Renn 2010)

Criteria level	Issue of conflict	Communication needs	Evaluation
1	Technical/science: Evidence, risk measurements and characterizations	Information transfer	Access to audience Comprehensibility Attention to public concerns Acknowledgment of framing problems Examples of instruments: Expert hearing, expert committees, expert consensus conferences, Delphi exercises
2	Trustworthiness	Dialogue with stakeholders and the public	Match between public expectations Openness to public demands Regular consultations Commonly agreed-upon procedures for crisis situations Examples of instruments: Stakeholder hearings, roundtables
3	Values and world views	Dialogue and mediation	Fair representation of all affected parties Voluntary agreement to obey rules of rational discourse Inclusion of best available expertise Clear mandate and legitimization Examples of instruments: Public hearings, surveys and focus groups, citizen advisory committees, citizen consensus conferences, citizen panels, mediation

Common instruments for debates on this level are expert hearings, expert committees, expert consensus conferences and Delphi exercises (Renn 2008; Aven and Renn 2010):

Expert hearings: Experts with different positions are asked to testify before the representatives of the organizing institution (most often a regulatory agency) or the deliberative panel. The organizers ask each expert specific questions and let them develop their lines of argument. Such hearings may also allow for open discussions among the experts. Hearings help provide a clearer picture of the variability of expert judgments and clarify the arguments supporting each position. Hearings do not provide consensus and may not resolve any conflict. They may, however, clarify the basis of the conflict or the different points of view.

Expert committees: Experts interact freely with each other, have time to learn from each other and are able to consult other experts if deemed necessary. They work independently of the agency or deliberative body to which they report.

Expert consensus conference: Experts are gathered in a workshop to discuss knowledge issues and to decide on a general standard to be applied throughout the world. The workshop may be organized in group sessions in order to prepare common standards or in plenary sessions to reach a common agreement.

Delphi exercises: A Delphi process is aimed at obtaining a wide range of opinions among a group of experts (Linstone and Turoff 2002). The process is commonly organized in four steps. In step 1, a questionnaire asks a group of scientists to assess the severity of a risk. The scientists provide their best assignments, possibly including some type of uncertainty interval to their answers. In step 2, the organizing team feeds back to each participant the scores of the whole group, including medians, standard deviation, and aggregated uncertainty intervals. Each individual is then asked to perform the same task again but now with the knowledge of the responses of all other participants. In step 3, this procedure is repeated until individuals do not change their assessments any further. In step 4, the organizer summarizes the results and articulates the conclusions. A variation of the classic Delphi method is the group Delphi (Webler et al. 1991). During a group Delphi, all participants meet face to face and make the assessments in randomly assigned small groups of three and four. The groups whose average scores deviate most from the median of all other groups are requested to defend their position in a plenary session. Then the small groups are reshuffled and conduct the same task again. This process is iterated three or four times until no further significant changes are made. The quality of Delphi outcomes depends strongly upon the accuracy and completeness of the expertise and information brought into the process.

The **second level** of risk debates is about trustworthiness and relates to risk management institutions. An example of such a debate is the dialogue between government, health agencies and public in relation to a pandemic and its mitigation. In such dialogues, trust can be gained if the risk management institution is competent, effective and open to public demands.

At the **third level**, we face different social values and cultural lifestyles, with impact upon risk management. In this case, neither technical expertise, scientific expertise, institutional competence or openness is adequate for risk communication. The communication requirements of the first and second level are insufficient to find a solution that is acceptable to all or most parties. The debate quickly turns into a political one between the right and the left, between industrialists and environmentalists and other value-driven groups.

There is a strong tendency for risk management agencies to reframe higher-level debates (conflicts) into lower-level ones: third-level debates are presented as first- or second-level debates and second-level debates as first level. A recent example is the discussion about measures to cope with the coronavirus, which is frequently said to be about science, when the discussion is really about giving weight to uncertainties and hence about values. This is an attempt to focus the discussion on science and technical issues in which the risk management agency is confident. Stakeholders who participate in the discourse are consequently forced to use first-level arguments to rationalize their value concerns. Unfortunately, risk managers and others often misunderstand this as 'irrationality' on the part of the stakeholders. Frustrated, many stakeholders then turn to direct action and protest. In the end, the result is disillusion and distrust (Aven and Renn 2010).

Major instruments at levels two and three are hearings, surveys and focus groups, roundtables, panels and mediation (Renn 2008; Aven and Renn 2010):

Hearings: Most regulatory regimes of the world require hearings with stakeholders or affected citizens under specific circumstances. Such hearings can serve a useful purpose if they are meant to give external stakeholders the opportunity to voice their opinion and arguments.

The idea of public hearings is that people who feel affected by a decision should be given an opportunity to make their concerns known to the authorities and, vice versa, to give the authorities the opportunity to explain their viewpoint to the public. Hearings provide opportunities for stakeholders to understand the position of the regulatory agencies or other direct players (such as industry). However, hearings have proven very ineffective for resolving conflicts or pacifying heated debates. On the contrary, hearings normally aggravate the tone of the conflict and lead to polarizations.

Surveys and focus groups: Surveys of the general public or special groups provides a mean to explore the concerns and worries of the addressed audience. Focus groups go one step further by exposing arguments to counter-arguments in a small group discussion setting (Krueger and Casey 2000). The moderator introduces a stimulus (e.g., statements about the risk) and lets members of the group react to the stimulus and to each other's statements. Focus groups provide more than data about people's positions and concerns; they also measure the strength and social resonance of each argument vis-à-vis counter-arguments.

Roundtables (advisory committees, stakeholder dialogues): Commonly the participants represent different social groups, such as employers, unions, professional associations and others. The advantage is that the ritual window-dressing activities (typical for classic hearings) can be overcome through a more strict working atmosphere. Essential for organizing a successful roundtable is the involvement of a professional moderator.

Citizen advisory committees: This type of committee is particularly popular in local and regional contexts, but there are also examples of advisory committees on a national level. Such committees can be very effective in detecting potential conflicts (early warning function) and getting the concerns of the consumers heard and reflected in the respective organizing institutions.

Citizen consensus conferences: This instrument is based on the belief that a group of non-committed and independent citizens is best to judge the acceptability or tolerability of technological risks (Andersen and Jaeger 1999). Some citizens (a group of non-experts) are invited to study a risk issue in detail and to provide the legal decision-maker, or an agency, with a recommendation at the end of the process. The citizens are usually recruited by self-selection, based on a specific protocol ensuring diversity in the group members.

Citizen panels, planning cells or citizen juries: Planning cells or citizen panels (juries) are groups of randomly selected citizens who are asked to compose a set of policy recommendations on a specific issue (Crosby 1995). The objective is to provide citizens with the opportunity for learning about the technical and political facets of the risk management options and to enable them to discuss and evaluate these options and their likely consequences according to their own set of values and preferences.

Mediation: Mediation and similar procedures rest on the assumption that stakeholders can find a common solution if they do not insist on positions but try to meet their crucial interests and underlying values. Under these circumstances, win–win solutions may be available that will please all parties. Mediation requires the involvement of a skilled and professional mediator.

Dialogue and mediation processes reveal the different positions and perspectives and can, in many cases, lead to an improved understanding among relevant stakeholders, increase awareness of and sensitivity to the dilemmas and concerns that are at stake and explore common ground for making the necessary trade-offs (Aven and Renn 2018). If these processes are well designed and conducted, they may lead to a common understanding of the problem and widespread support for a risk management solution.

Different types of risk communication models have been developed, which can help us target and tailor-make the risk communication. In the following, we present one of the most common and established models used.

7.2.2 The basic stages of risk information processes

The basic elements of risk communication processes are shown in Figure 7.1 (Bostrom et al 2018):

1   Exposure and attention:

- Exposure: What do risk communicators express, and how do they communicate? Which types of risks and which dissemination routes prevail in risk communications?
- Attention: How do messages attract people's attention? Important aspects are design and format features of risk messages.

2   Understanding and acceptance:

- Information processing: What governs understanding (e.g., mental models, expressions of risk and uncertainty)?

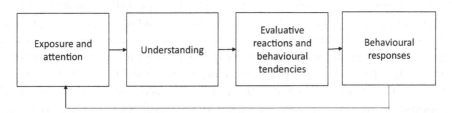

FIGURE 7.1 Basic stages of risk information processes (based on Bostrom et al. 2018)

- Acceptance of the message: Which messages are believed to be true? An influential factor is trust in sender.

3   Evaluative reactions and behavioral tendencies:

- Evaluative reactions to the message: Cognitive and emotional responses.
- Behavioral preferences: Which behavioral options are taken into account? What influences behavioral tendencies (e.g., perceived efficacy, culture, identity protection)?

4   Behavioral response:

- Behavior: Which contextual factors influence behavior (e.g., incentives, nudging)?
- Habit: Which factors foster retention and maintenance of behavior? Which factors foster change?

Exposure and attention

People typically attend to and remember information that supports their prior beliefs about a risk. A illustrative example is climate change risk, where people look for risk information sources that align with their underlying beliefs – the result being that those who oppose climate change mitigation become more strongly opposed and those who support climate change mitigation become more supportive.

To get attention and reach people, different types of warnings are used. For example, in the United States, current warning processes for risks such as tornadoes or flash floods include reverse 9–1–1 (in which warning systems autodial cellphones with warnings), social media posts, radio spots, TV banners and sirens. The aim is to capture attention and evoke an immediate response. To help people discern, understand and remember what to do, various design elements (e.g., visual, textual, graphical, numerical) can be used, including visual features such as color, motion, orientation and size, as well as visual design elements such as headers and subheadings.

A person's level of numeracy may strongly influence the attention to and interpretation of numbers and pictographs in risk messages. Low numerates seem to focus initially more on the pictographs than high numerates (Keller et al. 2014).

Understanding and acceptance

A main component of risk communication efforts is to provide an understanding of the relevant risks. As we know, uncertainty is a key element of risk, and probability is a common way of expressing uncertainty. Considerable research has been conducted within the risk communication field to provide guidance on how to best communicate risk and, in particular, the uncertainties and probabilities. It is a challenging task, as there are many perspectives and issues in this discussion, which also relate to the fundamentals of science in general as well as risk science. If the risk assessment and characterization lack a proper foundation, it is difficult for the communication to provide an understanding of the risks at hand; refer to the example of Section 7.1.1.

A fundamental requirement for any type of analysis and scientific work is that basic concepts be well defined and allow for a meaningful interpretation. If the analysts do not have a clear and solid understanding of what the risk assessment results mean, how is it then possible for others to understand the risk?

Returning to the risk conceptualization and characterization introduced in Chapters 2 and 3, (C,U)–(A,C,U) and (C',Q,K)–(A',C',Q,K), clarity is obtained on what risk means, what uncertainty in relation to risk means and how probability is used as a measure of uncertainty. According to this perspective, there are uncertainties about what type of events will occur and when, what the consequences will be given an event (what are the cause-effect relationships) and about frequentist probabilities specified in the risk assessment (these are examples of C'). In addition, there are uncertainties about beliefs and assumptions forming the knowledge K.

Risk assessment and characterization address these uncertainties by Q, typically using the pair of knowledge-based probability and strength of knowledge judgments. There are, however, limitations, as thoroughly discussed in Chapters 3 and 4. The analyst group could, for example, base their analysis on a model assumption which, surprisingly for the analysis team, turns out to be wrong.

Science produces characterizations like (C',Q,K) – (A',C',Q,K). The results are not facts but justified beliefs based on data, information, modeling, analysis, argumentation and so on. This fact represents a communication challenge, as the limitations of science, and the existence and acknowledgements of uncertainties, can be used to question research findings. An illustrative example is the cigarette industry. Cigarette manufacturers have consistently disputed evidence supporting the conclusion that smoking is dangerous by reference to the uncertainties and limitations of the analysis methods to prove causality. The use of appeals to uncertainty as an argumentative strategy to avert mitigative actions in societal controversies is so pervasive that it has been assigned a name: *scientific certainty argumentation methods*, or SCAMs for short (Freudenburg et al. 2008). People dislike uncertainties; they seek and prefer certain and unambiguous information that can guide them in their everyday decisions (Rabinovich and Morton 2012). Any discussion about uncertainties tends to be viewed as confusing, untrustworthy and disqualifying by the public (Frewer et al. 2003; Johnson and Slovic 1998); refer to discussion in Section 6.2.1.

Bassarak et al. (2017) show that controversial societal risks such as climate change are characterized by a psychological quality that they refer to as *disputed risk*, which reflects not only epistemic uncertainties but also evaluative uncertainty about how to form an opinion about the risk issue, covering aspects such as: the issue is perceived to be unsettled within the scientific community, scientific consensus is regarded as low and many diverse and controversial opinions exist concerning the issue.

The fact that science is about justified beliefs makes science vulnerable for misuse – by being selective in what information is highlighted and communicated. Actors may be tempted to present or frame the information in ways that favor their underlying conclusions, beliefs or motives. If objective knowledge on disputed risks does not exist, it is difficult for people to know if the risk communication conveys fair characterizations of the risks or is biased toward a particular view. Science is not represented by one voice, in particular for disputed risks. If there is a tendency to communicate only one view, science has become blurred with policy and politics; see discussion in Section 7.2.5.

As discussed in Chapter 6, messages that convey high uncertainty related to the hazards do not in general lead to high risk perception and a desire to enforce strict risk regulations. However, research has shown that messages that communicate high uncertainty, for example, in the prediction of climate change, are more persuasive for people who endorse a skeptical perspective on science than for people who hold a less skeptical view on science; see Rabinovich and Morton (2012) and Section 6.2.1.

Probability is the common tool used to express risk and uncertainties, and risk perception and communication research has given us considerable insights about how people interpret and react to probabilistic information. An interesting finding is that people, for example, understand a statement like a 30% chance of an event to be different than if the statement was expressed as a chance of 30 in 100. Another finding is that positive phrases of occurrences of events are interpreted differently than negative phrases. For example, there is a difference in saying that 95% of the medical operations are successful compared to expressing that 5% fail. Risk communicators need to take this type of insight into account when designing risk messages. The messages convey not only risk knowledge but also information about conversational implications such as encouragement or discouragement (Teigen 2012). Professional risk scientists should in general be less subject to interpretation issues as these than lay people. The difference defines what it means to be professional in this field. Qualitative probabilities are often preferred in risk communication, but it leads to interpretational problems, as people would understand the meaning of, for example, unlikely, likely and very likely differently.

The challenge of communicating probabilities cannot be seen as isolated from the risk characterization produced by the risk assessment. If probability numbers constitute a key dimension of the risk description, their message needs to be communicated. However, to be able to do this in a meaningful way, a proper foundation for the probabilistic analysis is required. A probability has scientifically different meanings (refer to Chapter 3), and if analysts and communicators neglect to address this, the result is likely to be a poor understanding of the risk.

People interpret risk messages and risk events through their mental models. A mental model is simplified representation (for example, an image) for the thought processes explaining how something works, for example, what causes an accident. Experts on a specific topic are expected to have better – more accurate – models than non-experts, as they possess a stronger knowledge about the topic. However, this is not in general true. A risk analyst may, for example, have a mental model based on risk being an inherent property of the activity considered objectively represented by a number equal to probability times loss, whereas a lay person has a mental model where uncertainty constitutes a key element of the risk concept. Then it can be discussed who has the best mental model; refer to Chapter 3.

Insights about the mental models of the recipients of risk messages is important for the proper design of these messages. A useful tool in this respect is to draft communication materials and empirically evaluate their effectiveness; see Section 7.3.

Evaluative reactions and behavioral tendencies

How people use and evaluate risk messages has been a topic of many studies. As expected, prior trust in the message source or sender is an important determinant of how the message is processed; see discussion in Section 6.2.1. Mental models represent another main determinant of an individual's evaluation of risk messages and of responses to risk events. These models also influence emotional reactions and behavioral responses (Böhm 2003; Böhm and Pfister 2000). Behavioral tendencies result from the emotional responses to a few dimensions of the message or situation, including agency, outcome desirability, certainty, fairness and coping potential (Ellsworth and Scherer 2003; Keller et al. 2012; Scherer 1999). The appraisal of these dimensions is based on the individual's mental model of the situation and differentiates between emotions such as fear, anger, sadness and disappointment. As discussed in Section 6.2.2, the emotions can have different effects on risk evaluations and behaviors, even if they share the same valence. For example, anger may reduce perceived risk, whereas fear amplifies perceived risk, even though both emotions have a negative valence (Lerner and Keltner 2001).

Research indicates that the mental representation of behavioral options – in particular, knowledge about them and judgments concerning their effectiveness and the ease with which they can be performed – plays a key role in forming risk behaviors and should consequently have an important place in risk communications (Bostrom et al. 2018). An example from politically relevant behaviors is provided by Bostrom et al. (2012), showing that perceived effectiveness of policy options was shown to be the strongest predictor of policy preferences in an international study on climate change risk perception and policy support.

A widely used concept in relation to this type of effectiveness is perceived efficacy, that is, a person's perceived ability to produce certain results. It is also commonly referred to group or collective efficacy, for example, in relation to health and environmental risks. Work indicates that messages conveying efficacy tend to increase health behaviors such as vaccination (e.g., O'Keefe 2013; Peters et al. 2013).

Risk is perceived differently depending on the cultural context, as was discussed in Section 6.2.3. Hence the same risk messages could be interpreted differently in, for example, two different countries or population groups. As highlighted by Bostrom et al. (2018), culture theories provide a basis for risk communicators to target specific groups and tailor risk communications to those groups to improve the effectiveness of the communications (Kahan et al. 2015; McKenzie-Mohr et al. 2012). The research is, however, not clear on the effectiveness of targeting and tailoring risk communications in this way (Bostrom et al. 2013, 2018; Hine et al. 2016; Myers et al. 2012). For health and environmental risk communications, there is evidence that targeting and tailoring are effective for some topics but not for others (e.g., global warming).

We would also like to draw attention to research on persuasion and attitude change. Factors that seem to enhance the persuasiveness of communication include the following (based on review in Renn 2008; Aven and Renn 2010):

- Attractiveness of information source: attractiveness relates to similarity of positions between source and receiver, likeability of the source and physical attraction
- Empathy or sympathy of the receiver for the source: this refers to the possibility of a receiver identifying with the source or its motivations
- Credibility of source: this relates to, for example, perceived competence, expertise, objectivity, impartiality and interest in the source
- Suspicion of honest motives: receivers do not observe or perceive any hidden agendas or motives behind the communication effort
- High social status or power of communication source: the effect of these two factors depends heavily upon the issue and the composition of the audience

All of these factors seem almost intuitively plausible. A communicator is likely to make a longer-lasting impression on the audience if the message appears accurate, honest and fair and if the communicator is likable, someone with whom the audience can easily identify. Psychological research has yielded some interesting, sometimes even counterintuitive, findings that relate specific aspects of message composition or personal style of communication to persuasive effect. Here are some examples (based on review in Renn 2008; Aven and Renn 2010):

- High-credibility sources, such as scientists or opinion leaders, produce more opinion change but no difference in message learning.
- Perceived expertise depends upon many factors, including status, education, perception of technical authority, age and social class.
- Stating explicitly the persuasive intent is usually more convincing than hiding such an intention and leaving it to the audience to make their own inferences.
- Perceived fairness and social status are both factors that can compensate for a lack of objectivity.
- Being explicit when drawing conclusions and presenting counterarguments has proven more effective than operating with implicit conclusions or presenting only one side of the story.

- The perception that the goals and motives of the source serve a common interest or refer to highly esteemed social values, such as protection of the environment or public health, enhances public confidence in the communicator but reinforces distrust if the task performance of the communicator is perceived as weak.
- The agreement to listen to disliked sources increases the probability of attitude change.

These insights are helpful to design communication programs and to train communicators for their task. However, it should be noted that many of these findings were accomplished in rather artificial laboratory environments and may to varying degrees be valid for a specific risk communication arena. The insights refer to changes in attitude, not behavior. Once attitudes toward a risk or an activity are formed, they generate a propensity to take action. However, a person's decision to take action depends upon many other factors than attitude, such as behavioral norms, values and situational circumstances. Thus the communication process can influence the receiver's behavior, but the multitude of sources, the plurality of transmitters and the presence of situational forces on personal behavior render it nearly impossible to measure or predict the effect of a single communication activity (Renn 2008; Aven and Renn 2010).

Behavioral response

The weak correlation between attitudes and behavior is a major challenge in risk communication that aims to change behavior (e.g., for emergency responses). It is difficult to change or modify attitudes through information but even more difficult to modify behavior. Research indicates that three factors are critical for increasing the chance of behavioral changes (based on review in Renn 2008; Aven and Renn 2010):

- continuous transmission of the same information even after a positive attitude has been formed toward taking action (need for constant reinforcement);
- unequivocal support of the most relevant information sources for the behavioral change advocated in the communication process (need for consistent and consensual information);
- adoption of the behavioral changes by highly esteemed reference groups or role models (social incentive for imitation).

Behavioral changes are rarely triggered by information alone. Rather, information may be effective only in conjunction with other social factors, such as social norms, roles and prestige (Renn 2008).

Research indicates also that risk communications that highlight effective action and efficacy promote protective behaviors, as discussed previously, as do fear appeals (Tannenbaum et al. 2015). Much behavior is susceptible to context effects. Bostrom et al. (2018) provides an example from insurance. If the default is that home insurance policies cover flood insurance unless one opts out, more people will have flood insurance than if they need to opt in to purchase it. The example illustrates how the choice architecture, or the *nudging*, can be used to influence risk decision-making (Thaler and Sunstein 2008).

Behavioural response is also strongly affected by habits. Many people continue to smoke despite the negative health effects. The science on this topic looks to ways of breaking the habit (Duhigg 2012). An approach is to try to reprogram the brain to respond to cues (with associated rewards) by creating new routines to respond to them. For example, the cue for a smoker is a chemical craving, and the reward is a good feeling. A new habit could be established by chewing a smoking cessation product in response to the cue. Over time, with repetition, the person will automatically reach for the smoking cessation product rather than a cigarette when the nicotine-craving feeling occurs (Bostrom et al. 2018).

Discussion of risk communication issues

The previous analysis shows how risk messages go through several processing stages – from exposure and attention via understanding and evaluation to behavioral intentions and action. Some messages may or may not attract attention, may be correctly interpreted or misinterpreted, may or may not be seen as trustworthy and may lead to sound or poor decisions, depending on different types of audience attributes.

Risk communication science highlights effective features of such messages (see the following box as an example) but also the importance of strategic framing and the politics of engagement (e.g., Fischhoff et al. 2011). As mentioned in Section 7.2.1, engaging stakeholders in deliberations and debates about risk may also constitute a function and goal of risk communications (Gregory et al. 2012; National Research Council 1989). Fischhoff (1995) refers to several development stages of risk communication and handling as observed over the last four or five decades – starting from the numbers (historical data, probabilities, etc.) to include explanations, context and engagement:

- All we have to do is get the numbers right
- All we have to do is tell them the numbers
- All we have to do is explain what we mean by the numbers
- All we have to do is show them that they've accepted similar risks
- All we have to do is show them that it's a good deal for them
- All we have to do is treat them nice
- All we have to do is make them partners
- All of the above

Risk communication could lead to consensus but also to sharpened disagreement, as it brings different stakeholders and actors to understand the depth of their differences in ways that exacerbate conflict (Bostrom et al. 2018).

Similar stages for the development of risk communication are described by Leiss (1996); see also Renn (2014):

- Phase 1 emphasizes the necessity of conveying probabilistic results to the general public and educating lay persons to acknowledge and accept the risk management practices of the respective institutions (model of message transfer).

- Phase 2 emphasizes persuasion and focuses on public relations efforts to convince people that some of their behavior was not acceptable (such as smoking) because of high risk levels, whereas public worries and concerns about many technological and environmental risks (such as nuclear installations, food additives and liquid gas tanks) were regarded as overcautious due to low risks (model of shared understanding about objective threats).
- Phase 3 of risk communication highlights a two-way communication process in which it is not only the members of the public who are to be engaged in a social learning process but also the risk managers.

REFLECTION

You are in a risk class, and the professor discusses the meaning of risk in a professional risk science context. The professor welcomes as many definitions of risk found in the literature as possible. The professor expresses that there are different perspectives and different needs, and all of these definitions are therefore relevant and of interest for risk analysts. Do you agree with this professor?

We do not. Risk science is about the most justified knowledge generated by the risk analysis field and discipline (refer to Appendix B), and over the years, many of these definitions have been rejected as not being suitable. The professor has a responsibility to bring forward this knowledge and not present all definitions as if they have the same justification and authority.

Effective warning messages contain five essential elements (Lindell and Brooks 2012; Mileti and Sorensen 2015; Bostrom et al. 2018):

- The source of the warning (so that recipients can assess the expertise and trustworthiness of the source).
- Identification of the hazard, including its certainty, severity and immediacy and the duration of dangerous conditions.
- The specific areas at risk (and safe areas, if not obvious), described in a way people can identify and understand.
- What to do (i.e., a protective action people see as effective and feasible), when and how to do it and what it will accomplish.
- Where to go for further information and assistance.

7.2.3 Openness and transparency

Many patients would like to receive a simple message from their physicians – not a detailed information package characterizing all aspects of the efficiency and risks related to the decease and the treatment. A confident and authoritative message from the physician can reduce patients' anxiety (Hoffrage and Garcia-Retamero 2018). The patients trust the physicians to provide guidance that is in the patient's best interest. Other patients, however, like to be fully informed. In any case, the patients should be given a choice. Physicians are increasingly encouraged to involve patients in the deci-sion-making rather than pursuing a paternalistic or authoritarian method in which they make the decisions for their patients (Hanson 2008). Many people today use internet and other sources to obtain information about their ailment, but much of this informa-tion is not science based. The desired result is an open and transparent risk communica-tion with sharing and discussion of risk-related information.

Many people today refuse to be 'educated' by experts or others. The paternalistic or authoritative approach is rejected. We observed this in relation to the nuclear indus-try, and we also see it today when it comes to many of the current global risk issues. Specific views are identified as the only legitimate ones, but many people reject the basic pillars that these views are based on despite the fact that the views are referred to as being founded on science. The problem is that the approach taken violates the fundamentals of science and also democracy. The issues have been politicized. If one view is considered the 'right one' with media supporting it, divergent perspectives and research are commonly removed from communication platforms or made more dif-ficult to trace. Media has, and often enforces, the ability to focus on topics, ideas and discussions that favor their own views and objectives. If alternatives are reported, they are typically attached with comments devaluating their perspectives. Such processes directly conflict with a policy based on openness and transparency – which is character-ized by an acknowledgment of the limitations of scientific knowledge and a need for a continuous 'battle' between different schools and camps for what the most warranted statements or justified beliefs are within the field of study (Bourdieu and Wacquant 1992). If this 'battle' is removed, there is no real science remaining but an unfortunate mixture of science and activism/politics.

Science often provides a comprehensive volume of findings reflecting a variety of results depending on perspectives and assumptions. For the purpose of a political cam-paign promoting a specific ('right') view, there is evidence available that can be used to serve that purpose. If the 'right' view is to be communicated, the message will be disturbed by science, allowing for uncertainties to be acknowledged and communicated. Institutions exist on the behalf of governments to make overall judgments about what the most justified knowledge available is, but these judgments can be contested, as they are to some degree value based. Simplifying the research message from multiple sources means that some research results could be completely ignored or judged of less impor-tance than other contributions. If the search for a 'right' view is strong, there is a risk that this process will be strongly biased. If media supports this view, they will, as discussed, contribute to either amplifying and attenuating the seriousness of the risks, depending on what the right view is. In the former situation, repeated cases and examples will be used to stress the risks, whereas in the latter situation, alternative insights will be ignored or given little attention; see discussion of the SARF model in Section 6.2.5.

As discussed by Bouder et al. (2015), there has been a change in recent years in the regulatory environment in many areas, in particular in relation to safety, health and environmental issues. The traditional elitist and expert-based approach has been replaced by a more transparent and open framework that ensures greater access to data, information and knowledge – a change to a new 'participatory-transparent' thinking with wider stakeholder participation and a trend toward enhancing transparency. People should have the legitimate right to know about the full background of regulatory decision-making. A key driver of this changing policy environment has been the lack of trust in regulators and industry, and a main contributing factor to this development has been the rise of 24-hour media and the internet. A variety of measures are commonly implemented by regulators in line with this policy, for example

- Maintaining dedicated web portals to, for example, provide public access to safety-related data
- Publishing recommendations from safety and risk assessments
- Public hearings
- Establishment of safety assessment and monitoring committees
- Setting up public workshops

Bouder et al. (2015) point to the fact that both policymakers and academics agree that transparency is in general something good – if adequately managed, it may improve the communication of benefits and risks. Measuring the improvements is, however, not straightforward, as discussed by Bouder et al. (2015).

The term 'transparency' is commonly interpreted to "entail conducting affairs in the open or subject to public scrutiny, or simply being honest" (Florini 1999; Birkinshaw 2006; Bouder et al. 2015). It is common to contrast transparency with opaque policies, where it is hard to discover who makes the decisions, what they are and what they are based on. Two types of transparency are often referred to (Coglianese 2009):

Fishbowl transparency, which captures the full disclosure of data and information without further explanations or contextualization, and

Reasoned transparency, which provides an interpretation and a message on the basis of the source data and information.

In practice, a mixture of these policies would normally be desired. Without reasoned transparency, people may have problems seeing what really matters, what the key information is. On the other hand, any interpretation would imply some level of value judgment, which could be subject to discussion.

7.2.4 Planning risk communication efforts and messages

Following the recommendations by Bier (2001a) (which are built on an extensive literature review), the planning of a risk communication to the public should provide

1   Clarification of legal requirements and organizational policies that could constrain the design of the risk communication message and/or format.
2   Clarification of the purpose of the risk communication.

3   Selection of a strategy that is appropriate for the purpose at hand; for example: simple, vivid messages to raise awareness of a hazard or risk; explanatory tools such as diagrams and analogies to educate people; and so on.

4   Identification of the characteristics of the audience(s) for the risk communication; for example: the audience's level of knowledge and education; the audience's mental models, attitudes and beliefs about the issue at hand; the audience's level of receptivity and openness to the ideas being communicated; the audience's concerns about the issue.

5   Identification of sources of audience information, such as focus groups, surveys, public information officers and articles and books describing the audience's views.

When it comes to designing risk communication messages, the recommendation from Bier (2001a) is to highlight trustworthiness with an open, responsible and caring attitude. Audience concerns should be listened to before attempting to impart new information.

Risk comparison is generally recommended as a useful risk communication tool but should be applied with care (Bier 2001a):

*   Make comparisons of the same risk at different times, comparisons with other causes of the same injury or disease and comparisons with unrelated risks, such as the risk of lightning.
*   Avoid comparisons with risks that are generally considered trivial, such as the risk of eating a number of tablespoons of peanut butter.
*   Perform pilot testing of the risk communication messages on a limited basis before using them more widely to ensure that they are easily understood and interpreted in the right way.

Additional recommendations are provided for some specific challenges related to audience comprehension of risk-related information (Bier 2001a):

*   For small probabilities, consider using graphical representations to illustrate how small a probability actually is.
*   For unfamiliar terms, or terms with unfamiliar meanings, give examples illustrating both what a term means and what it does not mean.
*   For explanations of complex phenomena, make use of explanatory tools such as diagrams and analogies to ensure that audience members develop accurate mental models of the phenomenon.
*   To avoid audience misconceptions or incorrect intuitions, acknowledge that the misconception or intuition is plausible (for example: since testing is a good thing, it is natural to believe that more testing is always better). Explain why the audience's view is inaccurate or incomplete (for example: testing effort on components that are not very important to risk can be wasted). Present a correct explanation and show why it avoids the flaws or weaknesses of the audience's original viewpoint (for example: doing better at focusing testing efforts on the

components that are most important to risk can improve safety while reducing the total testing effort).

Finally, Bier (2001a) provides some recommendations concerning strategies for enhancing stakeholder participation:

- Ensuring that there is a genuine commitment to the process before undertaking stakeholder participation.
- Clarifying the scope and objectives of any stakeholder participation (e.g., education, consultation, etc.) to all participants.
- Clarifying whether participants will in fact have a voice in the final decision.

A risk assessment provides a basis for charactering risk, and overall guidelines for how to form the characterizations are presented in Chapter 3. Clearly, there is a need to tailor-make the characterization based on the purpose of the risk assessment. Inspired by AIHC (1989) and Bier (2001b), we can formulate some common requirements when presenting risk assessment information and results:

- The presentation must be understandable and cover all relevant aspects as defined by the scope of the assessment.
- The presentation must show the applicability and usefulness of the assessment in relation to the decision-making situation considered.
- The presentation should provide a neutral characterization of risk (following the guidelines of Chapter 3) and treat contentious issues in a balanced way.
- The presentation should show traceability of all elements and steps of the risk assessment and characterization, from evidence through models to conclusions.
- Conclusions should be drawn so as to be relevant to the specific decision-making situation.
- The need for additional studies and research should be indicated.

In short, it is common to express that the assessment should be *relevant, timely* and *comprehensible* (AIHC 1992; Bier 2001b).

More specific guidelines for risk communication briefings point to the need for highlighting (see for example, Bloom et al. 1993; Bier 2001b)

- What is the issue?
- Who cares about it?
- What are the major stakeholders saying about it?
- The risk characterization (the seriousness of the risk, confidence in the results, uncertainties)
- Risk management options, related cost-benefit considerations

Many studies show that people, including decision-makers, have difficulties in understanding risk assessment results, expressed, for example, by means of probabilities. Risk analysts are trained in risk science, whereas decision-makers and lay people normally lack such training. Is the answer, then, to keep things very simple and avoid, for example, discussions of uncertainties? No, as risk is then not adequately described. Being a professional risk analyst means that one is able to communicate the results of the risk assessment in a way that is informative. As argued by Veland and Aven (2013), communicating about risks and uncertainties with managers and politicians is in general not a problem if properly conducted. Managers and politicians are in general well-equipped people who are able to relate to and deal with uncertainties and risk; in fact, these tasks are part of their job functions – to make decisions under risk and uncertainty. Managers and politicians quickly understand what the key issues are and what is at stake, if the risk professionals can do their job. The problem is, rather, that risk analysts are not able to report the risk and uncertainties and present them in an adequate way (Veland and Aven 2013). The present book aims at providing some help in providing the necessary clarity on the fundamentals of risk analysis, guiding risk professionals in these tasks.

This type of discussion is also relevant when discussing risk analysts presenting risk assessment results to lay people. People are faced with uncertainties in relation to car driving, security issues, potential pandemics and so on. They will realize that there are no numbers that can fully describe the risk in such situations, again provided risk analysts and scientists do their job properly, that is, build their work on risk science. Maybe it will take some time for all to understand the basic ideas of risk and a risk characterization, but the goal must be to increase public knowledge about the fundamentals of risk science to support risk communication.

7.3 HOW TO CONDUCT A RISK COMMUNICATION STUDY

In the following, we will provide an example of a risk communication study. It is based on a statistical survey, which is a common approach for such studies.

Suppose you work for the risk management office of a major university. Your role is to ensure that students are actively managing risk related to a variety of risk events, such as weather, terrorism, active shooters and campus assaults. Typically, students receive risk-related warnings from dedicated sources. For example, students may receive weather alerts from third-party apps on their phones, or they may receive alerts for campus-specific dangers directly from the university. Because the university wishes to provide standardized and consistent messages to all students, it has decided to send emergency alerts to all students using a dedicated application that is required to be loaded on students' electronic devices.

You are tasked with studying how the university community perceives and responds to the wording of these warnings. You enlist a group of test subjects who opt in to receive warning messages, then complete a survey.

The test messages involve two emergency scenarios: severe weather and a campus health alert. For each emergency scenario, two types of communication are tested: a short message with a link to more instructions and a long message requiring no additional information retrieval. Longer messages can convey more information but may be ignored due to the required reading time. Shorter messages can convey attention and urgency and are possibly more likely to be read but are not able to convey as much information. The two tests are described in the following:

Test 1: When a tornado (severe weather) is imminent, students are given a warning to move indoors. You have created two messages to test:

- W1: Alert: A tornado warning has been issued. Visit the emergency website for detailed instructions (linked here).
- W2: Alert: A tornado warning has been issued. A tornado has been sighted on the ground and you should take immediate action to take cover. Take shelter immediately. Stay away from windows, doors and walls that face the building's exterior.

Test 2: Campus health alert. When there is occurrence of health issues on campus, such as influenza, COVID-19 or mumps, the application provides a warning for students:

- H1: Alert: COVID-19 cases are rising. Visit the COVID-safety website for detailed instructions (linked here).
- H2: Alert: COVID-19 cases are rising. There are currently 19 cases on campus, 3 outbreaks, 20 students in quarantine and 12 students in isolation. Follow protective measures; get tested if you show symptoms.

You have decided to test two aspects of the risk communication: *access* and *clarity*. In addition, you would like to evaluate whether students are able to appropriately respond to the messages, measuring *response*. The survey measurements are as follows:

- **Access.** Whether the message was received: *Was the message received?* (1 = yes, 0 = no)
- **Clarity.** Whether there was ambiguity or confusion about the content of the message: *How clear was the message?* (1 = unclear, 2 = somewhat clear, 3 = very clear)
- **Response.** Whether receivers of the message acted in the intended manner after receiving the message. *Did you know how to respond?* (1 = no, 2 = yes, but I had questions, 3 = yes)

Confidence intervals are used to make an inference about the true population average scores for these questions. The results, using 50 responses, are shown in Table 7.2.

For the Access response, we use a confidence interval for the proportion of respondents who answered Yes (1). For example, in the Test 1 severe weather case for W1, the sample proportion is computed as:

$$\hat{p} = \frac{x}{n} = \frac{Number\ of\ respondents\ who\ answered\ Yes}{Sample\ size} = \frac{30}{50} = 0.6$$

The estimated standard error for the point estimator is:

$$Estimated\ Standard\ Error = \sqrt{\frac{\hat{p}(1-\hat{p})}{n}}$$

Then, the confidence interval for the proportion can be computed as:

$$\hat{p} \pm z_{\alpha/2}\sqrt{\frac{\hat{p}(1-\hat{p})}{n}}$$

For a 95% confidence interval, $z_{\alpha/2} = z_{0.05/2} = 1.96$; therefore, the confidence interval is computed as:

$$0.6 \pm 1.96\sqrt{\frac{0.6(1-0.6)}{50}} = 0.6 \pm 0.07$$

For the Clarity and Response measurements, we also use a confidence interval to compare the messages. The table uses the Microsoft Excel function named CONFIDENCE.T to compute the margin of error for confidence intervals. The function can be used with three arguments: CONFIDENCE.T (level of significance, standard deviation, sample size). Here, we use a 0.05 level of significance (corresponding to a 95% confidence interval), the computed sample standard deviation and a sample size of 50.

For all of the confidence intervals, the lower bound of the confidence interval can be computed as:

sample mean – margin of error,

The upper bound is:

sample mean + margin of error

For Test 1 (severe weather): The W1 confidence interval for Access was (0.53, 0.67), meaning that there was 95% confidence that the true population average Access level was within that range. Conversely, the W2 confidence interval measuring Access was (0.35, 0.49), meaning that there was 95% confidence that the true population average

TABLE 7.2 Sample results for severe weather and health scenario

	Sample data			Proportion descriptors			Mean descriptors		Confidence interval	
	Min	Max	Sample size	Yes responses	Sample proportion	Empirical standard error	Average	Empirical standard deviation	Lower bound	Upper bound
Test 1: Severe weather										
W1 (Short)										
Access	0	1	50	30	0.60	0.07			0.53	0.67
Clarity	1	3	50				1.90	0.40	1.79	2.01
Response	1	3	50				2.20	0.20	2.14	2.26
W2 (Long)										
Access	0	1	50	21	0.42	0.07			0.35	0.49
Clarity	3	3	50				2.90	0.50	2.76	3.04
Response	2	3	50				2.80	0.30	2.71	2.89
Test 2: Health										
H1 (Short)										
Access	0	1	50	12	0.24	0.06			0.18	0.30
Clarity	1	3	50				2.10	2.00	1.53	2.67
Response	1	3	50				1.7	2.00	1.13	2.27
H2 (Long)										
Access	0	1	50	15	0.30	0.06			0.24	0.36
Clarity	1	3	50				2.40	2.00	1.83	2.97
Response	1	3	50				2.50	2.00	1.93	3.07

Access level was within that range. The results show that the W1 (short) message was significantly more accessible compared to W2 (long). This can be seen by the lack of overlap in the confidence intervals; in other words, the lower bound for the W1 confidence interval is above the upper bound of the W2 confidence interval. Statistically, this means with 95% confidence, the true population proportions are not equal between W1 and W2. Similarly, the W2 (long) message has the highest scores for Clarity and Response, and there is no overlap in the confidence intervals. The W2 message is more clear and elicits a more appropriate response.

For Test 2 (campus health alert): The is overlap in confidence intervals for Access, Clarity and Response. For example, the W1 confidence interval for Access was (0.18, 0.30), while the W2 confidence interval was (0.24, 0.36). Therefore, the results are inconclusive; no tested message style performs significantly better.

There could be many reasons for the drastically different responses between the two tests. Aspects of health and safety related to viruses may be less understood by students. The notion of safety from severe weather can be physically seen and understood, as students have instantaneous feedback when assessing whether they are adequately protected. Students may also not clearly understand what behaviors adequately protect themselves from a virus, as students may have previously engaged in non-allowable behaviors and encountered no negative consequence. The scientific community itself may have disagreement on what behaviors are relatively safe vs. unsafe. Also, protecting oneself from a tornado typically involves protective action for the duration of the tornado warning, while protection from a virus requires days, weeks and even months of relatively more onerous action. Since it is relatively easier to protect oneself from severe weather, students may be more willing to engage with the warning messages.

While this demonstration served as an example of a risk communication survey, there are several important next steps for this example. There may be a need to evaluate whether simple messages are more adequate for some types of emergencies (such as severe weather) versus others. There may also be a need to test additional message styles, for example, using various fonts, sounds, spacing and colors. The university may also consider setting more intentional policies regarding how students should react to various warnings. For example, there may be a need to enforce various penalties for students engaging in unsafe behaviors after warnings have been issued. Conversely, the university may choose to implement policies that encourage or incentivize safe behaviors. Thus, risk communication cannot be seen in isolation but in sync with other policies, incentives and penalties.

7.4 PROBLEMS

Problem 7.1

Trevor in Section 7.1.1 showed a diagram with probability numbers for different loss of lives categories. Explain what these numbers mean and simulate that you communicate the results to the decision-makers of the company.

Problem 7.2

Do you consider the IPCC risk communication successful? Why/why not?

Problem 7.3

A risk researcher states that for the effective risk communication and use of risk assessments in practice, it is not outstandingly important to define with an exceptional accuracy the meaning of probability given all uncertainties related to obtained numbers. What is your response to this statement?

Problem 7.4

A vaccine is developed. How would you communicate that the vaccine is safe?

Problem 7.5

A new process plant is planned rather close to an area where people live. You are risk manager of the company operating the plant. How would you plan the risk communication in relation to these people? What main principles would you adopt?

Problem 7.6

Discuss to what degree you trust governmental communication about risk in the country you live. Why/why not?

Problem 7.7

Discuss to what degree risk communication should address potential surprises and the unforeseen.

Problem 7.8

Consider the issue of safety in your community, place of employment or campus. Select the most important or critical risk and create a risk warning message that will inform your community of the risk. Also, suggest how respondents should address the risk. For example, Section 7.3 considered addressing severe weather risk by instructing respondents to take shelter during severe weather. Share this risk warning message with your community and ask: 1) Was your message effective? 2) How can the warning be improved? Present your results to the class.

Problem 7.9

Consider efforts to simplify warnings for weather-related risk (Chief and Nagale 2020). See https://bit.ly/33VSEVi for an example. Will these proposed changes be effective? Explain.

Problem 7.10

Science is based on publications in scientific journals. The editors' handling and the review process are considered essential for ensuring the quality of the publications. The editor of a highly recognized journal with a scope addressing topics on societal safety and risks is also politically active, to a large extent related to subjects that the journal addresses. Do you think this represents a problem for the journal and science's standing in general in society as a producer of scientific knowledge?

Problem 7.11

Can a risk communication be successful if it is based on a traditional risk matrix approach? Why/why not?

Problem 7.12

Study the overview paper by Balog-Way et al. (2020). Identify one topic highlighted in the paper on risk communication research topics not addressed in the present chapter. Prepare a talk on this topic for your class.

KEY CHAPTER TAKEAWAYS

- Risk communication covers exchange or sharing of risk-related data, information and knowledge between and among different target groups such as analysts, experts, decision-makers, regulators, consumers, media and the general public.
- The main goal of risk communication is improving the understanding of the risks in order to make appropriate judgments and decisions.
- In general, risk communication requires an understanding of the target audience, a credible or trusted source and a properly formed message.
- Equally important is the quality of the risk analysis and risk characterizations that are used as a basis for communication.
- With few exceptions, such as proprietary information or that which may damage public security, the general recommendation is to apply an open, transparent and timely risk communication policy.
- The traditional elitist and expert-based approach has been replaced by a more transparent and open framework that ensures greater access to data, information and knowledge.
- The basic elements of risk communication processes are: exposure and attention, understanding and acceptance, evaluative reactions and behavioral tendencies and behavioral response.

PART IV

Risk management

CHAPTER 8

Basic theory of risk management

CHAPTER SUMMARY

This chapter discusses fundamental concepts, principles and approaches of risk management, including topics of risk governance. Risk management includes using and evaluating risk assessments; developing rules and processes under which risk management decisions are made; and deciding on the most appropriate decision, action, or investment for addressing risk. Risk governance refers to the application of governance principles for the handling of risk, linked to aspects like participation, accountability, effectiveness, coherence and proportionality.

Suppose a thorough risk assessment is performed and decision-makers are informed of the results. The challenge for the decision-makers is to recognize that these risk assessments do not in general justify a single most appropriate decision. The risk assessment has limitations – not all aspects of uncertainties are reflected, and there are other concerns and issues of importance for the decision-making not captured by the risk assessment. Even if we all agree on the results of the risk assessments, we do not necessarily agree on what risks we should accept or what the best decision is, as our values and priorities are different. There is a balance to be struck between creating value and protecting resources. This balance is not about science but about human values, ethics, management and politics. A key aspect in this regard is how much weight we give to uncertainties. In risk science, we refer to the *cautionary principle*, which is a guiding perspective for risk handling, expressing the idea that if the consequences of an activity could be serious and subject to uncertainties, then cautionary measures should be taken, or the activity should not be carried out. If you go for a hike in the mountains, it is wise to have extra clothes in case the weather should change. You give weight to the cautionary principle.

This chapter clarifies and motivates the rationale for the cautionary principle and how it relates to the *precautionary principle*, which is a special case of the cautionary principle, in the case that the uncertainties are 'scientific', meaning that we do not understand the phenomena involved. We are not able to develop accurate models of what is happening, for example, if we study the long-term effects on human health of a new type of chemical.

DOI: 10.4324/9781003156864-8

The chapter shows how the cautionary principle is closely linked to the key risk management strategies of vulnerability (robustness) and resilience management. These strategies are concerned with handling situations where a disturbance, change or failure has occurred, of known or unknown type, thus also meeting potential surprises and the unforeseen. It is emphasized that proper risk management gives due weight to resilience analysis.

Other risk strategies and policies of risk handling are also discussed, including *adaptive risk management*: different decision options are assessed, one is chosen and observations are made, learning is achieved and adjustments made. It can be seen as a way of implementing a cautionary approach.

In practice, the three major strategies, 1) risk-informed, using risk assessments; 2) cautionary/precautionary; and 3) discursive strategies for handling risk, are combined; the appropriate strategy is a mixture of these three main strategies. The discursive strategies and policies use measures to build confidence and trustworthiness through the reduction of uncertainties and ambiguities, clarification of facts, involvement of affected people, deliberation and accountability.

In addition to these three main categories of strategies and policies, it is also common to refer to 'risk-based requirement' – or just 'codes and standards' – strategies based on the use of codes and specific requirements that need to be met, which is applicable when the situations considered are simple – the phenomena and processes considered are well understood and accurate predictions can be made.

LEARNING OBJECTIVES

After studying this chapter, you will be able to explain

- the key concepts of risk management, governance and policy
- the basic strategies and policies for handling risk
- the main elements of the risk management process
- the importance of risk acceptance and tolerability judgments in risk handling
- the meaning and use of the precautionary principle
- the meaning and use of the cautionary principle
- the importance of vulnerability and resilience management
- the meaning and use of adaptive risk management
- the role of resilience in risk handling
- how risk strategies and policies depend on the type of risk problem

---

In this chapter, we first give some real-life examples illustrating important issues of risk management. Then we highlight some key definitions, covering inter alia risk management, risk governance and polices on risk. Next, we look into strategies for handling risk and relate these to different risk problems and issues. The following sections then discuss in more detail the strategy of using risk assessment, precautionary and cautionary thinking and management of vulnerabilities and resilience. We also give some reflections about ethical aspects in relation to risk handling. Key sources for the chapter are Abrahamsen et al. (2020), Aven (2010, 2016, 2019a, 2020d) and Aven and Renn (2010, 2018, 2019).

8.1 EXAMPLES ILLUSTRATING BASIC RISK MANAGEMENT ISSUES

8.1.1 Traveling by plane

We return again to the example of Chapter 3, where Anne and Amy consider a plane trip from Washington, DC, to San Francisco. Amy is hesitant because she views the trip as risky. Her judgment was to a large extent based on risk perception, her fear of flying. Amy acknowledges that a professional risk science judgment would express that the likelihood for the plane to crash is very low, and the knowledge supporting that judgment is strong.

So, what would be the conclusion for Amy? How should she handle the risk? Clearly there is no scientifically correct answer. It depends on how Amy gives weight to the different aspects or concerns. If she highlights her worries, she will not travel. However, if it is very important for her to travel to San Francisco, it may be that she is willing to accept the stress she feels by flying. She has to balance the pros and cons.

8.1.2 Nuclear industry

Germany has decided to phase out its nuclear power plants by the end of 2022 (Ethik-Kommission 2011). This decision was made following the 2011 Fukushima nuclear disaster. There are risks related to both potential nuclear accidents and nuclear waste. Judgments were made that the risks are unacceptable. Half of the German Ethics Commission, which paved the way for the German phase-out decision, argued that "Nuclear energy is not acceptable because of its catastrophic potential, independent of the probability of large accidents occurring and also independent of its economic benefit to society" (Aven and Renn 2018). They can be said to have given very strong weight to the cautionary principle. The other half argued using a cost-benefit type of reasoning: other means of electricity generation were feasible with almost the same benefit as nuclear power but with less risk (Renn 2015). Weight is also given to the cautionary principle in this case, although the argumentation is different. In most cases in life, trade-offs between different concerns must be made, and the cautionary principle then must be balanced against other attributes like costs and value generation.

8.1.3 Insurance

John has a nice house. He has implemented many measures to ensure that the probability of fire is very low. Yet he has purchased a fire insurance policy. A fire could occur, and he will not take the risk of losing everything. He gives weight to the cautionary principle.

8.1.4 Pandemic

Let us consider the early development of an epidemic, such as the coronavirus case in early 2020. We are faced with a potential for serious consequences, and there are scientific uncertainties – the precautionary principle may apply. Models providing accurate predictions are not available. The precautionary principle is interpreted as a guiding perspective for prudent risk handling when faced with such uncertainties. There is not really an alternative. Because of the uncertainties, science cannot alone lead us to the right decisions. Yet it should be used 'prudently', with care, embedded in a scientific approach. There is not a conflict between the precautionary principle and science. The precautionary principle is used to stimulate and justify research aiming at reducing the scientific uncertainties, and this is exactly what happened in the coronavirus case. While countries and regions mandated curfews and lockdowns, new knowledge about the virus and pandemic was gained, theoretical as well as empirical. An adaptive risk handling has been adopted all over the world, combining precautionary measures and scientific analysis. A pure science-based policy does not exist when the consequences are subject to scientific uncertainties. What can be said is that the policy is informed by science. The fact that different countries have come to different policies demonstrates this notion. However, key principles of risk science have been adopted.

Later in the development of the pandemic, a key topic was when and how to again open up key functions of society. Experts indicate that the risks in relation to the negative effects of shutdown are as least as large as the risks related to the virus. These types of statements lack, however, a strong rationale. The uncertainties are too large to make accurate estimates and predictions. Yet politicians have to make decisions. Then it is important to acknowledge that there is not a scientifically correct answer. Different weights on the uncertainties would lead to different conclusions. Different 'schools' (e.g., economists, health scientists, health bureaucrats) will provide different perspectives, but in the end, it is our politicians who need to balance all the different concerns and views and make the difficult decisions. How we confront the virus is mainly politics, not science. It is popular to state that we need to base these decisions on data and data analytics. Without reducing the importance of data, information and knowledge, we should be happy that the important decisions that politicians have to make are not mechanized or automated. They are too complicated to be prescribed by some algorithms. What is needed is broad deliberations of all relevant input, supported by experts, and that is exactly what our politicians have been elected to do.

8.1.5 Climate change

For the climate change issue, there is a strong knowledge base developed over the years, but there are also considerable uncertainties. IPCC summarizes this scientific knowledge base, which also captures judgments of uncertainty. Nonetheless, it is a common perspective that science has produced an unambiguous message, which points to only one path for action. However, what decisions to make is not about science, as science cannot prescribe what action to take. It can, however, provide input to such decisions, evaluating, for example, pros and cons of alternative arrangements and measures.

As risk analysts, we know that evidence is related to not only 'objective facts' but also concerns and beliefs that need to be taken into account in the risk handling. We know also that, as a basis for decision-making, value judgments are as important as evidence in the form of data, information and justified beliefs. Science balances 'confidence' (for example, stating that climate change is mainly man made) and 'humbleness' (reflecting that there are uncertainties). Risk science is critical for finding this balance, clarifying the meaning of the 'humbleness' part and pointing to the strengths as well as the limitations of the 'confidence' part.

In a scientific environment, critical issues are continuously debated, such as what the most warranted statements are. All statements and beliefs are looked into to decide whether they can be justified. This is the process of developing scientific knowledge over time. Climate change knowledge is not different; however, various stakeholders' strong interests and the political aspects make the discussion challenging. It is often hard to discern what is science and what is politics.

If we give weight to the cautionary and precautionary principles and conclude that climate change risk is unacceptable, it is a policy decision, not a scientific one. Science and the IPCC reports have informed us, but there is a leap from these findings to the decision that covers aspects related to how to weigh the uncertainties and what values are important and should be prioritized. Through the Paris Agreement, governments worldwide have in fact made such conclusions: the climate change risk needs to be reduced – strong actions are needed. Risk science can help in the process of selecting the most effective means, but the actual decisions as to how fast and in what way this is to be implemented are outside the scope of risk science.

8.1.6 Exploration of space

What is the aim of risk management in the context of space exploration? To reduce the risk of loss of human lives? No, of course not. If that had been the main goal, no manned space missions should have been initiated, as such missions would obviously involve high risks of loss of lives. The driving force for sending human beings in spacecraft is not safety and risk reduction but a genuine willingness to use resources to explore space and to expand into all possible niches. All "successful civilisations" have explored the surrounding areas and faced risks. The achievements are often astonishing, like sending humans to the moon: nearly 60 years ago (May 25, 1961), President John F. Kennedy presented a bold challenge before a joint session of Congress: send a man to the moon by the end of the decade. "No single space project in this period will be more impressive to mankind, or more important for the long-range exploration of space; and none will be so difficult or expensive to accomplish," Kennedy said.

Despite skeptics who thought it could not be accomplished, Kennedy's dream became a reality on July 20, 1969, when the Apollo 11 commander Neil Armstrong took "one small step for a man, one giant leap for mankind", leaving a dusty trail of footprints on the moon.

The space programs have not been without accidents and deaths, but most people would probably judge the success rate as extremely impressive. Although safety for the astronauts has always been a main concern, no one would expect that such a bold task could be realized without some accidents. Accidents are the price of exploration. It is not destiny but a consequence of the fact that the task is difficult and the risks are high.

Risk reduction enters the arena when a decision has been made about performing the activity: sending a human to the moon. The issue is then how to realize the project and ensure a sufficiently low risk for the astronauts. Different technological concepts may be considered in the process, and their suitability would be judged by reference to a number of attributes, including reliability and costs. One concept may have a large potential for risk reduction but could be rejected as it is too costly and would cause many years of delay to the project.

The spacecraft is then built and the space mission is to be accomplished. A number of factors, for example, related to the training and competence of the astronauts, would be critical for the successful completion of the mission. Risk reduction is a main objective of risk management at this stage of the project.

8.1.7 Investments of securities

An investor considers two investment strategies:

1   Invest $x million in securities of type a), which are characterized by high expected profit but also large uncertainties.
2   Invest the same amount in securities of type b), which are characterized by considerably lower expected profit but also a considerably lower level of uncertainties.

If risk management were all about reducing risk, strategy 2 would be favorable, but, of course, such a conclusion cannot be made in general. Depending on the investor's attitude toward risk and uncertainties, strategy 1 could be preferred. Risk reduction may be desirable but not without considering its value.

Striking the balance between Risk and Return

8.2 BASIC CONCEPTS AND TERMINOLOGY

Risk management refers to all activities used to address risk, such as avoiding, reducing, sharing and accepting risk. The terms *risk management* and *risk handling* are interchangeable. The term *risk governance* refers to the application of governance principles for the handling of risk. Such principles include aspects like participation, accountability, effectiveness, coherence, proportionality and subsidiarity (Aven and Renn 2018). Governance refers to the actions, processes, traditions and institutions by which authority is exercised and decisions are made and implemented. We interpret the term *risk management* in a broad way to also cover risk governance, but it should be noted that it is also common practice to use the *risk management* term in a more narrow way, which makes it important to explicitly refer to risk governance when we want to highlight aspects like participation, accountability and so on.

A risk strategy or policy is a plan of action on how to manage risk. Different types of strategies will be discussed in Section 8.3.

Risk assessments and cost-benefit types of analyses are examples of methods used to support risk management or handling. Risk communication can also be seen as an activity within risk management as here defined.

The medical profession addresses risk for patients who are susceptible to various health problems, such as diseases. The measures to address these diseases consist of two main categories: 1) Risk related to the occurrence of diseases and 2) risk related to the treatment of diseases given their presence. Examples of measures in the former category are vaccines and research to discover and eliminate dangerous viruses. Examples of measures in the latter category are medical operations and the use of medicine.

Using the terminology introduced in Chapter 2, these two risk contributions relate to (A,U) and (C,U|A), respectively. The importance of proper management of (A,U) is obvious, as if we can avoid or eliminate a dangerous virus, nothing would be better. However, there will always be risk sources and events occurring that threaten health; hence, we also need to properly handle the (C,U|A), that is, the vulnerability, reducing the effects of the virus or any other threat (event).

In general terms, we speak about risk prevention and risk mitigation:

- Risk prevention is commonly understood to mean avoidance of relevant risk sources (for example, eliminating dangerous viruses) or interception of the risk source pathway to undesirable consequences such that none of the targets is affected by the risk sources (for example, early containment of dangerous viruses). More generally and using the terminology of Chapters 2 and 3, risk prevention is about reducing the risk contribution as defined by (A,U), for example, by eliminating some events A' or reducing probabilities P(A') for some identified events A'.
- Risk mitigation refers to the reduction of vulnerability given an event A.

A model for risk handling is shown in Figure 8.1, from the identification of the issue or problem, to the decision using risk assessments and other analyses. The risk issue is often formulated as a task of choosing among a set of decision alternatives or options.

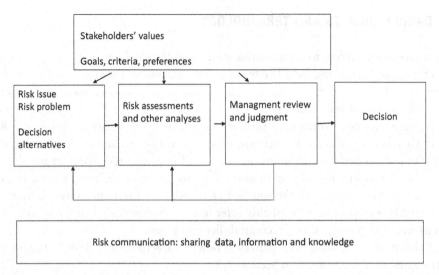

FIGURE 8.1 Model of the risk management process using risk assessments and other analyses to support decision-making (based on Aven 2012b)

For examples of such issues, see Section 8.1. The number of alternatives evaluated needs to be practically manageable and thus not too many. Choosing alternatives is typically established through an integrated process between managers and analysts. Then, risk assessments are performed to understand the risks involved for the alternatives considered. Other studies could also be conducted, for example, analyses of people's risk perception and cost-benefit types of analyses that relate the risks to costs and benefits. The generation of the risk information is often supplemented with decision analysis tools, which are systematic approaches to organizing the pros and cons of a decision alternative; see Chapter 9. These tools differ with respect to the extent to which one is willing to make the factors in the problem explicitly comparable.

The risk assessments and the other analyses are used as input to the decision-making process. The decision-makers conduct a management review and judgment (MRJ) and then make a decision; refer to Section 4.2.3. The MRJ is formally defined as the process of summarizing, interpreting and deliberating over the results of risk and other assessments, as well as other relevant issues (not covered by the assessments) in order to make a decision. The MRJ is based on the recognition that all assessments have limitations: there are aspects not fully captured by the assessments. The decision-maker needs to take into account *all* aspects of importance for the decision, not only those addressed by the assessments. Most assessments are based on some specific assumptions and beliefs, and decision-makers must also consider the validity of these assumptions and beliefs. Stakeholders' values, goals, criteria and preferences provide essential input to the MRJ but could also strongly influence the definition of the risk issue or problem, as well as the assessments, in particular when performing analyses combining risks, costs and benefits. Risk communication is central in all features of this process for sharing of data, information and knowledge between experts, analysts, decision-makers and other stakeholders.

8.3 MAIN RISK HANDLING STRATEGIES AND POLICIES

The basic risk management strategies or policies are:

1 Use of risk assessment (for short, being risk informed)
2 Giving weight to the cautionary and precautionary principles. Vulnerability (robustness) and resilience management are key instruments
3 Discursive

In addition to these strategies, we also speak about a 'risk-based requirement' strategy or 'use of codes and standards'. It is applicable when the situations considered are simple – the phenomena and processes addressed are well understood and accurate predictions can be made. An example is the specific requirements (for example, related to speed limits) used for how to design and operate car traffic safety. The experience is extensive, and numerous risk assessments have been conducted. If the requirements are met, no further risk assessments are required for the risks covered by the requirements. The other basic strategies 1 and 2 have largely been replaced by a 'risk-based requirement' strategy. However, when considerable changes are discussed, for example, using new technology, a strategy based on 1 and 2 is required.

The appropriate strategy in practice would typically be a mixture of the basic strategies 1–3. What strategy to give weight to depends on the risk issue or problem – the context – as shown in Table 8.1. We distinguish between the following main types of risk problems:

- *Simple risk problem*: the phenomena and processes considered are well understood and accurate predictions can be made. Minor uncertainties (examples: many health issues and transportation risks).
- *Uncertainty*: accurate predictions cannot be made, for example, as a result of lack of understanding of underlying phenomena or complexity.
- *Difference in values*: These differences relates to the consequences C and the uncertainties U.

TABLE 8.1 Risk strategy and related factors and instruments for different types of risk problems

Risk problem	Risk strategy/policy	Influencing factors	Typical instruments used
Simple risk problem: the phenomena and processes considered are well understood, and accurate predictions can be made. Minor uncertainties (examples: many health issues and transportation risks)	'Risk requirements' (codes and standards) Risk assessment informed		Codes and standards (based on considerable experience and many risk assessment of similar activities) Statistical analysis and traditional risk assessments

(Continued)

TABLE 8.1 (Continued)

Risk problem	Risk strategy/ policy	Influencing factors	Typical instruments used
Moderate levels of uncertainties (a potential for severe consequences, example: nuclear industry)	Risk informed using risk assessment Cautionary	Relevancy and amount of data, available models	• Informed by risk assessments, combining data and models • Reducing risk related to risk sources and events, for example, by avoidance or substitution • Reducing vulner-abilities, strength-ening resilience (highlighting for example, contain-ment, redundancy, diversification, etc.)
High levels of uncertainties (a potential for severe consequences) (example: climate change risk, terrorism)	Risk informed using risk assessment Cautionary, precautionary	• Lack of understanding of underlying phenomena • Complexity	• Informed by risk assessments, which provide broad risk characterizations, highlighting knowl-edge aspects and uncertainties • Reducing risk related to risk sources and events, for example, by avoidance or substitution • Weight on reduc-ing vulnerabilities, strengthening resilience (highlight-ing for example, containment, redundancy, diver-sification, etc.)
Value differences (a potential for severe consequences) (examples: climate change risk, nuclear industry)	Discursive The discursive strategy uses measures to build confidence and trustworthiness through the reduc-tion of uncertainties and ambiguities, clarification of facts, involvement of affected people, deliberation and accountability	• Priorities • Risk attitude (appetite) • Ubiquity • Persistence • Reversibility • Delayed effects • Potential for mobilization	Political processes, participatory discourse; competing arguments; beliefs and values are openly discussed

For the Uncertainty category, it is also relevant to distinguish between uncertainties related to risk sources (agents) and events (A) and the related effects (C), sometimes referred as the consequences of the risk-absorbing system (Renn 2008). Different strategies are appropriate for meeting uncertainties about A relative to C. When it comes to A, the obvious strategy is to avoid, if possible, the activity that has the potential of leading to A; refer to Section 8.1.2. Substitution of hazardous chemicals is an example of this strategy, in which the dangerous chemical is replaced by non-dangerous or less dangerous ones. Another strategy is to perform research aiming at improving the understanding of how A can occur and in this way be able to avoid A or reduce its probability of occurrence. This strategy applies to many diseases. For the consequences C given the occurrence of A, the strategies would aim at reducing vulnerabilities and strengthening resilience through measures such as

- containment (for example, avoiding a virus's spread to other areas)
- redundancy (there is a back-up in case of a failure of a component)
- strengthening of the immune system (concerning health: exercise regularly, eat healthy food, get enough sleep, etc.)
- diversification (mix completely different types of strategies, for example, investments)
- design of systems with flexible response options (in case of some disturbance or stress, avoid that response being rigid with no adaptation to the specific situation considered)
- the improvement of conditions for emergency management (for example, by elevating exercises and drills)

Concerning the difference in values, there are different views on how much weight to give to uncertainties relative to the potential benefits, particularly in political decision-making processes. Differences in risk attitude or 'risk appetite' can often explain the different views. Some people are more willing than others to take on risky activities in pursuit of values. Today, many people and parties give strong priority to measures to protect the environment: if an activity threatens the environment, it is rejected. Other people and parties have their focus on the values that these activities can generate and, as long as the risks are not too high, they support the activities.

Many factors related to the consequences at stake influences differences in values, including (WBGU 2000; IRGC 2005):

1 Ubiquity, the geographic dispersion of potential damage;
2 Persistence, the temporal extension of potential damage;
3 Reversibility, the possibility of restoring the situation to the state before the damage occurred;
4 Delayed effect, which characterizes a long period of latency between the initial event and the actual impact of the damage;
5 Potential for mobilization, that is, violation of individual, social or cultural interests and values, generating social conflicts and psychological reactions in individuals and groups who feel afflicted by the risk consequences.

From many perspectives, climate change risk scores high on all of these factors.

The different types of risk problems call for different risk-management strategies, as shown in Table 8.1. Simplified, we can say that the following strategies apply:

- *Simple:* codes and standards and risk informed (accurate predictions using risk assessments)
- *Uncertainty:* risk informed (highlighting knowledge and lack of knowledge assessments) and cautionary/precautionary
- *Differences in values:* discursive

Smoking and many health risk issues are examples of simple risk problems. Climate change risk is an example of both Uncertainty and Difference in values. The nuclear industry is subject to moderate levels of uncertainties, but rather high level of differences in values.

Alternative ways of classifying risk management strategies involve risk avoidance, reduction, transfer and acceptance (retention). While avoiding risk means either selecting a path that avoids from the risk (e.g., by abandoning the development of a specific technology) or taking action in order to fully eliminate the risk, risk transfer deals with ways of passing the risk on to a third party, for example, through insurance. Risk retention essentially means making an informed decision to do nothing about the risk and to take full responsibility both for the decision and any consequences occurring thereafter (Aven and Renn 2010).

Risk reduction can be accomplished by many different means, depending on the type of activity considered. In safety context, it is common to refer to the as low as reasonably practicable (ALARP) principle (see Section 8.6 and Chapter 9). According to the ALARP principle, an identified measure should be implemented unless it can be documented that there is an unreasonable disparity ('gross disproportion') between costs/disadvantages and benefits.

ALARP assessments require that appropriate measures be generated. As a rule, in a risk assessment context, suggestions for measures always arise, but often a systematic approach for the generation of these is lacking. In many cases, the measures lack ambition to make large changes to the risk profiles. A possible way to approach this problem is to apply the following principles (Aven and Renn 2010):

- On the basis of existing solutions (base case), identify measures that can reduce the risk by, for example, 10%, 50- and 90%.
- Specify solutions and measures that can contribute to reaching these levels.

The solutions and measures must then be assessed prior to making a decision on possible implementation. Although expressed by numbers when it comes to ambition (10%, 50% and 90%), the assessments need not express risk on a precise scale. The important point here is to generate measures with a certain magnitude of risk-reducing effect.

The examples of Section 8.1 clearly demonstrate that risk management is not about an isolated risk reduction process but about balancing different concerns, for example, costs and improved safety for people and the environment. Willingness to take risk is required to create values. Based on the space example of Section 8.1.7, we distinguish between

1   risk linked to the realization of the activity
2   risk linked to constraints such as concept selection and technology
3   risk linked to 'free variables' within the constraints of 1 and 2. These free variables are often related to human and organizational factors.

Using this classification, risk reduction is not the main objective in case 1, but it can be for 2 and 3. Increased activity in general gives higher risks, so it would be meaningless to say that risk reduction is the main objective in the case of 1. When it comes to 2 and 3, risk reduction is, however, an objective, but there will always be other concerns to be balanced. This is illustrated by, for example, the adoption of the ALARP principle.

Risk management is to a large extent about finding the proper balance between development and protection; see Figure 8.2. Protection is supported by the cautionary and precautionary principles and related strategies. Development, on the other hand, is promoted by taking and accepting some risks in pursuit of values. Risk assessments and other types of analyses provide input to the process of obtaining this balance, more specifically in relation to what option to select, the acceptance of activities and systems and so on.

Aiming at such a balance means that we do not talk about risk in isolation. We have to look at what the alternatives are: the costs, benefits and risks of each, and find the overall best one. The risk associated with the alternative selected is by definition acceptable. If you consider two investment strategies, 1 or 2, and choose 1, it means that you accept the risk related to 1.

However, to protect values, for example, to protect the lives and health of employees working in hazardous industries, it is common to express more explicitly what acceptable (tolerable) or unacceptable (intolerable) risks are – independent from the benefits that are associated with the activity. Risks that cannot be compensated for by benefits, regardless of how plentiful they may be, are commonly referred to as *inviolate* or *categorical*. The risks alone are enough to ban the activity (Josephson 2002; Aven and Renn 2018).

FIGURE 8.2 Risk management as a balancing act (based on Aven 2014)

Many principles have been identified as critical for the practice of risk management in society and organizations (see, e.g., ISO 2018a). Examples include the active leadership by top management, customization of the risk management to the context, the integration of the risk management in other organizational activities, dynamic risk management, risk management based on system thinking and continuous improvement through learning and experience. A philosophy of 'management by objectives' is often applied, the idea being to for all activities to specify objectives and then develop strategies and measures to meet these. It is a strongly debated philosophy, with some obvious strengths but also some severe weaknesses, as thoroughly discussed in the literature, particularly in the quality management discourse (e.g., Deming 2000; Bergman and Klefsjö 2003). The focus on objectives easily leads to a compliance regime, in which the main driver becomes meeting a number of task goals, without really improving overall performance.

We remember from Chapter 2 that risk captures the idea of a potential for undesirable consequences of the activity considered. The actual consequences could, however, be positive, and this points to the challenge of risk management discussed previously: avoiding undesirable consequences but at the same time also seeking positive outcomes. Risk management alone cannot ensure this. This management needs to be placed in a broader context of the management or governance of the activity considered, with the many visions and objectives that could here be relevant. Think, for example, about an enterprise with its business goals or a government with some political ambitions and programs. Success in relation to such objectives and goals clearly extends beyond the risk handling. What risk management and risk science can offer is suitable concepts, principles, approaches, methods and models for understanding, assessing, communicating and handling risk. Other means are as important as those provided by the risk field. For example, the enterprise may benefit from knowledge from performance analysis and quality management, and the government may obtain valuable insights from general policy analysis. It is not always easy to define sharp boundaries regarding what is risk management and what it is not. Think about the antifragility concept of Taleb (2012). The basic idea of this concept is that some types of stress, risk and uncertainties need to be welcomed – 'loved' – in order to improve and become better over time. It captures the same idea that we all know from training: some pressure and stress is needed to improve the form over time. The antifragility concept can be viewed as a principle for improving performance over time, but it can also be considered a principle for risk management when having a special focus on developments over time. Referring to the (C,U)–(C',Q,K) terminology for the risk concept and its description, the antifragility concept is about management of (C,U) taking into account the dynamics of the situation. Most setups and approaches for understanding and managing risk are based on a static perspective, in the sense that the risk about the future activity does not incorporate the dynamics of the situation with events occurring leading to changes in the thinking and new goals and performance targets. For the enterprise, a failure at some point in time can be an opportunity to rethink the organizational goals and from a longer perspective be decisive for its survivability and profitability. Most of us have experienced the same thing: a failure or mistake can be the start for something new and better.

This discussion relates to *adaptive management* and *adaptive risk management*: different decision options are assessed, one is chosen and observations are made; learning is achieved and adjustments made (Bjerga and Aven 2015). The key is to learn while proceeding, as exemplified by the pandemic example of Section 8.1.4.

Risk changes, often quickly, and a dynamic risk management approach is essential to be able to anticipate, detect, acknowledge and respond to these changes in an appropriate and timely manner (ISO 2018a). Depending on changes in uncertainties and knowledge, the risk management strategy could change; refer to Table 8.1. Again we refer to the COVID-19 risk handling as an illustration. A potential serious threat was emerging early in 2020, and a precautionary strategy was needed.

Systems thinking can be understood as a way of seeing the whole, highlighting the interrelationships between the interactions between the components of the system (Senge 1990; Langdalen et al. 2020). The idea of systems thinking is frequently referred to in accident analysis, organizational theories and quality discourse (see, e.g., Deming 2000; Hollnagel et al. 2006; Leveson 2011). It is considered essential for meeting complexity; see discussion in Section 8.7. Langdalen et al. (2020) provides some illustrating examples of the importance of systems thinking. A key point highlighted is that to evaluate the effects of a safety measure, it is not sufficient to consider the measure in isolation. System thinking is required, as resources are limited – spending resources on some safety measures may imply reduced resources for other safety measures (Abrahamsen et al. 2018).

REFLECTIONS

A risk analyst argues that to use expected values to guide decision-makers is to apply systems thinking. Do you agree?

Yes, it can be, as the expected value under certain conditions approximate well the average value when considering a large number of activities – the 'total system'; refer to Section 3.1.1. However as thoroughly discussed in Chapters 3 and 9, it is necessary to see beyond expected values in most cases to adequately describe risk and support risk decision-making.

In the following, we will look closer into the main risk management strategies and policies 1) risk informed and 2) giving weight to the cautionary and precautionary principles, where vulnerability (robustness) and resilience management are key instruments. The discursive strategy (3) is discussed in Chapter 7.

8.4 RISK INFORMED

The risk-informed strategy makes use of formal risk assessments to support decision-making. On a macro level, the process generated is as illustrated in Figure 8.1. A more detailed description of the process covers the following steps

1   Establish the context, which means, for example, to define the objectives of the risk assessment and to organize the work. Often a set of decision alternatives are identified.
2   Perform the risk assessment.
3   Perform a management review and judgment (MRJ).
4   Handle the risk (decision-making).

Findings from analyses other than risk assessments, such as cost-benefit type of analysis, also provide input to the MRJ.

The consequences addressed in the risk assessment are commonly related to quantities such as loss of lives, injuries, environmental damage and economic parameters. A number of classification systems exists for the consequences, some also integrating different dimensions. An example is provided in Table 8.2 for social concerns. See also Problem 8.1.

TABLE 8.2 Consequence categories related to social concerns (based on Aven and Renn 2010, p. 104)

Category	Ranking	Definition
Significant positive	+3	• Major beneficial improvement to human health. • Large-scale benefits to individual livelihoods (e.g., large-scale employment). • Major improvements to community facilities/utilities. • Notable impact on the wider economy (e.g., extensive use of local supplies).
Modest positive	+2	• Moderate beneficial improvement to human health. • Medium benefits to individual livelihoods (e.g., employment impacts). • Modest improvements to community infrastructure/utilities. • Moderate impact on the wider economy (e.g., some local sourcing of supplies).
Limited positive	+1	• Some beneficial improvement to human health. • Some benefits to individual livelihoods (e.g., additional employment opportunities). • Limited improvements to community facilities/utilities (e.g., no discernible improvement). • Some impact on the wider economy (e.g., limited local procurement).
None	0	• No impact on human health. • No impact on livelihoods. • No impact on community facilities/utilities. • No impact on the wider economy.
Limited negative	−1	• Limited impact on human health and well-being (e.g., occasional dust, odors, traffic noise). • Limited impact on the livelihoods of individuals (e.g., isolated incidents related to ethnic tensions and some restrictions on access to income source).

Category	Ranking	Definition
		• Limited impact on access to community facilities and utilities (e.g., access to cultural centers restricted to a limited extent). • Sparse impact on the wider economy, at a local, regional and national level (e.g., limited procurement).
Moderate negative	–2	• Modest impact on human health and well-being (e.g., noise, light, odor, dust, injuries to individuals). • Moderate impact on individual livelihoods (e.g., restricted access to income source). • Medium impact on access to community facilities and utilities (e.g., access to utilities restricted for long periods [weeks] of time). • Moderate impact on the wider economy, at a local, regional and/or national scale (e.g., only moderate levels of employment and supplies sourced within the area). • Potential breach of company social policy and/or legislation.
Significant negative	–3	• Emergency situation with harmful consequences to human health (e.g., fatalities). • Disastrous consequences on the livelihoods of individuals (e.g., curtailment of access to primary income sources). • Calamitous consequences on those seeking to access community facilities and utilities (e.g., resettlement of large numbers [1,000s] of households). • Disastrous consequences on the economy (e.g., all employment and suppliers within the area). • Breach of company social policy and/or legislation.

Examples of other types of criteria for classifying the consequences (Aven and Renn 2010) include:

- *Effectiveness:* does the decision alternative achieve the desired effect?
- *Efficiency:* does the decision alternative achieve the desired effect with the least resource consumption?
- *Minimization of external side effects:* does the decision alternative infringe upon other valuable goods, benefits or services, such as competitiveness, public health, environmental quality, social cohesion and so on?
- *Sustainability:* does the decision alternative contribute to sustainability? Does it assist in sustaining vital ecological functions, economic prosperity and social cohesion?
- *Fairness:* does the decision alternative burden the subjects of regulation in a fair and equitable manner?
- *Political and legal implementation:* is the decision alternative compatible with legal requirements and political programs?
- *Ethical acceptability:* is the decision alternative morally acceptable?
- *Public acceptance:* will the decision alternative be accepted by those individuals who are affected by it? Are there cultural preferences or symbolic connotations that have a strong influence on how the risks are perceived?

"Surprisingly enough, this is in the range of acceptable risk"

Assessing decision alternatives against these criteria may create conflicting messages and results. An alternative that proves effective may turn out to be inefficient or unfair to those who will be burdened. Other alternatives may be sustainable but not accepted by the public or some stakeholders.

Combining such consequences categories with assignments of probability (typically using some intervals) and strength of knowledge judgments, a risk characterization is established as discussed in Chapter 3. The point made here is that the consequences considered could show huge variations depending on the situation and the purpose of the risk assessment.

Risk assessments are commonly used together with some risk acceptance criteria or tolerability limits. Examples of such criteria and limits are presented in Table 8.3. The basic motivations for using such criteria or limits are summarized in Table 8.4.

TABLE 8.3 Examples of risk acceptance criteria based on risk metrics

Risk metric	Explanation	Purpose	Examples of limit
Individual risk	Probability of death for a fixed but arbitrary selected person within a specific group	Ensure a maximum risk for each person in this group	$<10^{-5}$ for a period of one year
Societal risk using an F–N curve	A curve showing the probability (F) of an accident having at least N fatalities	Ensure a maximum societal risk	P(number of fatalities $> x$) $< C/x^{\alpha}$ where C and α are constants which provide position and steepness of the curve
Environmental based on restitution time	Probability distribution for the restitution time T for an ecosystem	Ensure a maximum environmental risk related to accidents	$P(T > t) < 0.05/t$ (t in years)
Investment risk value-at-risk (VaR)	A specific quantile in the probability distribution of the loss of an investment	Ensure a maximum risk	VaR at 99% level

TABLE 8.4 Summary of arguments used for applying risk acceptance criteria in relation to risk assessments (based on Abrahamsen et al. 2020)

Actors	Person or other value (environmental, assets)	Risk producer (for example, a company)	Authorities
Argumentation	Contribute to the key aspects of risk being addressed		
	Contribute to ensure a maximum risk level (for example, a minimum safety level for people due to accidents)	Contribute to simplify risk management processes Contribute to ensure transparence and consistency in the risk management	Contribute to ensure that societal expectations are met Contribute to simplify risk management and regulation processes

TABLE 8.5 An overview of key challenges and problems by using risk acceptance criteria for ensuring a maximum risk level (based on Abrahamsen et al. 2020)

Problem, challenge	Comments
a1) The risk metric used can to varying degrees be informative for describing risk	For example, the VaR metric does not reflect the magnitude of the loss exceeding the VaR number Strength of knowledge judgments are not taken into account Thus, risk is not 'controlled', only an aspect of risk
a2) The risk assessments are to varying degrees taking into account uncertainties and limitations in the supporting knowledge	Uncertainties and limitations in the knowledge are not reflected by the risk acceptance criteria approach when comparing the risk assessment numbers with the probabilistic-based criteria
a3) Quantitative risk acceptance criteria often lead to 'mechanistic' decision-making – the need for management review and judgment is undermined	It is easy to interpret the criteria as expressing prescriptions for how to act, without acknowledging the limitations of the approach
a4) It is often difficult to see the implications of a specific criterion before having analyzed the activity	If the knowledge basis is rather weak, the result of a criterion could be that good arrangements and measures are rejected despite the fact that they perform well taking into account all aspects (pros and cons) of these arrangements and measures
a5) Quantitative criteria often result in unfortunate adaptations	Such adaptations are well known from the quality management literature. The assessments and focus are unilaterally directed towards meeting the criteria, and other aspects are overlooked or neglected

By introducing the risk acceptance criteria, the idea is to ensure some maximum risk level for values such as life and health that need to be protected. Alternative arrangements and measures will be considered in the decision-making process, and different concerns must be balanced. Then such criteria will provide constraints for this balance with respect to these values. There are, however, a number of challenges and problems associated with the use of risk acceptance criteria or tolerability limits for this purpose, as summarized in Table 8.5. The criteria build on some risk metrics, which may to varying degree reflect risk. Issues of uncertainties and knowledge strength are to a large extent ignored. A specific criterion level can be difficult to justify. The approach leads to a prescriptive use of risk assessment neglecting the importance of management review and judgment. A compliance focus on the project level is often the result of using such criteria, instead of searching for the overall best solutions for the whole system or an activity considered, also highlighting potential improvements over time.

The contribution to transparency (refer to Table 8.4) lies in the formulation of explicit risk acceptance criteria that can be compared with explicit and documented risk analysis results. The final decision may involve subjective judgments, especially when using semi-quantitative and qualitative risk acceptance criteria, but the risk assessment process is well defined and the principles and logic of the decision-making are clear. With regard to transparency, however, it is often a problem that the risk criteria are not precisely defined. They are commonly expressed by probabilities, but it is not made clear what a probability of, for example, 10^{-4} means in the context used, in particular how it relates to uncertainty. The management review and judgment process involves a certain loss of transparency, but this loss must be balanced against the need to see the results of the risk assessments in a larger context that takes into account their limitations and weaknesses. The arguments for having a MRJ are strong – from a professional risk science perspective, it is difficult to justify decision-making processes that do not see beyond the risk assessment results when concluding what to do in relation to risk.

The contribution to consistency (refer to Table 8.4) lies primarily in the fact that the same type of criteria can form the basis for all similar risk decisions. Simple, clear and efficient planning processes are desirable, and the use of criteria and requirements are tools to achieve this. Checking that specific numerical criteria are met is easy to implement, and deviations and problem areas will be clearly identified. Risk acceptance criteria can remove the need to make case-by-case evaluations.

It can, however, be a challenge to balance the need for firmness and uniformity on the one hand and adaptability and flexibility on the other. Applying the same criterion for many activities may lead to unreasonable requirements for some of these activities, as they have some specific features. The need to break down the problem and simplify it must be balanced against the overall objectives and ambitions. A project can achieve success and a significant reduction in risk for this project but also contribute to increasing the overall risk for the organization considered. The specification of criteria and requirements at a lower activity or system level must be executed with care. The whole must not disappear in the work of meeting detailed criteria and requirements.

There are management strategies and methods that try to balance these considerations: the need for decomposition and simplicity with the need to see the whole and the overall goals. Such strategies and methods do not emphasize predefined numerical criteria and requirements as strongly as in the philosophy described previously. Instead, the focus is on the overall objectives, what the alternative solutions are, the understanding of these and how they can be improved and what they entail in terms of all the pros and cons.

Society, through regulations and political statements, may have some expectations and requirements for how to perform the activities, for example, about ensuring a specific safety level. The risk acceptance criteria could be an instrument for ensuring such expectations and requirements. However, care has to be shown, as imposing constraints on how actors are to perform the activity could have unfortunate implications. For example, if a government expresses an expectation that a specific industry in that country should be world leading on safety issues, a possible way of implementing this could be to adjust the risk acceptance criteria accordingly. However, the industry may consider such criteria 'problematic' restrictions in the work of finding the best solutions and measures, violating key business economic principles.

Proper risk management means that risk is not sacrificed at the expense of other interests and concerns. The use of risk acceptance criteria is generally considered a means for ensuring that this does not happen. It can, however, be discussed to what degree such criteria are more efficient for this purpose than other approaches as long as management is truly committed to prudently handling all relevant risks. There are industrial examples showing that the use of risk acceptance criteria in some cases actually led to less focus on risk and safety improvement processes (Aven and Vinnem 2007).

Given all these challenges with using risk acceptance criteria, the conclusion is that if one would like to apply such criteria, they should be used with care. As thoroughly discussed in Chapter 3, strength of knowledge judgments should always accompany probabilistic risk metrics and thus need to be considered in the risk acceptance discussion. Moreover, the use of such criteria can easily lead to the wrong focus: namely meeting the criteria rather than finding the best possible solutions and measures, taking into account the limitations of the analysis, uncertainties not reflected by the analysis and other concerns important for decision-making. As strongly highlighted by the risk analysis science, a risk decision should be risk informed, not risk based. There is always a need for management review and judgment.

In the following, an adjusted approach is suggested using risk acceptance criteria in combination with strength of knowledge judgments (Aven 2013); see also Table 8.6:

1   If risk is found acceptable according to probability with large margins, the risk is judged as acceptable unless the strength of knowledge is weak (in this case, the probability-based approach should not be given much weight).
2   If risk is found acceptable according to probability, and the strength of knowledge is strong, the risk is judged as acceptable.
3   If risk is found acceptable according to probability with moderate or small margins, and the strength of knowledge is not strong, the risk is judged as unacceptable, and measures are required to reduce risk.
4   If risk is found unacceptable according to probability, the risk is judged as unacceptable, and measures are required to reduce risk.

TABLE 8.6 Adjusted procedure for use of risk acceptance criteria in view of considerations of the strength of knowledge (based on Aven 2013)

Probability-based justification	Above limits	Unacceptable risk	Unacceptable risk	Unacceptable risk
	Small margin below	Unacceptable risk	Unacceptable risk	Acceptable risk
	Large margins	Further considerations needed	Acceptable risk	Acceptable risk
		Poor	Medium	Strong
		Strength of knowledge		

8.5 PRECAUTIONARY PRINCIPLE

The precautionary principle emerged as a concept in the 1970s–80s in German environmental law, and it was incorporated into international law following the North Sea Conference on the Protection of the North Sea (NSC 1987). In the declaration from this conference, we can read:

> Accepting that, in order to protect the North Sea from possibly damaging effects of the most dangerous substances, a precautionary approach is necessary which may require action to control inputs of such substances even before a causal link has been established by absolutely clear scientific evidence.
>
> (NSC 1987)

We find a similar formulation in the well-known 1992 Rio Declaration (Freestone and Hey 1996):

> In order to protect the environment, the precautionary approach shall be widely applied by States according to their capabilities. Where there are threats of serious or irreversible damage, lack of full scientific certainty shall not be used as a reason for postponing cost-effective measures to prevent environmental degradation.

Since then, the principle has permeated most international environmental conventions, and the principle is also a cornerstone of European Union (EU) regulations and law, stating that regularity actions may be taken in situations where potentially hazardous agents might induce harm to humans or the environment, even if conclusive evidence about the potential harmful effects is not (yet) available (EU 2002, 2017).

Real-life situations created a need for a principle to protect values (related to the environment and humans' health and lives) in the face of uncertainties and risks. The formulation of this principle was, however, not straightforward; it has created debate and still does. Some of the issues discussed are:

- What are the criteria for invoking the principle? When do we have 'scientific uncertainties'?
- Is the idea adequately reflected by a 'principle'? Is the precautionary principle a decision rule? A decision rule can be seen as a logical statement of the type "if [condition], then [decision]". As formulated by Peterson (2007), "A decision rule simply tells decision-makers what to do, given what they believe about a particular problem and what they seek to achieve".
- Is the principle irrational in the sense that it leads to inconsistent decisions?
- Does the principle promote a risk-averse attitude, hampering development and innovation?

Risk science has now discussed these issues for over 30 years as both a generic, fundamental risk science and an applied risk science challenge (refer to Appendix B). The applied part relates to the challenges of using the precautionary principle in specific settings, for example, implementing the principle in EU regulations and law on or related to, for example, climate change, food and/or chemicals. The discussion in this book focuses on generic, fundamental risk science. Considerable work has been conducted to clarify the meaning of the principle. The scientific literature includes, for example, many definitions of the principle beyond those referred to previously. At any point in time, there can be a discussion on what the most warranted judgments (most justified beliefs) on this topic are. We will highlight the following two definitions of the principle based on work conducted by the Society for Risk Analysis (SRA 2015):

- An ethical principle expressing that if the consequences of an activity could be serious and subject to scientific uncertainties, then precautionary measures should be taken, or the activity should not be carried out.
- A principle expressing that regularity actions may be taken in situations where potentially hazardous agents might induce harm to humans or the environment, even if conclusive evidence about the potential harmful effects is not (yet) available.

The SRA documents do not, however, explain what 'scientific uncertainty' and related statements like 'conclusive evidence not yet available' mean in this setting. The issues have been subject to considerable discussion in the literature; see, for example, Sandin (1999), Stirling (1998, 2007), Aven (2011c), Cox (2011), North (2011) and Vlek (2011). Aven (2011c) argues that it is sensible to relate scientific uncertainties to the difficulty of establishing an accurate prediction model for the consequences considered. If such a model cannot be established, the precautionary principle can be invoked. Following this thinking, it is clear that the principle is not to be interpreted as a decision rule but as a guiding perspective for risk handling, a perspective which is considered expedient, prudent or advantageous (Aven 2020d, 2020a, p. 179). Judgments are needed to decide when the uncertainties are scientific and the principle can be invoked.

Risk scientists have performed detailed studies of the case that a decision rule-based interpretation of the principle is adopted. It is shown, for example, that the use of the principle in this case leads to inconsistencies (Peterson 2006, 2017; Stefánsson 2019).

However, recent work points to the fact that the conditions applied to ensure these results are founded on comparisons of likelihood judgments (Aven 2019a, 2020a). For example, one such condition states that "If one act is more likely to give rise to a fatal outcome than another, then the latter should be preferred to the former, given that both fatal outcomes are equally undesirable" (Peterson 2006). However, in the case of large uncertainties, such judgments cannot be justified. An assessor (which could be the decision-maker) may judge an event A to be more likely than an event B, but the decision-maker should not give much weight to this when the judgment has a poor basis.

We see how risk science discussions lead to new knowledge concerning the understanding of the principle. Considerable criticism has been raised against the principle, and this has led to further discussion and new insights. A good example, in addition to those mentioned previously, is the work by Sandin et al. (2002), which defends the precautionary principle against five specified charges. One of these charges is that the principle marginalizes science. As discussed by Sandin et al. (2002) (see also Aven 2011c), this charge can be rather easily refuted. The point is that, for the situations addressed, science is not able to provide clear answers because of the scientific uncertainties. The principle is to be considered a risk management strategy in the case of weak knowledge about the activities considered, for example, generated by the operation or use of a specific system or product. The measures to be taken could include holding back the activity until more scientific knowledge is available. As such, the principle stimulates science and scientific work rather than marginalizing it. The pandemic case of Section 8.1.4 provides an illustration of this discussion.

If a company would like to introduce a new product into the market, the basic idea of the precautionary principle is that the company has the burden of proof of showing that the product is safe and the negative risks associated with its use are acceptable. Risk science provides knowledge about how to make judgments about what is safe and acceptable risk in such a context. Probability theory and Bayesian analysis are shown to be useful instruments in this regard by calculating the probability of the product having hazardous effects given all available evidence. Acceptance of the product can only be given if this probability is small and the evidence strong.

Risk science shows how these ideas relate to fundamental theory of statistical inference and hypothesis testing. According to statistical principles, the null hypothesis is that the product is acceptable, and strong evidence must be provided to show harmfulness that would warrant restricting the product. Attention is mainly paid to the error of type I: to wrongly reject a true null hypothesis, that is, conclude that the product is hazardous when that is in fact not the case. However, following precautionary thinking, considerable weight also has to be placed on the error of type II: to not conclude that the product is hazardous when that is actually the situation. Adopting a strict precautionary approach, there is initially "a red light which is only switched to green when there is convincing evidence of harmlessness" (Trouwborst 2016). The burden of proof is reversed. The traditional perspective stresses the costs (interpreted in a wide sense) of erroneously taken protective measures, whereas the precautionary perspective stresses the costs of an erroneous lack of protective measures. Risk science discusses the tensions and need for balance between these two concerns. There is no scientific optimal formula. The issue is mainly about values, priorities, ethics, management and politics. It is about balancing development and protection. Too much protection hampers development, and vice versa. The result is a difference between different political parties, countries and cultures; see discussions in HSE UK (2001), Wiener and Rogers (2002), Sunstein (2005) and Wilson et al. (2006).

Risk science provides knowledge about aspects to consider when finding this balance and what principles, approaches and methods can be used to support the communication and decision-making. Extensive theories exist for this purpose, such as decision analysis (see Chapter 9). Risk science explains that some principles, approaches and methods promote protection; others promote development. The precautionary principle belongs to the former category, whereas cost-benefit type of analysis supports the latter, with its expected value focus, placing little weight on uncertainties and risks.

Risk science clarifies and explains what risk is and the challenges related to measuring its magnitude when the activities considered are subject to large uncertainties – all important knowledge for understanding the rationale for the precautionary principle. There is no way to calculate meaningful risk numbers in this type of situation – obviating the precautionary principle. What a sufficient level of scientific uncertainty is for invoking the precautionary principle can be strongly influenced by risk perceptions. Climate change uncertainties and risk are a good illustration of this issue.

REFLECTION

Think about a medical treatment for which the side effects are highly uncertain. Should the conclusion then be to apply the precautionary principle and not use the treatment?

It depends on the situation. If a patient is close to dying because of the disease, using the treatment could still be of interest. If the disease is not so serious, one would be more cautious using the treatment.

8.6 CAUTIONARY PRINCIPLE

The precautionary principle is well known among scientists, politicians and the general public. However, the closely related and broader cautionary principle is not so often referred to. This principle expresses that if the consequences of an activity could be serious and subject to uncertainties, then cautionary measures should be taken, or the activity should not be carried out. Thus, the key difference between the cautionary and the precautionary principle is that the latter refers to 'scientific uncertainties', whereas the former just refers to 'uncertainties'. In none of the examples discussed in Section 8.1.1–8.1.3 do we face scientific uncertainties, and, hence, the precautionary principle does not apply. As scientific uncertainties are a type of uncertainty, the precautionary principle is a special case of the cautionary principle.

The level of caution adopted needs to be balanced with other concerns such as costs. However, all industries would introduce some minimum requirements to protect people and the environment, and these requirements can be considered justified by reference to the cautionary principle.

Illustrations of some policies supported by the cautionary principle related to industrial safety:

- robust design solutions, such that deviations from normal conditions do not lead to hazardous situations and accidents;
- design for flexibility, meaning that it is possible to encounter a new situation and adapt to changes in the frame conditions;
- implementation of safety barriers to reduce the negative consequences of hazardous situations if they should occur, for example, a fire;
- improvement of the performance of barriers by using redundancy, maintenance/testing and so on;
- quality control/quality assurance;
- the ALARP principle, saying that the risk should be reduced to a level which is as low as reasonably practicable.

The cautionary principle provides guidance on how to handle risk. Arguing as in the previous section, it is not a decision rule interpreted in a narrow sense, which maps an observation to an action. There is not necessarily an observation to consider when addressing risk. What we are facing are potential observations, events and their effects. Like the precautionary principle, the cautionary principle is understood as a guiding perspective for risk handling, a perspective which is considered expedient, prudent or advantageous.

In relation to risk, there are two main types of concern: the need to create values on the one hand and protection on the other. The cautionary principle is of the latter type. It gives weight to the uncertainties. It has a role in notifying people and society in relation to protection against potential hazards and threats with serious consequences.

In society, there is a continuous 'battle' between development on the one side and protection on the other. This battle is rooted in differences in values and priorities but also in scientific and analytical argumentations. For example, public administration is strongly guided by the use of cost-benefit type of analysis (CBA), which means that risk and uncertainty considerations are given little attention beyond expected values. Hence the creating concern is highlighted more than protection. Following a cost-benefit type of analysis, nuclear industry in a country would normally be 'justified'.

The cautionary principle has a role to play in relation to this type of consideration and management. A warning is in place – there is a potential for serious consequences, and there are uncertainties. The development tools have spoken – now it is time for the protection side to highlight important aspects for the decision-makers to adequately balance the various concerns.

As a guiding perspective for risk handling, there are several aspects that the cautionary principle seeks to highlight. First, it points to the need for actions when the consequences C can be extreme, as in the nuclear example of Section 8.1.2 or the insurance example of Section 8.1.3. Caution is needed when the potential for such Cs exists. Related uncertainty and likelihood judgments affect the degree of caution.

Second, the cautionary principle points to actions when the consequences are sensitive to how the activity is realized. Consider a car in which the driver considers passing another car on a rather narrow road. The driver may abandon passing or choose to carry it out, increasing concentration and awareness when passing the car. The driver gives weight to the cautionary principle. Lack of awareness can, for example, easily lead to severe consequences.

The same type of sensitivity argumentation can be used for the nuclear example. These examples also illustrate a third aspect, captured by the saying 'better safe than sorry'. It is considered wise to be cautious, even when it does not seem necessary, to avoid problems, failures and losses later. If one is not cautious, one may later regret it. If you go for a hike in the mountains, it is wise to have extra clothes in case the weather should change. The calculated risk reduction of having lifeboats on a ship or offshore installation may be low but justified by references to the cautionary principle. Events may occur making the lifeboats important for saving lives.

In statistics, there are two types of errors: false negative and false positive. In science, it is generally considered more important to avoid false positives than false negatives, as discussed in the previous section. We will avoid concluding that a substance has a positive effect when that is not the case. This is in line with the cautionary principle. Without a strategy for avoiding false positives, the consequences could be serious. We (society) will not introduce a new substance if we are not sufficiently confident that it is safe. The point of departure is that the substance is risky, and the producer must demonstrate that it is safe. In this sense, the burden of proof is reversed. Society does not have to prove that the substance is dangerous.

As the last and fifth aspect, the cautionary principle highlights the case of scientific uncertainties, and it is then referred to as the precautionary principle.

It is illustrative to relate the cautionary principle to risk. As defined in Chapter 2, risk can be viewed as the combination of the consequences C of an activity and related uncertainties U, denoted (C,U). Describing or characterizing the risk, we are led to (C',Q,K), where C' are the specified consequences, Q a measure (in a wide sense) of uncertainty and K the knowledge supporting this measure; refer to Chapter 3. This representation of risk covers, as special cases, most other commonly used conceptualizations of risk. Without loss of generality, we can also write risk as (A,C,U), where A represents events (changes, hazards, threats), which can lead to some consequences C. The risk characterization can then be reformulated as (A',C',Q,K).

From this basis, we see that the cautionary principle applies when risk is judged high in the following ways:

1   There is a potential for C values that are extreme (highly negative).
2   There is a potential for serious C values if the activity is not cautiously executed – C is very sensitive to how the activity is implemented.
3   There is a potential for serious C values if something unlikely, surprising or unforeseen should happen (for example, an event A not anticipated or a new type of event A).
4   Weak knowledge about the consequences C of a specific type of activity, for example, about the effect of the use of a specific drug. There is a potential for extreme C values.
5   There is a potential for extreme C values, and these are subject to scientific uncertainties.

Some of these criteria are closely linked and overlapping, for example, 4 and 5. What constitutes a serious or extreme C is a judgment call, as illustrated by the examples in Sections 8.1.1–8.1.3. The consequences may relate to different types of values, such as life, health, the environment and economic quantities. Interpreting risk as previously, the cautionary principle reflects a type of risk aversion, in the sense of disliking the risk (the negative consequences and the related uncertainties). See Problem 8.12.

Due considerations need to be given to potential surprises, although their risk contribution is by definition difficult to measure or describe. As resilience measures are to a large extent motivated by meeting potential surprises and the unforeseen, it is also a challenge to adequately characterize the effect of resilience measures. Surprising scenarios will always occur in complex systems, and traditional risk management approaches using risk assessment struggle to provide suitable analysis perspectives and solutions to meet the risks (Turner and Pidgeon 1997; Hollnagel et al. 2006; Aven and Ylönen 2018). Measuring the benefit of investing in resilience is difficult. Such an investment can contribute to avoiding the occurrence of a major accident, although the effect on calculated probability and risk numbers could be relatively small.

The cautionary principle highlights the need for further evaluation of the assessments: Is the knowledge supporting the judgment strong enough? Could there be a potential for surprises relative to current knowledge? The assessors may consider the situation to be characterized by rather strong knowledge, yet the cautionary principle stimulates both better analysis and resilient measures. The key point made is that the analysis could have limitations in accurately reflecting the real world, and surprises can occur relative to current knowledge. This justifies robustness and resilience-based measures, for example, implementation of safety barriers, different layers of protection ('defense-in-depth'), redundancy, diversification, the ALARP principle and so on. Current industry practice is based on this thinking and these types of measures – the cautionary principle is commonly adopted.

A remark is warranted concerning the ALARP principle. It is a principle designed to support protection, as the basic idea is that a measure that will reduce risk should be implemented (HSE UK 2001; Jones-Lee and Aven 2011; Ale et al. 2015). Only in the case that it can be demonstrated that the costs are grossly disproportionate to the benefits gained does the measure not need to be implemented. Thus, it can be said to highlight cautionary thinking. However, in practice, we often see that the gross disproportion criterion is verified by using cost-benefit types of analyses (Aven and Renn 2018). These analyses are based on expected values and hence do not give much weight to risk and uncertainties. The result is that ALARP becomes a principle highlighting development more than protection. Alternative ways of verifying the disproportionate criterion are therefore suggested; see Section 9.5.1. The essential point is that if cautionary thinking is to be given weight, measures should also be considered when the computed expected net present value is negative, provided that these measures can contribute to a reduction in risks and uncertainties and a strengthening of the robustness and resilience of relevant systems.

Whether the uncertainties are scientific is a critical issue in many cases. Think about cautionary measures implemented in airports all over the world. We do not know when an attack will occur, but, certainly, if no such measures had been implemented, extreme events would have been the result. The cautionary principle is applicable and has played a key role in the way this problem has been dealt with. The cautionary principle has not been specifically referred to – as it is not broadly known – but the previous discussion has shown that it is in fact a main perspective adopted for handling the risk. Using the broader concept of the cautionary principle instead of the precautionary principle, we can avoid unnecessary discussion about what type of uncertainties we face, and focus can be placed on action and how to manage the risk. However, for some specific situations, it may be important to clarify whether the uncertainties are in fact scientific so that the appropriate measures are taken. In relation to the approval of new products, the concept of scientific uncertainties is important to ensure proper qualification processes. Clarifying when we face scientific uncertainties is also important in relation to other contexts, such as climate change, when more knowledge and science can reduce the scientific uncertainties and, in this way, clarify the issues and better distinguish between discussions about uncertainties and discussions about values.

8.7 VULNERABILITY AND RESILIENCE MANAGEMENT

There are a variety of hazards and threats that require study, such as illnesses, fires, hurricanes, flooding events, cyber attacks and terrorist attacks. The topics of vulnerability and resilience management intend to address these hazards and threats. If you get seriously ill, you will receive medical treatment, and hopefully you will recover. If a fire occurs in your house, you rely on alarms and fire extinguishers and help from the professional fire services. These are events that occur from time to time, and measures have been developed to address them. Barriers at different stages of the scenario developments are implemented; see the bow tie in Figure 4.4. Assessments can evaluate the effectiveness of the preparedness and in particular the effectiveness of the barriers intending to reduce the consequences of the events. The performance of the barriers would highlight aspects like:

- functionality: the ability of the barrier to function as intended assuming its presence
- integrity (reliability): the ability to be in place and intact when needed
- robustness/resilience: the ability to withstand and recover from disturbances and situations that are somewhat different from the normal state

Consider the example of ventilators used to help the breathing process for patients during the COVID-19 pandemic of 2020. We are concerned about the quality of the ventilators in supporting the breathing process (functionality) and that they actually work when needed (integrity, reliability). In addition, we would like the ventilator to maintain or at least quickly resume its function in case of a sudden unplanned power failure.

On a national level, we can similarly think about the role of the ventilators for the treatment of patients. When planning for a possible pandemic of this type, health agencies and hospitals would consider purchasing, storing and maintaining ventilators in case of an emergency with acute need for such ventilators. Clearly a high number of available ventilators would increase the resilience in relation to such a situation and thus reduce the vulnerabilities. However, there is a cost associated with the ventilators, and before the pandemic, these costs would need to be balanced against risk. While advances in 3D printing and other aspects of ingenuity can help respond quickly to sudden and unforeseen needs, even these aspects require some amount of resources and planning. Investments into resilience thus cannot be seen as isolated from risk considerations.

With hindsight, it is easy to conclude that more ventilators should have been available when the situation occurred, but there are many 'demands' for reduced vulnerabilities and improved resilience, and the resources are limited. It becomes a management and political task to prioritize. Risk assessments and risk science in general provide support for such prioritizations.

The example illustrates the importance of seeing vulnerability and resilience management in relation to risk. This does not mean, however, that vulnerability and resilience cannot at all be managed without detailed assessments of what hazards and threats that can occur. If you exercise regularly, do not smoke, eat healthy food and sleep enough, your resilience is in general strengthened in case of potential diseases, whether the known or unknown type. This is the great attraction of resilience management. We can improve resilience without knowing exactly what type of threats that can occur and express their likelihoods as needed in traditional risk assessments. In situations with large uncertainties, this is important, as risk assessments then are not able to produce reliable predictions of what will happen. It is of special relevance for complex systems, where it is acknowledged that surprises will occur.

Another illustrating example of the importance of vulnerability management is cybersecurity. A cyber system may suffer from a number of vulnerabilities, in the sense that if is exposed to a specific type of threat, it fails. Focusing on vulnerability and resilience management, the aim is to eliminate these vulnerabilities or reduce the consequences if they occur. The experts can, for example, improve the design/architecture of the system and how it is operated. The basic idea is that if there is a vulnerability present, it will be exposed at some time. Of course, in practice, there will always be some need for considerations of risk, as the improvements may have strong economic implications, and prioritizations may be necessary. Yet vulnerability and resilience management would systematically look for ways of making the program or system better from a vulnerability and resilience point of view.

The arrangements and measures used to meet the vulnerabilities and enhance resilience depend on the type of application considered. Many people today are particularly focused on the vulnerability and resilience management in relation to climate change. Whatever the reasons for the changes, the message is that we need to be prepared. The implications of the change could be many and difficult to predict. Thus, we need to address the vulnerabilities and resilience. Studying the literature on this topic, we find a huge number of contributions studying and providing guidance on how to best conduct this management. The challenges are, however, considerable, as many policies considered would affect people's lives to a large extent. Providing scientific support for many of these policies is also difficult, as the uncertainties about what is actually happening in relation to climate change and what the implications will be are large. The result is that the handling of the problem quickly becomes a mixture of science and politics.

Think about flooding in a specific region. To reduce the vulnerabilities and strengthen resilience in case of such an event, many arrangements and measures can be considered. Science can assess the effect of such arrangements and measures for different flooding scenarios. It is, however, difficult to make reliable predictions about the occurrence of flooding and its impacts. Hence the decision-making concerning these arrangements and measures will to a large extent be political, using available scientific knowledge, but also building on other factors, such as risk perception and value judgments.

The level of vulnerability and resilience for a system or organization is linked to the ability to sustain or restore its basic functionality following a stressor. It is commonly stated that a resilient system has the ability to (Hollnagel et al. 2006):

1   respond to regular and irregular threats in a robust yet flexible (adaptive) manner;
2   monitor what is going on, including its own performance;
3   anticipate risk events and opportunities;
4   learn from experience.

Through a mix of alertness, quick detection and early response, the failures can be avoided. Elements 1–3 are also highlighted in other theories, including high-reliability organizations and the concept of collective mindfulness, with its five principles: preoccupation with failure, reluctance to simplify, sensitivity to operations, commitment to resilience and deference to expertise (Weick and Sutcliffe 2007). There is a vast amount of literature (see, e.g., Weick et al. 1999; Le Coze 2019; Hopkin 2010) providing arguments for organizations to organize their efforts in line with these principles in order to obtain high performance (high reliability) and effectively manage risks, the unforeseen and potential surprises.

Resilience management is largely motivated by the need to cope with the complexity of modern socio-technical systems; complexities that foster surprises and risks. Traditional risk assessments are based on causal chains and event analysis, failure reporting and risk assessments, calculating historical data-based probabilities. This approach has strong limitations in analyzing complex systems, as it treats the system as being composed of components with linear interactions, using methods like fault trees and event trees, and has mainly a historical failure data perspective. These problems are addressed in resilience engineering, which argues for more appropriate models and methods for such systems; see Hollnagel et al. (2006) and discussion in Sections 5.5 and 5.6.

8.8 ETHICS

Ethics is the doctrine of morality: the norms, values and attitudes that we follow or should follow. There are different ethical theories, and they are classified in different ways. Two main categories are teleological ethics and deontological ethics. In teleological ethics, the rightness of actions is determined by the action's relation to the result of the action. The ethics of consequences is the most important teleological ethics. In the ethics of consequences, the rightness of an action is determined solely by the consequences of the action. Utilitarianism is a form of consequentialism. This theory claims that an action is right if and only if it leads to the greatest amount of benefit (good consequences), with all individuals taken into account.

In deontological ethics (duty ethics), the moral value of actions is assessed on the basis of rules, duties, principles and so on. It is not only the results of the action that determine whether it is morally right but also the ethical principles behind it. Examples of other ethical theories are virtue ethics (the moral quality of an action must be judged by the virtues of the mind and intention of the person and not by the extent to which the action is in line with norms and rules or by the consequences of the action) and discourse ethics (equal and rational participants must find out to common ethical standards through open and informed discussion).

Formulating a maximum level of risk can be said to be a principle rooted in the ethics of duty. That everyone who works at a facility must be ensured a certain minimum level of safety is such a principle. It is considered ethically right, because every life has an intrinsic value. However, it is not obvious what this level should be and how it should be formulated. We know that measuring risk is challenging. Employees can receive salaries and other benefits that take into account the risks. The economic, social and cultural contexts can be different. The use of quantitative risk acceptance criteria can be seen as a practical method of ensuring such a maximum risk level, but, as thoroughly discussed in Section 8.2, there are a number of challenges associated with the use of such criteria for this purpose. A main point here is that judgments will always be made in relation to the implications of specifying the level of the criteria. Aspects of consequence ethics are included. The focus here is on people, but the reasoning is also relevant when it comes to the environment. For material and economic values, it is more difficult to argue on the basis of ethics of duty.

Let us then look at the ALARP principle. The principle states that a measure that can improve safety must, as a general rule, be implemented; only in the event that one is able to document that the costs are grossly disproportionate in relation to the benefit can one refrain from implementing the measure. The principle can be said to be justified both by ethics of duty and by ethics of consequences. It is not only about the consequences of the measure. Safety is a priority; the cautionary and precautionary principles are emphasized. These principles can also be justified on the basis of the ethics of duty: one is faced with situations where values are at stake, there is uncertainty related to what the consequences will be and measures must be implemented or the activity not carried out. In a practical risk management context, however, these principles must also be balanced with other considerations, especially value creation and development. Again, aspects of consequence ethics are included.

A closer look at different types of risk acceptance criteria and related types of principles shows that they are justified by both ethics of duty and ethics of consequences. This can be said in general about most if not all risk management strategies. The strategies are based on some fundamental ideas and principles, but their implementation cannot be made without looking at the consequences. Since the consequences in a risk context are unknown when decisions are made, assessments of these consequences must be made, and uncertainty must be addressed. How one does this becomes a topic for evaluation and ethical considerations. The ethics of duty is also central here. For example, let us look at the problem of using the expected present value as a measure of the consequences as in cost-benefit analysis (see Section 9.3). This criterion is based on a principle of maximum socio-economic benefit and can thus also be justified on the basis of duty ethics, even though it primarily reflects the ethics of consequences to assess and balance all gains and losses. However, strong arguments can be made against this criterion as an overall measure of the goodness of a measure or a solution, as will be discussed in Chapter 9. A key point is that the expected present value places little emphasis on risk and uncertainty aspects. Consequently, principles that emphasize these aspects need to be highlighted – such as the cautionary and precautionary principles.

Other ethical theories can also be mentioned here. Since the consequences are unknown and we must limit ourselves to giving assessments of these based on knowledge and probabilities, it is also possible to argue that we must look at the underlying virtues and intentions behind the management, and then we are into virtue ethics.

We see how balancing of different concerns is done in risk management, where these considerations are supported in different ways by the various ethical theories. No ethical theory provides answers to the right action in a risk management context. The ethical theories cannot be directly used as guidelines for what constitutes good risk management. Awareness of these theories is, however, important in clarifying the dilemmas and tensions that we face in risk management.

Finally, a comment on equity issues. There are many examples where equity issues have been overlooked or ignored, in relation to both time (e.g., future generations) and social groups (e.g., exporting hazards to developing countries). The way risk is commonly characterized, using losses and probabilities, and also the use of cost-benefit analyses, normally does not highlight such issues of distribution. Ethical considerations may, however, require regulatory action, even if the activity in total is cost effective. For example, concentrating hazardous facilities in poor countries may be seen as a violation of equity, even if this provides revenues to these countries. Using national resources to build hospitals for the political elite, while the rest of the population is left with poor health care, is another example where equity considerations require regulatory interventions. These are just some examples showing that interventions are justified in cases where desirable societal goals are not met from either an economic or an ethical point of view. In practice, the issues are less obvious than in these two examples, but the two examples clearly show the need for correction (Aven and Renn 2018).

For further reading on this topic, we refer to Hansson (2003), Aven (2007), Ersdal and Aven (2008) and Vanem (2012).

8.9 PROBLEMS

Problem 8.1

In risk assessment, it is common to use consequence categories that integrate values related to human life and health, environment and assets. Find an example from the internet. What do you think about this practice?

Problem 8.2

It is common to use risk matrices based on probability and consequences that basically prescribe what is the decision to make. If you search on the internet, you will find many of these, typically with red color showing that the risk is unacceptable, green that it is acceptable and yellow that the risk should be reduced. What do you think about the use of such matrices?

Problem 8.3

Find examples of F-N curves on the internet which point to acceptance, ALARP region and unacceptable risk. Do you find the use of such curves useful for managing societal risks? Why/why not?

Problem 8.4

What does ALARA mean, and is there a difference between ALARP and ALARA?

Problem 8.5

In an industry, the operating company is by regulation required to specify risk acceptance criteria when using risk assessments to protect human lives and health and the environment. Argue why the company would normally specify 'weak criteria'.

Problem 8.6

Refer to the previous problem. Is there is a logic in that the company itself specifies the risk acceptance criteria? Why/why not?

Problem 8.7

Use the internet to explain the meaning of *hedging* in a risk management context.

Problem 8.8

The cautionary principle is introduced to protect values. The concerns are negative or undesirable consequences. A natural question to ask is whether it is possible and meaningful to define and use an opposite type of principle, highlighting wanted, desirable and positive consequences. In Aven (2019a), the following definition is suggested: If the consequences of an activity could be highly positive and subject to uncertainties, then the activity should be carried out and supporting (stimulating) measures should be taken. The principle could be referred to as a type of 'development principle' or inspired by the 'anti-fragility' concept of Nassim Taleb (Taleb 2012; Aven 2015d), an 'anti-cautionary principle'.

Review the five criteria of Section 8.6 and suggest how these can be adjusted to be in line with this 'development principle'.

Problem 8.9

We consider the swine flu in 2009. Vaccination was here offered to people despite considerable uncertainties about the efficiency and possible side effects of the vaccine. How was the precautionary principle then applied?

Problem 8.10

Provide an example where containment is an efficient risk management strategy.

Problem 8.11

This chapter presents and discusses fundamental and generic risk management strategies and policies. For specific applications, more detailed aims and principles of the management can be derived. For example, in enterprise risk management (ERM), it is common to refer to the following objectives (COSO 2020)

- Aligning risk appetite and strategy – Management considers the entity's risk appetite in evaluating strategic alternatives, setting related objectives, and developing mechanisms to manage related risks.
- Enhancing risk response decisions – ERM provides the rigor to identify and select among alternative risk responses – risk avoidance, reduction, sharing, and acceptance.
- Reducing operational surprises and losses – Entities gain enhanced capability to identify potential events and establish responses, reducing surprises and associated costs or losses.
- Identifying and managing multiple and cross-enterprise risks – Every enterprise faces a myriad of risks affecting different parts of the organization, and enterprise risk management facilitates effective response to the interrelated impacts, and integrated responses to multiple risks.
- Seizing opportunities – By considering a full range of potential events, management is positioned to identify and proactively realize opportunities.
- Improving deployment of capital – Obtaining robust risk information allows management to effectively assess overall capital needs and enhance capital allocation.

(COSO 2020)

Do you find these objectives in line with the presentation of risk management in this chapter?

Problem 8.12

Check the definition of the concept of risk aversion in Appendix A. Is the technical definition of risk aversion commonly used in economic contexts useful when discussing the scope and rationale of the cautionary principle and precautionary principle?

Problem 8.13

'De minimis risk' is a basic concept and principle of risk management, which states that risks which are sufficiently small can be ignored. Do you consider this principle meaningful and useful? Why/why not?

Problem 8.14

Use asbestos to motivate the precautionary principle.

Problem 8.15

In 2020, the world faced an extreme crisis as a result of the coronavirus. Thus, it can be argued that the risk handling failed. How did it fail?

Problem 8.16

The 2020 coronavirus pandemic involved risk management failures (as studied in the previous question). However, one can argue that the learning from the risk management failures have improved the ability to be resilient to future pandemics. How?

KEY TAKEAWAYS

- There are three main categories of strategies and policies for handling risk: 1) risk-informed, using risk assessments, 2) cautionary/precautionary and 3) discursive strategies. In practice, the appropriate strategy is a mixture of these three main strategies.
- In addition to these three main categories of strategies and policies, it is also common to refer to 'codes and standards' – which is applicable when the situations considered are simple – the phenomena and processes considered are well understood and accurate predictions can be made.
- When using risk assessments and other analyses, management review and judgment is needed.
- Risk acceptance or tolerability limits should be used with care.
- Risk management is about balancing protection and development.
- Protection is supported by the cautionary and precautionary principles and related measures that can improve the robustness and resilience of relevant systems.
- Development, on the other hand, is promoted by cost-benefit type of analysis, as these tools are expected to be value based and, hence, give little weight to uncertainties and risks.
- When making decisions in relation to important societal issues, one could argue that the ideal is to be informed by all relevant evidence from all relevant stakeholders. Science-based decision-making is often referred to, but it is more accurate to refer to evidence- and knowledge-informed decision-making, as evidence and knowledge can be more or less strong and also erroneous in some cases. The beliefs can be based on assumptions that may turn out to be wrong. Hence, decision-makers also need to address these limitations and uncertainties related to the knowledge basis. In addition, there could be different values related to the different concerns, which could strongly influence the decision-making.

CHAPTER 9

Methods for balancing different concerns

Decision and cost-benefit analysis

CHAPTER SUMMARY

This chapter presents approaches and methods that can be used to integrate various concerns and compare decision options where risk is an issue. For example, consider a person deciding whether to invest in a new bicycle helmet. This decision may involve many concerns; often cost is a distinguishing one. But risk related to a potential accident also needs to be taken into account. The helmet reduces the risk, but is the reduction sufficient to justify the cost of the helmet and the inconvenience of using the helmet? To make a decision, the person needs to balance these concerns and issues. The chapter will tell you about approaches and methods that can be used for this purpose. In particular, it will explain how to transform all concerns, including risks, into one unit (monetary values or a utility function), and it will discuss problems related to making such a transformation.

These approaches and methods are known as decision analyses. Decision analysis covers the following basic elements, in line with Figure 8.1: 1) identify the problem and develop goals; 2) generate decision alternatives, assess these using a decision analysis method and rank the decision alternatives; and 3) perform a management review and judgment, and make a decision.

There are different decision analysis methods. This chapter looks into multi-attribute analysis, cost-effectiveness analysis, cost-benefit analysis and expected utility theory.

A multi-attribute analysis presents the consequences of the various decision alternatives separately for the various attributes of interest. There is no attempt made to transform all the different attributes in one comparable unit, such as in dollars. It is the decision-maker that has to apply weights to the different attributes.

Cost-effectiveness analysis evaluates the effectiveness of a measure (for example, a safety improvement measure or an investment in a business context) by computing the expected cost of the measure divided by the expected gain obtained by this measure, for example, measured by the expected number of saved lives (in a safety context) or expected increase in profit (in a business context).

A cost-benefit analysis computes the expected net present value of a project or measure. Following this approach, all benefits and costs of the project are transformed to a common scale, typically money. If this value is positive, the project is recommended.

DOI: 10.4324/9781003156864-9

In safety applications, the value of a statistical life is used to transform the expected number of fatalities to monetary units by multiplying VSL and this expected number. VSL is defined as the maximum value the decision-maker is willing to pay to reduce the expected number of fatalities by 1. Similar values can be defined in other settings.

The expected utility approach is based on computing the expected utility for a project, where a utility function is specified for all possible outcomes of the project. The project is considered attractive if the expected utility is positive.

All of these approaches and methods are to be seen as decision support tools, not prescriptions for actions. As with all tools, the decision analysis tools have weaknesses and limitations. It is the decision-maker's responsibility to consider issues such as: 1) assumptions made; 2) limitations of data and information used as inputs to these decision-analysis tools; and 3) how to use the output of these tools in coordination with personal, organizational or societal values and other factors important for the decisions to be made. A management review and judgment (MRJ) is always needed. In general, a multi-attribute analysis is recommended, as it does not force the assessors to integrate attributes and concerns more than found suitable for the situation considered. It also adequately acknowledges the importance of MRJ.

LEARNING OBJECTIVES

After studying this chapter, you will be able to explain

- what a cost-effectiveness analysis is
- what a cost-benefit analysis is
- the meaning of VSL
- the meaning of ICAF
- what a decision analysis is
- what a multi-attribute analysis is
- the difference between descriptive and normative decision analysis
- what a multi-attribute analysis is
- the strengths and limitations of these analyses
- how risk is reflected by these analyses
- what a decision diagram is

First in this chapter, some examples are presented pointing to issues of importance for the discussion. Then, we review cost-effectiveness analysis, cost-benefit analysis, expected utility theory and multi-attribute analysis. A discussion follows where we look closer into the strengths and weaknesses (limitations) of these methods. We also consider how to implement the ALARP principle. The chapter is partly based on Aven (2010, 2012b, 2015b).

9.1 EXAMPLES ILLUSTRATING BASIC ISSUES

9.1.1 Casinos and other situations where expected values provide a useful decision rule

If you run a casino, you can make decisions related to fees on the basis of expected values. For all games, the frequentist probabilities are known, and you can predict the costs with accuracy. From this, you can determine what the appropriate fees should be. For example, if a game has a probability of winning 100 equal to 0.0001, the expected value is 0.01 (100 · 0.0001); that is, the average cost for the casino would be 0.01 per game in the long run. Thus, the fee needs to exceed 0.01 per game in order to make a profit from this particular game (neglecting all other facility costs).

An insurance company would normally be well informed by using expected values, as these values would estimate how much it would need to pay in claims when averaged out over many similar incidents with customers. Similarly, there are many cases from transportation and health where expected values would provide accurate estimates of populations and hence can guide the decision-making. For example, hospital capacity can be designed based on the expected number of people who will get some specific diseases over a specified period of time. These expected numbers would approximate the actual number of patients if the time interval considered is long. However, the actual number of patients may fluctuate in a weekly or monthly basis.

9.1.2 John purchases fire insurance

If we change the perspective from a company or society to an individual, the expected value would not be very informative. Think, for example, about the insurance example of Section 8.1.3, where John has purchased a fire insurance policy. He may calculate a probability of fire equal to 0.00001 on a yearly basis and estimate the value of the house as $100,000, leading to an expected cost value equal to 1 (100,000 · 0.00001). However, this value is not helpful for John's decision concerning the purchase of the insurance. The average cost is not relevant for John. What matters is that there is risk related to a potential fire, and that risk is not properly described by an expected value. John is willing to pay considerably more than the expected value for the insurance. Economists and psychologists refer to John as *risk averse*. Basically, it means that he dislikes the negative outcomes more than the weight given by the expected value.

9.1.3 Daniel Bernoulli: the need for seeing beyond expected values

The need for seeing beyond the expected values in decision-making situations goes back to Daniel Bernoulli (1700–1782) more than 250 years ago; refer to discussion in Section 3.3.1. In 1738, the *Papers of the Imperial Academy of Sciences in St. Petersburg* carried an essay with this central theme: "the value of an item must not be based on its price, but rather on the utility that it yields" (Bernstein 1996). The author was Daniel Bernoulli, a Swiss mathematician who then was 38 years old. Bernoulli's

St. Petersburg paper begins with a paragraph that presents the thesis that he aims to attack (Bernstein 1996):

> Ever since mathematicians first began to study the measurement of risk, there has been general agreement on the following proposition: Expected values are computed by multiplying each possible gain by the number of ways it can occur, and dividing the sum of these products by the total number of cases.

Bernoulli finds this thesis unsound as a description of how people in real life go about making decisions, because it addresses only gains (prices) and probabilities and not the utility of the gain. Usefulness and satisfaction need to be taken into account. According to Bernoulli, rational decision-makers will try to maximize expected utility rather than expected values. The attitude to risk and uncertainties varies from person to person. And that is a good thing. Bernstein (1996, p. 105) writes:

> If everyone valued every risk in precisely the same way, many risky opportunities would be passed up. . . . Where one sees sunshine, the other sees a thunderstorm. Without the venturesome, the world would turn a lot more slowly. Think of what life would be like if everyone were phobic about lightning, flying in airplanes, or investing in start-up companies. We are indeed fortunate that human beings differ in their appetite for risk.

Modern decision analysis is based on Bernoulli's thinking.

Bernoulli provides this interesting example in his article: two men, each worth 100 ducats (approximately $4,000), decide to play a game where the expectation is the same for both players, based on tossing coins, in which the probability of winning or losing is 50–50. Each man bets 50 ducats on the throw, which means that the probability of ending up worth 150 ducats or of ending up worth only 50 ducats is the same. Would a rational player then play such a game? The expected value is 100 ducats for each player, whether they decide to play or not. However, most people would find this play unattractive. Losing 50 ducats hurts more than winning 50 ducats. There is an asymmetry in the utilities. The best decision would be to refuse to play the game.

9.1.4 Scrap in place or complete removal of plant

A chemical process plant is to be decommissioned. The plant is old, and the company that owns the plant would like to scrap and cover the plant in place. People who live close to the plant, environmentalists and some of the political parties are skeptical to this plan. They fear pollution and damage to the environment. Large amounts of chemicals have been used in the plant processes. The company therefore opens considerations of other alternatives than scrapping in place. One alternative is a removal and disposal approach:

All materials are removed from the plant area and to the extent possible reused, recycled and disposed of. A major operation is conducted related to lifting and transport of a huge plant 'component'. The lifting and transport are difficult, and there is concern about the operation resulting in a failure with loss of lives and injuries. There are large uncertainties related to the strength of the component materials – if the lifting operation is commenced, it could be stopped at an early stage because it cannot be completed successfully. A considerable cost is associated with this initial phase of the operation. The costs associated with full removal are considerable.

The company is large and multinational with activities in several countries. Due to the tax regime, the state will pay a major part of the cost associated with the removal and disposal. Nevertheless, the company makes the final decision on which cessation alternative to be implemented. The authorities, through the supervisory bodies, see to it that laws and regulations are met. The company seeks a dialogue with these bodies to ensure that the parties have a common understanding of the regulations' requirements.

The question is now what principles, what perspective, should be adopted to choose the best alternative. More specifically, considerations need to be made on

- How formalized should the decision-making be?
- How should risk and uncertainty be taken into account?
- Should risk acceptance criteria be defined?
- Should the ALARP principle and cost-benefit type of analyses be adopted?
- Should one attempt to use utility functions to weight values and preferences?
- How can one ensure stakeholder involvement (environmental organizations, neighbors, politicians, agencies)?

The company decides that its decision is to be based on an overall consideration of technical feasibility, costs, accident risk, environmental aspects and effects on public opinion. More formal decision-making processes using cost-benefit analysis and expected utility theory were also considered but not used in the study; see Problem 9.6.

The studies were carried out by recognized consultants. Of the results obtained, we briefly look into the cost and accident risk analyses.

The predicted cost of the scrapping in place alternative is $10 million, with a 90% prediction interval given by +/–$5 million. This means that the analysts that have done the assessment are 90% confident (interpreted using knowledge-based probabilities) that the cost would be within the interval $[5,15] million. For the removal and disposal alternative, the corresponding numbers are 100 and [50,150], thus substantially larger costs.

When it comes to accident risk, the most concern is related to the removal and disposal alternative. Focus is here placed on the risk related to an unsuccessful operation; what are the consequences? Risk assessments are conducted, and they conclude that there are large uncertainties related to whether the lifting operation can be executed without losing the component. Unproven techniques have to be used for the operation, and there are large uncertainties in the quality of the component materials. These uncertainties can be reduced by detailed analysis and planned measures. The 'remaining uncertainties' in relation to the event 'the lifting operation is successful' is expressed by a probability 1/20. Compared to typical uncertainty levels for ensuring technical feasibility of industrial projects, an unreliability of 1/20 is considerably higher than what is normally accepted. This is, however, a unique type of operation, and it is difficult to make good comparisons. The strength of knowledge supporting this judgment is considered rather weak.

The operation does place personnel at risk, but the level is not considered much higher than what is typical for large industrial projects. The transport of the component is not seen to be a safety problem with the planned measures implemented.

Following the plans for scrapping the plant, there will be no environmental problem – all chemicals will be removed. Measurements will be carried out to ensure that there is no pollution present.

Several environmental organizations and residents who live in the neighborhood of the plant are skeptical about the conclusions of the company related to the environmental impacts. How can one be sure that all chemicals are removed? They refer to the bad reputation this company has from similar activities internationally and the fact that it could be technically difficult to ensure that no 'surprises' occur in the future if the company implements its plans.

The political parties have different views on this issue. All express that the company must remove all chemicals so that people can feel that they are safe, but there are different opinions on whether this means that the removal and disposal alternative should be chosen.

The company makes an overall evaluation of all inputs, the various studies and statements from a number of groups and the dialogue with the supervisory bodies; see Table 9.1.

The company concludes that it finds the scrapping in place alternative the best one. As there is no safety and environmental problem with this alternative, as seen from the company perspective, the additional cost of the removal and disposal alternative cannot be justified. The company is convinced that the procedures for removing all chemicals would work efficiently – measurements will be carried out to ensure that there is no

TABLE 9.1 The company's comparison of the two decision alternatives with respect to the attributes of costs, safety, environmental issues and public view. The question marks indicate uncertainty. The company sees no safety problems for alternative 1 but acknowledges some issues related to the lifting operation. It acknowledges, however, that there are some pollution issues related to alternative 1

Attribute	Alternative 1: Scrap in place	Alternative 2: Removal and disposal
Costs	$ 10 +/- 5	$ 100 +/- 50
Safety	+++	+++ ???
Environmental issues	+++ ??	+++
Public view	👎	👍 ?

pollution – but respects that others, in particular the people that live close to the plant, are concerned. The company recognizes the importance of this problem but cannot see that it justifies the rather extreme increased costs the removal and disposal alternative would imply. If this alternative is chosen, one could spend a substantial amount of money ($10–30 million) and risk not succeeding at all. The company is not concerned about the decision on scrapping in place damaging its reputation, as it has been open about all facts and judgments made.

Whether the chosen alternative would satisfy the requirements set by the authorities would depend on the documentation that the company can provide. It turned out in this case that the supervisory bodies required more studies to reduce the uncertainty related to the environmental impacts of scrapping in place. The final outcome would then largely be determined by the supervisory bodies' consideration of this uncertainty, and that consideration could of course be influenced by other stakeholders, such as environmentalists. There are seldom sharp limits to acceptability, and then the issue and discussion in itself will give an impression that there is a significant uncertainty on the environmental impacts.

Given the new documentation, and some additional measures to be implemented to reduce uncertainty, the supervisory bodies found the chosen alternative to satisfy the requirements set by the authorities.

Not all environmentalists and residents living close to the facility were happy about this conclusion, but they could not reverse it. Protests were used to physically stop the operations, but after a short delay, the facility was scrapped and covered in place. So far, no pollution has been identified.

This example illustrates what is sometimes referred to as an acceptable risk problem. It typically involves experts, the public, politicians and other interested parties such as environmentalists. Making decisions in such a context is hard, as the benefits of the activities could be unclear or are disputed or are not shared; the potential consequences are large, the uncertainties are large; the advantages and disadvantages do not fall upon the same group or in the same time-frame; decisions are seen to be forced upon smaller groups by a higher or distant authority; and experts and others discuss the severity of the hazards and risks.

Extensive political conflict, complexity of a problem and media coverage may strengthen the effects of these factors. Under these circumstances, decisions may not be accepted by society, and the position of authorities and the experts who advise them may become disputed.

Often the experts are seen as acting on behalf and under control of an interested party, producing results and advice that this party wants to hear and see. In some cases, the same experts are seen to be in the 'camp' of the other party, no matter how 'objective' they try to be in establishing the facts and formulating their findings; refer to discussions in Chapters 6 and 7.

REFLECTION

How do you think the previous analysis would be different than shown in Table 9.1 if conducted by another stakeholder?

The neighbors, environmentalists and some political parties would probably also have specified considerable uncertainties for alternative 1 on the attribute of safety – not trusting the company's promise that all chemicals would be removed and thus not posing a threat to the people who live close to the area.

9.2 COST-EFFECTIVENESS ANALYSIS

A risk-reducing measure, titled Measure 1, is considered. It costs 0.1. We may think of one unit of cost as $1 million. The measure reduces the assigned probability of a fatality from 2/100 to 1/100. These probabilities relate to a period of 10 years. The probability of two or more fatalities is negligible and can be ignored. To simplify the analysis, suppose that there are no other factors to consider. Should the measure be implemented?

Clearly, this question cannot in general be answered, as the conclusion depends on how the decision-maker balances cost and safety. What is the value of a probability of a fatality of 1/100 compared to 2/100? We cannot give a general answer, but we can compute a cost-effectiveness ratio or index, expressing cost per expected life saved, which gives a reference and a link between cost and safety.

We see that the index in this case is $0.1/[(2/100) - (1/100)]$, which is equal to 10. It is common to refer to this number as ICAF – the implied cost of averting a fatality.

Using this index, the conclusion is thus that if the decision-maker finds 10 (million dollars) too high to be justified, the analyst should not implement the measure. Otherwise, the decision-maker should implement the measure.

In practice, the ICAF varies considerably for measures that are implemented, demonstrating that there are many factors justifying a measure that extend beyond the expected cost-effectiveness ratio here considered, refer discussion in Section 9.5.

More formally, the cost-effectiveness index of a measure can be written $E[Z]/E[B]$, where Z denotes the cost of the measure and B the benefits, for example, expressed through the number of saved lives.

9.3 COST-BENEFIT ANALYSIS

Consider the problem discussed in the previous section. Another way of performing the analysis is to express a cost value of a 'statistical life' – commonly referred as the VSL, expressing the maximum value the decision-maker is willing to pay to reduce the expected number of fatalities by 1 (Robinson and Hammitt 2015; Viscusi 2018). Alternatively, VSL/100 can be expressed as the maximum the decision-maker is willing to pay to reduce the expected number of fatalities by 0.01.

Thus, if VSL has been specified as higher than or equal to 10, Measure 1 can be justified but not if VSL is specified as lower than 10.

Assume that a VSL has been determined. Then the total expected value by implementing this measure can be written, taking into account the cost savings of the measure due to a reduced expected number of fatalities:

- $0.1 + VSL [(2/100) – (1/100)] = –0.1 + VSL/100.$

We see that for this expected value to be positive, VSL needs to exceed 10. If for example, VSL = 15, the expected value is positive, and the measure is recommended, whereas if VSL = 5, the expected value is negative and the measure is not considered attractive.

In practice, we need to take into account time and the discounting of money, reflecting the idea that $1,000 today is worth more than $1,000 next year. The point is that if you have $1,000 today, you can invest this money and earn some interest on that amount in order to have a larger value next year. There is rate of return, a discount rate, r, such that the value of $1,000 next year equals $1,000/(1 + r)$ today. Looking two years ahead, the value of $1,000 is $1,000/(1 + r)$ next year and $1,000/(1 + r)^2$ today. Following this argument, we derive the net present value of a project or investment:

$$NPV = \sum_{t=0}^{T} \frac{X_t}{(1+r)^t}$$

where T is the time period considered (normally expressed in years), X_t is equal to the cash flow at year t, $t = 0, 1, 2, \ldots, T$, and r is the discount rate.

In practice, cash flows are in general uncertain and are replaced by their expected values $E[X_t]$. A project is considered attractive if the expected cost-benefit is positive, that is, $E[NPV] > 0$. What discount rate to use has been subject to considerable discussion, and different approaches have been suggested, including the use of the so-called capital asset pricing model (CAPM) (Copeland and Weston 1988). A key point when used in a cost-benefit analysis is that the discount rate expresses an attitude related to future rate of returns, and its determination is thus equally about strategy and policy as analysis and empirical records. See Problem 9.8 for further discussion of the discount rate.

9.4 EXPECTED UTILITY ANALYSIS

The expected utility approach to decision-making is based on computing the expected utility of the consequences C' for different decision alternatives and choosing the alternative with the highest expected utility. If u denotes the utility function, the approach is based on computation of $E[u(C')]$ for the various decision alternatives.

The following example illustrate the ideas of the method.

In the example, the possible consequences for two alternatives are (2,X) and (1,X), where the first component represents the benefit and X represents the number of fatalities due to a potential accident, which is either 1 or 0. Now, what is the utility value of each of these consequences (outcomes)?

According to the utility approach, the best consequences or outcome, (2,0), is given a utility value 1. The worst outcome would be (1,1), and it is assigned a utility value 0. It remains to assign utility values to the outcomes (1,0) and (2,1). Consider the standard of balls in an urn, with u being the proportion of balls that are white. Let a ball be drawn at random; if the ball is white, the outcome (2,0) results; otherwise, the outcome is (1,1). We refer to this lottery as '(2,0) with a chance of u'. Now, how does '(2,0) with a chance of u' compare to achieving the outcome (1,0) with certainty? If u = 1, it is clearly better than (1,0); if u = 0, it is worse. If u increases, the gamble gets better. Hence, there must be a value of u such that you are indifferent between '(2,0) with a chance of u' and a certain (1,0); call this number u_0. Were u > u_0, the urn gamble would improve and be better than (1,0); with u < u_0, it would be worse. The value u_0 is the utility value of the outcome (1,0). Similarly, we assign a value to (2,1), say, u_1.

As a numerical example, we may think of $u_0 = 0.90$ and $u_1 = 0.10$, reflecting that we consider a life to have a high value relative to the gain difference.

Now, according to the utility-based approach, a decision maximizing the expected utility should be chosen.

We suppose $P(X = 1) = 0.02$ for alternative 1 and 0.01 for alternative 2. It follows that the expected utility for the first alternative is equal to

$$1 \cdot P(X = 0) + u_1 \cdot P(X = 1) = 1.0 \cdot 0.98 + 0.1 \cdot 0.02 = 0.982,$$

whereas for the other alternative

$$u_0 \cdot P(X = 0) + 0 \cdot P(X = 1) = 0.9 \cdot 0.99 + 0 \cdot 0.01 = 0.891,$$

The conclusion is that the first alternative is to be preferred. Observe that the expected values computed previously are in fact equal to the probability of obtaining the best outcome, namely a gain of 2 and no fatalities. To see this, note that for the first alternative, the consequence (2,0) can be obtained in two ways, either if X = 0 or if X = 1 and we draw a white ball in the lottery. Thus, by the law of total probability, the desired results follow for the first alternative Analogously, we establish the result for the other alternative.

We conclude that maximizing the expected gain would produce the highest probability of the outcome (2,0) and, as the alternative is the worst, (1,1), we have established that maximizing the expected utility value gives the best decision. This is an important result. Based on the requirements of consistent (coherent) comparisons for events and for consequences, we are led to the inevitability of using the expected utility as a criterion for choosing decisions among a set of alternatives.

The expected utility approach applies to only one decision-maker. See Problem 9.9.

9.4.1 Health risk example

In this section, we study a decision problem related to the health risk example studied in Problem 4.12. We test a patient when there are indications that this patient has a certain disease. Let X be 1 or 0 according to whether the test gives positive or negative response. Furthermore, let θ be the true condition of the patient, 'the state of nature', which is defined as 2 if the patient is seriously ill, 1 if the patient is moderately ill and 0 if the patient is not ill at all.

Suppose analysis and testing of this patient at the hospital gives updated posterior probabilities of θ = 2,1 and 0 equal to 0.3, 0.7 and 0.0, respectively. Two possible medical treatments are considered: d_1, which would be favorable if θ = 1, and d_2, which would be favorable if θ = 2. The expected proportion of normal life expectancy given θ and d_i is shown in Table 9.2. We see that if θ = 2, then treatment d_1 would give a life expectancy of 10% relative to normal life expectancy.

From these expectations, a utility function can be established, reflecting the preferences of the patient or, alternatively, the physician.

Let us look at how we can do the elicitation of the utility function for the patient. The starting point for establishing the utility values are 0 and 1, corresponding to immediate death and normal life, respectively. We then ask the patient to compare an expected life length without the operation of, say, 15 years with a thought-constructed operation having a mortality of x%; however, if the operation is successful, the patient would enjoy a normal life with an expectancy of 30 years. Note that this is an exercise not directly linked to the medical treatment that this patient is to undertake. The patient is asked the minimum probability of success from the operation needed to undergo the

TABLE 9.2 Expected proportion of normal life expectancy

Decision	$\theta = 2$	$\theta = 1$
d_1	10%	80%
d_2	50%	50%

operation. Say it is 90%. Then this number is the utility value related to a proportion of life expectancy of 50%. Obviously, this probability would be higher than 50%, as the patient is 'guaranteed' 15 years of life with no operation, while the operation could lead to death.

Other utility values are established in a similar way, and we arrive at the utility function shown in Table 9.3.

In Figure 9.1, a decision tree for the decision problem is presented. The tree grows horizontally from left to right. Beginning from the left, there are two decision alternatives represented by two branches of the tree. This describes the decision structure, but at the end of the terminal branches (each of which is either d_1 or d_2), we add two others, one labeled θ_1, the other θ_2, corresponding to the health state of the patient. The points where the branches split into other branches are called nodes: decision nodes and random nodes depending on whether the branches refer to possible decision choice alternatives or event/state uncertainties, respectively. These two types of nodes are represented by a square and a circle, respectively.

According to the expected utility approach, the decision maximizing expected utility should be chosen. We find that the expected utility for the two decisions is given by

$$E[u_1] = 0.40 \cdot 0.3 + 0.95 \cdot 0.7 = 0.785$$
$$E[u_2] = 0.90 \cdot 0.3 + 0.90 \cdot 0.7 = 0.900.$$

Thus, decision d_2 should be chosen. Of course, this is what the mathematics says. The analysis is based on simplifications of the real world, and it is based on the preferences of the patient only. The costs involved are not reflected. The physician must also take this into account, if relevant, when establishing the utility function. In most cases, patients and physicians agree on which treatment to undertake, but conflicts could

TABLE 9.3 Utility function for the two decision alternatives

Decision	$\theta = 2$	$\theta = 1$
d_1	0.40	0.95
d_2	0.90	0.90

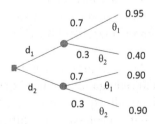

FIGURE 9.1 Decision tree for the decision problem summarized in Table 9.3 (based on Aven 2012b)

occur. The use of utilities provides a tool for communication of values but does not solve the difficult problem of dealing with different preferences between patients and physicians.

A utility-based approach ensures coherency in medical decision-making. Viewing the totality of activities in medicine, there is a strong need for using resources effectively and obtain 'optimal' results. The introduction of utility functions could be somewhat standardized to ease the assessment, as a number of examples can be generated.

Clearly, the previous analysis would also provide decision support without deriving the utility function. The numbers in Table 9.2 are very informative as such. It provides valuable insights and a good decision basis. If we do not introduce a utility function, we would evaluate the predictions and assessed uncertainties but give no numerical utility value on the possible outcomes. As a patient, this could be satisfactory, as coherency in decision-making is not critical. For the hospital and society, coherency is, however, an issue. The need for management review and judgment is also strong. The results of the analyses must be placed in a larger context.

9.5 MULTI-ATTRIBUTE ANALYSIS: DISCUSSION

The three approaches described in Sections 9.2–9.4 are all quantitative and can provide useful decision support by integrating costs, risks and benefits. In practice, qualitative and semi-quantitative approaches are also commonly used, as demonstrated by the examples in Section 9.1.4. Compared to quantitative approaches, these qualitative approaches often provide less structure and traceability in the analysis process, challenging the common goals of being transparent and consistent. Clearly, transparency and consistency are easier to achieve when using quantitative and explicit criteria for decision-making, as there are well-defined rules for what to do depending on the results of the analysis. One of the main motivations for using cost-benefit analysis is to ensure that the resources available are used effectively, for example, avoiding spending a disproportionate amount of money in one area or sector for improving safety compared to others.

However, the quantification and related rules may result in important aspects of the decision-making process being misrepresented or neglected. Many decision-making processes involving risk are not trivial – they cannot be replaced by formulas and numbers. Assessments and management of knowledge and uncertainty are demanding and contain elements that cannot be easily measured. As argued in Chapter 3, there is a need for seeing beyond probabilities and expected values to make prudent risk decisions. There is always a need for a management review and judgment process that takes into account the limitations of the analysis and add concerns and issues not reflected in the formal analysis process. An MRJ process will involve a certain loss of transparency, but this loss must be balanced against the need to see the results of the analysis in a larger context that considers all aspects of the decision-making. The arguments for having an MRJ are strong – it is professionally difficult to justify decision-making processes that do not consider the limitations of the approach used. Transparency is attractive up to some limit; we need to acknowledge the importance of MRJ and specify what it covers. This helps to clarify what are professional and scientific issues and what are more managerial and political matters.

As discussed in Section 8.4, when it comes to consistency, we need to balance the need for firmness and uniformity on the one hand and adaptability and flexibility on the other. It may be the case that the ICAF and the value of a statistical life are higher for one type of activity than another, but there may be good reasons for this. We may, for example, think about two industries, one where there is potential for an accident with extremely serious consequences and one where only relative minor types of accidents can occur. In this case, ICAF and VSL may differ. An MRJ may compromise consistency in that the decision-making process is influenced by considerations that go beyond those of the formal analysis, but, on the other hand, the MRJ will help ensure the necessary adaptation and flexibility for the specific situation considered.

The qualitative and semi-quantitative approach outlined in the example of Section 9.1.4 is often referred to as a *multi-attribute analysis*. The analysis is a decision support tool addressing the consequences of the various decision alternatives separately for the various attributes of interest. For each decision alternative, attention is given to attributes such as investment costs, operational costs, safety, environmental issues and so on. For some attributes, quantitative analyses may be used, while for others, such as political and social aspects, qualitative analyses are typically adopted. In a multi-attribute analysis, there is in general no attempt to transform all the different attributes in one comparable unit. It is the decision-maker who has to weigh the different attributes. The trade-offs are made implicitly.

A multi-attribute analysis includes other approaches discussed in this chapter, as these approaches can be seen as providing input knowledge on specific attributes. In general, a multi-attribute analysis is recommended, as it does not force the assessors to integrate attributes and concerns more than found suitable for the situation considered. It also adequately acknowledges the importance of MRJ.

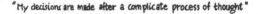
"My decisions are made after a complicate process of thought"

In Table 9.4, we have summarized the assessed strengths and weaknesses of the different approaches discussed in this chapter: multi-attribute analysis, cost-effectiveness analysis, cost-benefit analysis and the expected utility approach. The summary is based on the following set of ideal goals of such an analysis:

1 It does not only see one project in isolation but relates it to the unit's (for example, a government's or a company's) total portfolio of projects
2 It reflects the specific features of the project(s) considered (flexibility)
3 To the extent that it is practical, all relevant issues are taken into account
4 The value of time is taken into account ($1,000 for you today is worth more for you than $1,000 next year)
5 The approach is transparent, showing the importance and weight of all factors of importance for the decision-making
6 The approach contributes to consistency; the same procedure and logic are applied for all projects
7 It is objective, not depending on who conducts the analysis
8 It adequately reflects risk and uncertainties
9 It is relatively simple to perform
10 It allows for 'broad decision-making processes' (for example, political processes)

TABLE 9.4 A summary of the assessed strengths and weaknesses of multi-attribute analysis, cost-effectiveness analysis, cost-benefit analysis and expected utility approach.

Analysis approach criterion	Multi-attribute analysis, qualitative, semi-quantitative and quantitative analysis of different attributes	Cost-effectiveness analysis based on $E[Z]/E[B]$, where Z denotes the cost of the measure and B the benefits of the measure, for example, expressed by the number of saved lives	Cost-benefit analysis, based on expected net present values $E[NPV]$	Subjective expected utility, based on expected utility $E[u(C')]$, where C' is the consequences and u the utility function
General	The decision-maker must weigh the different concerns and attributes based on the analysis and using MRJ	$E[Z]/E[B]$ is computed to compare options with respect to cost-effectiveness	$E[NPV]$ is computed to compare options with respect to cost-benefit. A project is considered attractive if $E[NPV] > 0$	$E[u(C')]$ is computed to compare options with respect to expected utility. A project is considered attractive if $E[u(C')] > 0$

1 Portfolio perspective	To some degree	Yes	Yes	To some degree
2 Flexibility	Yes	To some degree	To some degree	To some degree
3 All-inclusive (comprehensive-ness)	Yes	No	No	No
4 Reflect the value of time	To some degree	To some degree	Yes	To some degree
5 Transparent	To some degree	Yes	Yes	Yes
6 Consistency	To some degree	Yes	Yes	Yes
7 Objective	To some degree	To some degree	To some degree	Subjective
8 Reflect risk and uncertainties	Yes	To limited degree	To limited degree	To some degree
9 Simple	Yes	To some degree	To some degree	No
10 Broad decision-making processes	Yes	To some degree	To some degree	No

The cost-effectiveness and cost-benefit analyses take a portfolio perspective by focusing on expected values. Hence, risk and uncertainties are only to a limited degree reflected, as shown in Table 9.4. We refer to thorough discussion in Section 3.3.1 concerning expected values to represent risk and uncertainties. It is also valid for costs and benefits. The main issues with expected values are not revealing the spectrum of potential consequences and not reflecting the strength of the knowledge supporting the probability judgments.

The quantitative approaches score higher on transparency and consistency than the multi-attribute analysis, as discussed previously, but lower on flexibility and comprehensiveness. The expected utility approach is difficult to carry out, but it allows for important aspects of uncertainties and risk to be reflected, although not to the same extent as multi-attribute analysis, as the expected utility approach is based on probabilities, not reflecting the strength of the knowledge supporting these probabilities.

The theoretical basis of the expected utility approach for guiding risk decisions is acknowledged but also its limitations. According to the expected utility theory, a rational decision-maker is assumed to be indifferent to the level of uncertainty beyond its effect on the outcome subjective probability distribution. Whether the probability is founded on a strong or weak knowledge basis is not relevant. Clearly, this is a limitation of this approach. If the knowledge supporting a specific analysis is poor, it would be meaningless to take action just on the basis of a conclusion expressing that the expected utility is positive.

The subjective expected utility approach has a strong rationale and appeal, but this approach is purely subjective and does not provide any guidance for collective decision-makers on how to use their resources in an optimal way; refer to Problem 9.9. The approach is also difficult to use in practice, with its very demanding way of specifying utilities. All the other approaches allow for comparisons with reference values to ensure some level of proportionality in use of resources. The expected utility approach is more subjective, as it incorporates the decision-makers' value judgments and priorities. Computing, for example, the index $E[Z]/E[B]$ is more science based than computing $E[u(C')]$, which is to a large extent about how to weight different concerns. When making a judgment about the index $E[Z]/E[B]$ being high or low, value judgments are taken into account, but such value judgments are not included in the formula itself as they are for the expected utility $E[u(C')]$.

The expected utility approach requires that preferences be specified for all consequences, which is a difficult task in practice and, more importantly, not necessarily something that management would like to do. Refer to the example of Section 9.4, where utilities were established for cost and loss of lives. Specifying a value of a life is required in order to use this type of approach. This value is related to an arbitrary person in the population, not a specific individual, of course. In a cost-benefit analysis, the value of a statistical life is needed, which is conceptually not the same as the number used in the utility approach. In practice, these numbers could be the same, but the point is that the utility approach requires values (utilities) to be assessed for all consequences C', whereas for the cost-benefit approach, the value judgments to be made by the decision-makers relate to the probabilities and expected values related to C' and not the value of C'. Thus, in the cost-benefit case, we assess values in a 'world' constructed by the analysts, not the real world as in the utility-based approach. Usually it is much easier to relate to this constructed world, as we can make use of appropriate summarizing performance measures.

A cost-benefit analysis requires the specification of the VSL, not the value of a life. These are not interchangeable terms. We should acknowledge that a life has in principle an infinite value – there should be no amount of money that a person would find sufficient to compensate for the loss of one's child, and society (or a company) should not accept a loss of a life with certainty to gain a certain amount of money. On the other hand, the VSL has a finite value, reflecting that decisions need to be made balancing benefits and risks for loss of lives. The value of a statistical life is a decision support tool. Now we are to make a decision influencing the future; then by specifying a VSL number, it is possible to obtain an appropriate balance between benefits and risk. When the future is here, we would focus on the value of life and not the value of a statistical life. For example, if a person becomes ill, the money used to help the person would not be determined by reference to the VSL but to the value of the life. For this person and their closest family, the value of the life is infinite. What we refer to here is the value of loss that we are willing to accept, given that this benefit is present. What we are willing to pay to obtain a benefit is something else. How much should society be willing to pay to save a (statistical) life? In a cost-benefit analysis, the focus is usually on this willingness to pay rather than willingness to accept. This is, however, not an obvious approach, as it means a standpoint with respect to what the starting point is. For example, does the public have a right to a risk-free life, or does industry have a right to cause a certain amount of risk? In the former case, the public should be compensated (using willingness to accept values) by a company wanting to generate risk. In the latter case, the public should compensate (using willingness to pay values) a company keeping its risk level below the maximum limit (Bedford and Cooke 2001, p. 282).

The term *rationality* has been thoroughly discussed in the literature. Here we use it in a broad sense, expressing that if we adopt some rules that our statements or actions should conform to, we act in a way that is consistent with them – we act rationally. As there are many ways of defining rules, this means that whether a behavior is rational will depend on the rules adopted. The rules of the subjective utility theory are such a set of rules, but it follows from our definition of rationality that people who do not abide by the percepts of this theory are irrational; they may have perfectly sensible rules of their own which they are following most rationally. Consequently, if one adopts the structure for decision-making presented in this book, you would behave rationally, according to the rules set by this structure.

We emphasize that we work in a normative setting, saying how people should structure their decisions. We know from research that people are not always rational in this sense. A decision-maker would in many cases not seek to optimize and maximize utility but rather look for a course of action that is satisfactory. This idea, which is often referred to as a *bounded rationality*, is just one out of many ways of characterizing how people make decisions in practice. See Problem 9.14.

A decision, and a decision-making process, may be regarded as good by some parties and bad by others. Consider an industrial development. One particular development concept could be considered good for the company but not so attractive for the society as a whole, as it could mean a rather high environmental risk compared to another development alternative. Decisions need, however, to be made, and proper considerations need to be taken for all relevant parties. But such considerations are not easily transformed to a mathematical formula and explicit trade-offs. In many cases and in particular when dealing with societal risk problems, we believe that more can be gained by deliberations for people to exchange views, consider evidence, negotiate and attempt to persuade each other. Deliberation that captures part of the meaning of democracy and contributes to making decisions more legitimate is also a part of our decision framework, although not explicitly shown in Figure 8.1.

The tools we have discussed here for structuring the decision-making process and providing decision support can also be used for decisions made by a group. Decision analyses, which reflect someone's preferences, would give insights to be used as a basis for further discussion within the group. Formulating the problem as a decision problem and applying formal decision analysis as a vehicle for discussions among the interested parties provides the participants with a clearer understanding of the issues involved and why different members of the group prefer different actions. Instead of trying to establish consensus on the trade-off weights, the decision implications of different weights could be traced. Usually, then, a shared view emerges what to do (rather than what the weights ought to be).

The earlier risk science literature (e.g., Kaplan and Garrick 1981; Paté-Cornell 1996) has strongly supported a quantitative approach, although also addressing assumptions and the background knowledge of the assessments. Today the knowledge aspects of risk can be viewed as more strongly highlighted compared to the standards in earlier works. For example, when subjective probabilities are used to measure or describe uncertainties and risk today, recommendations are also given for including judgments of the strength of the knowledge supporting these. This cannot be done mathematically and is not covered by quantitative risk theory. Incorporating knowledge and its strength as basic elements of risk characterization, the awareness of these aspects of risk is increased, and changes in knowledge are more easily reflected in risk characterizations.

Also, the potential for surprises is highly relevant for risk management, but it is not an aspect of a pure mathematical perspective. In risk science today, it is a major area for research and development. Such events represent a challenge in risk management not only because of limits of imagination but also unwillingness to think outside the box and anticipate potential problems (Paté-Cornell and Cox 2014).

9.5.1 Implementing the ALARP principle

The ALARP principle states that measures that can improve safety should in general be implemented; only in the case that it can be demonstrated that the costs are grossly disproportionate to the benefits gained can one refrain from implementing the measures. The challenge with the principle and this approach is that it presumes the existence of an underlying driving force for producing measures with ALARP in mind. Without such measures, risk reduction will not be obtained. Once one has identified a measure, the question becomes how to verify ALARP and a possible mismatch between costs and benefits. Reference is then often made to cost-benefit analyses and calculations of expected net present values. Such analyses in their classical form are largely based on expected values and will consequently to a limited degree reflect risks and uncertainties. The ALARP principle recognizes the need for balancing the pros and cons of a measure, but it can be argued that protection is the primary consideration the ALARP principle seeks to support – and measures that promote protection and safety should normally be implemented – only in the event that one is able to document a gross disproportion should the measure not be implemented (Ale et al. 2015; Aven 2020a). However, by using cost-benefit analyses based on expected values, the focus of the criterion is in reality shifted away from protection to measures that promote development and growth; refer to discussion in Section 8.6. To meet this critique, various alternative approaches for verifying ALARP have been suggested, including the following when discussing the need for a risk reducing measure (Aven and Vinnem 2007):

- If the costs are small: implement the measure if it is considered to have a positive effect in relation to relevant objectives or other reference values.

- If the costs are large: make an assessment of all relevant pros and cons of the measure. If the expected present value (or similar indices) can be meaningfully calculated, implement the measure, provided this value is positive.
- Also, consider implementing the measure if it generates a considerable positive effect on the risk and/or other conditions, for example:

   ○ Reducing uncertainty or strengthening knowledge
   ○ Strengthening the robustness in the case of hazards/threats and/or strengthening the resilience

This procedure acknowledges the importance of fundamental safety principles, such as robustness and resilience, to meet uncertainties and potential surprises.

In the same spirit, a seven-stage process has been suggested for evaluating ALARP (Aven and Vinnem 2007)

1 Perform a crude multi-attribute analysis of the benefits and burdens of the various alternatives, addressing all attributes of interest for decision-making, including the risk and robustness/resilience. The analysis would typically be qualitative and its conclusions summarized in a matrix such as that shown in Table 9.1. From this crude analysis, a decision can be made to eliminate some alternatives and include new ones for further detailing and analysis. Frequently, such crude analyses give the necessary platform for choosing one appropriate alternative. When considering a set of possible risk-reducing measures, a qualitative analysis in many cases provides a sufficient basis for identifying which measures to implement, as these measures are in accordance with good engineering or operational practice. Also, many measures can be quickly eliminated as the qualitative analysis reveals that the burdens are much more dominant than the benefits. If the costs are small, the ALARP principle would imply that an identified measure improving the safety should be implemented.

2 Determine the need for further analyses to give a better basis for concluding which alternative(s) to choose. This may include various types of risk assessments.

3 Compute, if meaningful, cost-effectiveness or cost-benefit types of metrics to provide information about the performance of various alternatives and measures. Sensitivity analyses should be performed to see the effects of varying key parameters, such as the value of a statistical life. Often the conclusions are rather straightforward when calculating indices such as the expected cost per expected number of saved lives. If a conclusion about gross disproportion is not clear, then these measures and alternatives are candidates for implementation. Clearly, if a risk-reducing measure has a positive expected net present value, it should be implemented. Crude calculations of expected net present values, ignoring difficult judgments about valuation of possible loss of lives and damage to the environment, will often be sufficient to conclude whether this criterion could justify the implementation of a measure.

4 Conduct an analysis of uncertainties in the underlying phenomena and processes. The key issue is to assess the degree for which the predicted values in the analysis

(for example, the expected costs) will deviate from the actual values (for example, the costs). A strength of knowledge assessment provides useful input to this analysis. The alternatives are assessed with respect to their robustness/resilience, in particular their ability to cope with surprising events occurring.

5   Conduct an analysis of manageability. To what extent is it possible to control and reduce the uncertainties and thereby arrive at the desired outcome? Some risks are more manageable than others in the sense that there is a greater potential to reduce risk. An alternative can have a relatively large calculated risk under certain conditions, but the manageability could be good and could result in a far better outcome than expected.

6   Conduct an analysis of other factors such as risk perception and reputation whenever relevant, although it may be difficult to describe how these factors would affect the standard indices used in economy and risk analysis to measure performance.

7   Conduct a total evaluation of the results of the analyses to summarize the pros and cons of the various alternatives, where considerations of the constraints and limitations of the analyses are also taken into account. A risk-reducing measure may not be justified by reference to the cost-benefit type of analysis, but if it contributes strongly to increased robustness/resilience, it may still be recommended for implementation.

REFLECTION

Is the assessment based on steps 1–7 restricted to ALARP processes?

No, the process can also be used in other contexts where decisions are to be made under uncertainty. The process can be seen as a multi-attribute analysis based on both qualitative and quantitative input.

ALARP judgments will necessarily have to be flexible to account for a variety of considerations. Many good safety measures can be implemented cheaply in the early stages of a project, but the costs could become grossly disproportionate later. Effective use of ALARP thus requires that ALARP be carefully planned in relation to important cost-driving decisions.

9.6 PROBLEMS

Problem 9.1

You consider investing in one of two projects, a) or b). The costs of the projects are 10 and 20, respectively. The rewards are given by the probability distributions of Table 9.5. Compute related cost-effectiveness indices. Should you then choose the project with the highest index?

TABLE 9.5 Probability distributions for projects a) and b)

Profit value	100	200	1000
Project a)	0.45	0.50	0.05
Project b)	0.60	0.30	0.10

Problem 9.2

Define the concept of VSL. Say VSL = 30. Explain how this value is used in CBAs.

Problem 9.3

Define the implied cost of averting a fatality. A company has defined a value of a statistical life equal to 20. What does it mean, for example, when ICAF = 10? Explain the link between VSL and ICAF.

Problem 9.4

Explain how a cost-effectiveness analysis based on ICAF calculations can contribute to identifying suitable safety measures. Discuss the limitations/weaknesses of this method.

Problem 9.5

In his paper from 1738 referred to in Section 9.1.3, Daniel Bernoulli presented and discussed the St. Petersburg paradox. This problem was first suggested by his cousin, Nicolaus Bernoulli. It is based on a casino game where a fair coin is tossed successively until the moment that heads appears for the first time. The casino payoff is $2 (ducats) if heads comes up on the first toss, $4 (ducats) if heads turns up for the first time in the second toss and so on. In general, the payoff is 2^n. Thus, with each additional toss, the payoff is doubled. How much should the casino require the player to stake such that, over the long run, the game will not be a losing endeavor for the casino (Tijms 2007; Bernstein 1996; Aven 2010)?

Problem 9.6

Return to the chemical process plant example of Section 9.1.4. In risk management, the use of a standard cost-benefit analysis based on expected net present value as well as the expected utility approach were discussed, but they were not found suitable for the situation considered. Provide some reflections why this was the case.

Problem 9.7

Look into a business book to explain the concept of internal rate of return (IRR). How is the IRR related to the NPV value?

Problem 9.8

How is it possible to better reflect risks and uncertainties using the concept of NPV? Study the finance literature and explain the difference between systematic and non-systematic risk. Use a simple example with two projects to illustrate the idea of diversification.

Problem 9.9

The expected utility approach is established for an individual decision-maker. Arrow (1951) proved that it is impossible to establish a method for group decision-making that is both rational and democratic, based on some reasonable conditions that he felt should be fulfilled. Study this result by Arrow. Outline the basic ideas.

Problem 9.10

How could uncertainty be reflected in a cost-effectiveness analysis?

Problem 9.11

A risk measure costs $2 million and the expected number of saved lives is 0.01. Compute E[NPV]. Make assumptions if required.

Problem 9.12

In many countries, a three-regime safety regulation regime is adopted in industries: A distinction is made between intolerable risk (I), ALARP region (II) and acceptable (negligible) risk (III). Discuss the suitability of this approach.

Problem 9.13

Think about the health example of Section 9.4.1. Suppose the probabilities are based on a large and relevant database. Hence, the assessor would judge the related strength of knowledge as strong. The challenge of using the expected utility approach would thus be mainly linked to the elicitation and use of the utility function. We argued in Section 9.4.1 that an individual may find it difficult to derive the utility function and instead base the decision-making on a qualitative evaluation of the information provided by Table 9.2 on the expected proportion of normal life expectancy for the two decision alternatives. For the hospital and society, we argued that a utility function could provide a useful instrument for ensuring coherency in decision-making. But why not use a cost-benefit type of analysis instead, as the hospital and society deal with a large number of patients?

Problem 9.14

Considerable scientific work has been made to present alternatives to the expected utility theory (rank-dependent utility theory, prospect theory). A main motivation for this work is the fact that people do not act according to the expected utility theory. Do you find that type of argument convincing? Why/why not? Point to some of the basic ideas of these alternative theories.

Problem 9.15

This chapter reflects on the differences and similarities between VSL and the value of a single human life. Explain the following: 1) How are these terms defined? 2) How are these terms similar to each other? 3) How are these terms different from each other?

Problem 9.16

Search your local news to find a risk topic being discussed. For example, consider the building of oil pipelines, airline safety, drug policies, counter-terrorism and health-related policies (e.g., COVID-19). Create a presentation that explains the following: 1) What is the risk issue? 2) Who are the relevant decision-makers and stakeholders? 3) What are the most appropriate decision analysis tools for this risk topic and why? 4) Conduct a hypothetical decision analysis example that could be used for this risk topic. After your presentation, poll your class to determine the most appropriate risk measure.

KEY TAKEAWAYS

- Cost-effectiveness analysis evaluates the ratio E[cost]/E[benefits].
- A cost-benefit analysis computes E[NPV]. If this value is positive, the project is recommended.
- The value of a statistical life is defined as the maximum value the decision-maker is willing to pay to reduce the expected number of fatalities by 1.
- The expected utility approach computes E[u(C')], where C' is the specified consequences of a decision alternative and u is a related utility function. The project alternative is attractive if the expected value is positive.
- A multi-attribute analysis presents the consequences of the various decision alternatives separately for the various attributes of interest. The decision-maker has to weigh the different attributes.
- A procedure is outlined for how to implement the ALARP principle, which extends beyond the use of cost-benefit type of analyses.
- All of these approaches and methods are to be seen as decision support tools, not prescriptions for actions. A management review and judgment is always needed. In general, we recommend the use of a multi-attribute analysis with input from different tools, including the other analyses discussed in this chapter.

Tackling practical risk problems and issues

General knowledge and experience

CHAPTER SUMMARY

Chapter 10 discusses how individuals and organizations can address challenges in tackling practical risk problems and issues, integrating theories and methods from the other categories of topics (parts). Special attention is devoted to the added value of risk analysis (covering risk assessment, risk communication and risk management) and risk science relative to the contributions from other fields and sciences. Organizational capacity (human resources, knowledge, etc.) needed for achieving high quality risk analysis is a key focus area. This chapter focuses on challenges related to three main activities:

- Understanding risk and risk problems using risk analysis and risk science competences, recognizing that it is not sufficient to rely on knowledge gained through standards and applied certifications.
- Conducting pre-analysis and framing, which relate to inter alia the objectives of the analysis and use of standards.
- Treating and communicating uncertainties, which is a main challenge in using risk analysis in practice.

LEARNING OBJECTIVES

After studying this chapter, you will be able to explain

- the importance of having strong and broad risk science competencies for solving risk problems
- the importance of proper framing of the risk analysis issue or problem
- the importance of proper treatment and communication of uncertainties in risk analysis

---

These topics have been selected based on our and other risk scholars' experiences with pitfalls and shortcomings observed in practice with analyses supporting risk

DOI: 10.4324/9781003156864-10

management decisions. An important source for the present work is the efforts made by the SRA Applied Risk Management Specialty Group (ARMSG 2019).

10.1 COMPETENCES

This section discusses types of competences that are needed to perform a risk analysis. Our key message is this: For many types of risk analyses, it is essential that the analyst team cover competences in many if not all of the main topics of risk analysis and risk science. Furthermore, to be able to work in such a team, it is also important that the team include analysts with a broad knowledge basis in risk analysis and risk science, not only specialists on each topic.

Consider the following example: A team is asked to guide the government on a strategy or policy to deal with climate change risk. The team would need to be multi-disciplinary, covering experts on climate change, economy, policy and so on in addition to risk analysis and risk science. Now, what kind of competence is required for the risk analysis and science? Here are some areas for which risk analysis and risk science can contribute:

1   conceptualize and understand climate change risk
2   characterize climate change risk
3   represent and express uncertainties
4   methods for assessing climate change risk
5   communicate climate change risk
6   understand perception aspects related to climate change
7   strategies for handling climate change risk

An expert in risk communication or risk assessment would only be able to give guidance on some of these areas. But these areas are closely linked. To communicate the risk, it is essential to also have insights about how risk should be understood, conceptualized and characterized. To take a concrete example: Risk communication provides guidance on communication issues related to probability but has to a limited degree based this guidance on a distinction between frequentist and subjective probability, with scientifically precise definitions as shown in Chapter 3. From a risk analysis and risk science perspective, such a distinction is essential to explain what the risk assessment results express and how they are to be understood. From this perspective, it is also important to highlight the strength of knowledge supporting the probabilities. However, this topic is not addressed in risk communication theory. As a last example, think about the difference between risk perception and professional risk judgments. From a risk analysis and science point of view, it is important to understand what this difference is about, in particular in relation to risk communication, as discussed in Chapters 6 and 7. This issue is also addressed by the risk communication field but seldom with a risk science foundation, clarifying underlying concepts on risk, uncertainty and knowledge. Conversely, experts on risk assessment could lack understanding of fundamental principles of communication.

In a team composed of experts with different competences, these challenges can be solved by integrating the insights from all relevant areas, but this integration would benefit from a core common knowledge base related to risk analysis and risk science. With an understanding of the basics of the different risk topics, it is more likely that this integration would work as intended, being able to communicate the climate risk in a way that is solid from both a risk science and a communication perspective.

In the same way, an expert on risk management and governance, without an understanding of the fundamentals of risk assessment, would struggle to provide strong guidance on risk strategies and policies. To illustrate, think about a case where a policy is to be formulated where risk assessment forms a basic pillar for the policy. Then it is essential to understand what the risk assessment actually delivers and what the limitations are, in particular related to uncertainties. We can also argue the opposite way. In order to provide guidance on risk assessment, it is important to have some risk management knowledge to adequately support the decision-making and dialogue with different stakeholders.

Again, it is possible to develop teams where experts supplement each other, but the interaction would obviously benefit from the team members having some insights beyond their own specialty. Many of the topics of risk analysis and risk science are conceptually challenging – it may take considerable time and efforts to become acquainted with a new area not previously studied or worked on.

As another example, think about a risk manager who guides the management of a company on issues related to risk. Examples of issues include

- Informing management about risks
- Clarifying for management what the scope and limitations of risk assessments and other type of analysis, like cost-benefit analyses, are
- Conduct and interpret risk perception surveys among employees, customers and other stakeholders
- Communicating risk-related knowledge to different stakeholders
- Clarifying for management what the possible types of strategies and policies are that can be implemented to handle the risk
- Clarifying for management what the possible instruments are that can be used to follow up a risk-handling strategy

Being able to provide such guidance requires broad competence in risk analysis and risk science. This competence can involve coordination with colleagues of various domain expertise and risk science subjects, but as discussed for the government example, these expertise topics are highly integrated and need a unified risk analysis approach. It is therefore important to have experts with knowledge on risk analysis and risk science, who can obtain a holistic perspective of what concepts, principles, approaches, methods and models are suitable for understanding and handling the issue at hand.

Whatever the issue is, climate change or enterprise risk, the risk scientist needs to understand the risk problem to be able to provide adequate guidance. It is not possible to have the same depth as, for example, natural scientists or economists on specific climate change or business topics, but some level of insight is required to understand what the key questions to be asked are and why. As a risk analyst or scientist, you are included in the team, as you possess essential knowledge on risk analysis and risk science, not because of an expectation that you alone are to add something to the climate change or enterprise risk issue. The work is interdisciplinary. Risk science is just one of several competencies needed. The content of this book aims at providing an overview of topics that should be covered by risk analysis and risk science, as a minimum. As a risk analyst or scientist, you are expected to provide input and guidance related to the specific issue discussed, founded on the best risk knowledge available across all domains and application areas, on the concepts, principles, approaches, methods and models for understanding, assessing, characterizing, communicating and handling risk. This is what risk analysis and risk science are about; refer to Appendix B.

A natural scientist, an engineer or an economist can take a course in risk analysis and risk science tailor-made to their area of interest, but they would not be experts in risk analysis or risk science. Such a competence would require more extensive coursework, containing topics of generic and fundamental risk science knowledge across all types of problems and domains. Many of the common risk assessment methods are rather easy to carry out, but if the applications are not sufficiently founded on risk science, the result can easily lead to poor use and misguidance. An example is the common use of risk matrices in risk assessments (see discussion in Section 3.3.2).

Due to the newness of the risk science field, it is a challenge that few degree programs are offered, leaving a market to different types of applied certification arrangements and schemes, for example, based on international standards. However, these standards are to a large extent based on consensus-oriented processes and not science; refer to discussion in Aven and Ylönen (2019) and Section 10.2. Hence, care must be shown in equating competence on risk science with knowledge on the substance of a specific standard. For example, the well-known ISO standard 31000 on risk management (ISO 2018) provides highly problematic guidance on risk conceptualization and characterization. Probability is a key concept for describing risk but is not meaningfully explained in this standard, and there is no reference to the need for highlighting the strength of knowledge supporting the probabilities.

Different roles require special competencies, such as a risk assessor working in a food agency or a risk manager in a company. It is important in all functions to have broad risk science knowledge, as discussed previously, but many other qualifications are important in such roles, including ability to communicate and collaborate with others. For a manager, generic management competencies are needed, in the same way as a risk assessor needs generic analysis and research competencies. For the career, it is commonly seen as a good rule to first become a specialist in a specific area before spreading out to the larger system (Greenberg 2017, p. 304). At later stages, more interests are likely to be devoted to broader issues, requiring knowledge across disciplines. This could mean extending the interests from some topics of risk analysis and risk science to all of the main topics, as well as knowledge related to the applications considered. This would allow for scientists and analysts from different fields to work effectively together, improving the understanding of the problem considered and contributing to its solving. As Greenberg (2017) mentions, it is scientists trained in this way who would be able to advise elected officials and their staff and the media.

<u>REFLECTION</u>

You are a risk scientist. Is your work then multi-/interdisciplinary?

It could be if you work on an applied problem, for example, related to health or climate change risk. But if you work on fundamental issues of risk analysis, for example, how to best assess uncertainties in risk assessments, your work is not multi-/interdisciplinary – there is only one field, discipline and science involved, namely risk analysis (risk science). See Appendix B.

10.2 PRE-ANALYSIS AND FRAMING

This section discusses the importance of pre-analysis and framing, which cover clarifications of the problem or issue addressed, identification of the main actors (stakeholders) and premises and constraints for the analysis to be conducted. The pre-analysis and framing specifically outline the scope of the analysis, which includes objectives, decisions, decision alternatives, stakeholders, outcomes and deliveries, approaches and methods, risk and decision criteria, resources required, responsibilities, time management and relationship with other projects.

The clarification of the problem or issue includes specifying the aim of the analysis. There are many potential uses of risk analysis, including

- Support decision-making on choice of alternatives and measures
- Demonstrate that an activity is safe
- Empower stakeholders with risk-related knowledge
- Reduce concerns and increase trust and confidence

Clarity on the aim is essential, as the choice of analysis approach and methods may strongly depend on why the analysis is conducted. All decisions to be supported by the risk analysis should be identified, including a list of relevant decision alternatives. If the aim of a risk analysis is to increase stakeholders' understanding about the risk related to an activity, the analysis can be seen as a basis for selecting suitable actions. The decision is about what to do next given the findings of the analysis. In general, we seek 'decision-focused' risk analysis, but care must be shown in interpreting this too narrowly. Risk analysis can also be used as an instrument for defining objectives. Think of scientists who are exploring an unknown substance. Based on a risk assessment, they are able to characterize the risks and make judgments about the severity of these risks. From these results, the decision-makers may conclude what objectives should be formulated for further handling of the substance.

To clarify the problem or issue addressed, it is essential to take into account all relevant perspectives and parties. For example, when analyzing the risks related to a new process plant to be developed, expert knowledge is needed as well as judgments and perceptions from neighbors who are affected by the plant. Safety agencies and politicians are also relevant stakeholders.

From a practical risk analysis point of view, it is important that the analysis be positioned appropriately in the relevant organization chart, describing both the strategic and tactical plans. In an enterprise, it is critical that the analysis be in line with its main objectives and visions, such that the analysis for each stage obtains appropriate funding and support. The risk analysis timeline must effectively support specific points in decision-making. Results of the risk analysis must be available when the relevant decision are to be made. There is often tension between the risk analysts and decision-makers when it comes to timing, as the risk analysts would like to wait for relevant knowledge (including data and information) to be available, whereas the decision-makers and project development need input from the analysts even when such knowledge is not available to ensure progress and efficiency of the project. A balance is to be found. The risk analysis needs to be fully and effectively engaged with the decision-makers, for example, when formulating the scope and timeline of the risk analysis.

The risk analysis requires use of conventions and procedural rules, offering structure for conducting activities, whether it concerns risk management for all the activities in an enterprise or a specific risk assessment to be conducted for a critical operation. Examples include scientific protocols (how to do scientific work) and technical standards. Although these standards are typically voluntary in use – they are just to be considered guidance documents that offer advice, suggestions and recommendations – they strongly influence the practice of risk analysis. In general, standards are said to contribute to uniformity and coherent coordination of performance (Timmermans and Esptein 2010; Blind and Mangelsdorf 2016). They represent a key element in current risk and safety regulation schemes, which is based on functional requirements, allowing for alternative arrangements and solutions to meet these requirements. The standards constitute one way of meeting the requirements. In this way, the standards can be seen as a system of compliance (Blind and Mangelsdorf 2016).

There are many types of standards in the risk and safety fields, some linked to specific applications; others are more generic, such as ISO 31000 on risk management (ISO 2018), which provides general principles and guidelines on risk management for use by any public, private or community enterprise, association, group or individual. According to ISO, the use of these standards will result in an increased likelihood of identifying threats and opportunities and achieving desirable outcomes, as well as more effective allocation and use of resources. It also points to the benefit for organizations of being able to compare their practices with internationally recognized benchmarks.

ISO is a global standard-setting body, a kind of umbrella organization, consisting of national standards bodies. ISO defines a standard as a "document, established by a consensus of subject matter experts and approved by a recognized body that provides guidance on the design, use or performance of materials, products, processes, services, systems or persons" (ISO 2020a). The development of an ISO standard is based on market needs. Independent technical experts nominated by the respective national committees will form a technical committee that is responsible for a specific subject area, such as risk management. These experts start drafting a standard proposal that is shared for comments. The standards are developed through a multi-stakeholder process; all stakeholders can participate in commenting on the draft proposal. There is a voting process that indicates whether consensus is achieved. If there is a disagreement, the draft will be modified further, until it is accepted (ISO 2020b).

We believe we now have a consensus
on the risk standard

The use of standards has both benefits and limitations. Some of the benefits are pointed to previously. There are also many problems associated with using standards; see discussion in Aven and Ylönen (2019) which focuses on ISO 31000. The paper argues that the use of standards based on consensus-based processes may lead to the application of inadequate principles and methods. The standards favor compromise and the lowest common denominator of available options at the expense of scientific quality. Experience indicates that it is difficult to influence the thinking supporting the standards. Contemporary risk science is not adequately reflected.

In general, standards can be an obstacle to incorporating new needs, ideas and developments, such as those emerging from most recent research. The standards may also lack flexibility to adapt to specific needs for various organizations, industries and other perspectives related to risk analysis and management. This can be understood by regarding standards as institutionalized schemes and practices, which become powerful and not easily refuted, innovated or replaced by other practices once they have been placed into use.

It can be argued that the aim of the standards is to standardize well-established technologies, not new ideas. Standards are regularly reviewed and updated, which provides opportunities to introduce recent developments. Yet experience shows and the discussion in Aven and Ylönen (2018) concerning ISO 31000 demonstrates that such developments are not easily included. This is a critical issue that needs to be addressed in the following years as industry momentum for risk science grows.

The ISO standards are developed through a multi-stakeholder process, and ISO highlights that they are established by consensus. However, this selling point is problematic because it may mislead potential users. It indicates that all relevant parties find the standard acceptable. This is not the case. Risk science has, for example, raised serious concerns about some of the main aspects of ISO 31000; see Aven and Ylönen (2019) and Problem 10.1. The authors of this book do not find the ISO 31000 standard acceptable from a scientific point of view. If we compare the ISO 31000 standard with the SRA Glossary and guidelines developed by the Society for Risk Analysis, there are several conflicting perspectives. Consensus is thus not established if the reference is the broader community of professional societies and organizations working with risk. Within the formal processes of ISO, it can be argued that the processes are consensus based, but consensus is only obtained because the processes are limited to some stakeholders and power is exercised.

Even if consent on the content of a standard is achieved, it is relevant also to look at how the consensus is gained. It can be questioned whether consensus is based on real acceptance as right, beneficial and good for business, health, environment and society (normative acceptance) or if the acceptance is more pragmatically driven. The ISO approach has a basis in both types of acceptance, but we will argue that the pragmatic dimension is the dominant one (Aven and Ylönen 2019). If the normative acceptance had been the most important perspective, more weight would have had to be given to, for example, the scientific quality of the standards. The world-leading experts in the field should necessarily have been a part of the stakeholders involved in the process. However, such a process would have been very difficult to carry out in practice, if consensus is seen as an overriding principle for the development of the standards.

In our view, and in line with Aven and Ylönen (2019), the ideal of consensus-building processes in standard developing needs to be challenged. Rather, the ideal should be high quality, as judged by the risk science community. Scientific risk organizations need to take greater responsibility as knowledge organizations and seek to influence the risk field on what represents high-quality risk analysis. The Society for Risk Analysis has taken this challenge seriously and has developed several documents providing guidance on risk terminology and fundamental principles (SRA 2015, 2017b). The SRA's vision is to be the world-leading authority on risk science and have impact globally. We welcome this development, as the risk analysis and management applications need stronger scientific-based guidance, as clearly shown by the ISO 31000 example.

Regulators for different areas should give increased support to scientific organizations to build the organizational capacity to meet such a responsibility. At the same time, the risk science community should increase its participation in standardization activities like ISO. It should build liaisons with the standardization organizations to influence the content and quality of the standards. There are processes initiated in this direction, but there is a long way to go.

10.2.1 Example: definition of the scope

You are a risk manager in a company operating a process plant. A major accident has occurred in a similar plant managed by a competing company. The executive board of your company asks you to conduct a risk assessment of your company's plant to understand if any related risk issues exist. Let us outline a possible scope for such an assessment.

First, we consider the objectives of the assessment. You may formulate the objectives in this way:

- Review the knowledge gained by the accident, with a special focus on issues of relevancy for this plant. These issues may relate to process-oriented issues but also risk assessment methodology (for example, that the accident revealed weaknesses in the methods used to analyze risk).
- Perform a risk assessment of the plant in view of the updated knowledge. Point in particular to changes in the risk characterizations as a result of the new knowledge.

The risk assessment is based on established tools used in the company, possibly improved to take into account the new knowledge. Risk is to be characterized on the basis of principles, as presented in Chapter 3. The assessment is to support a decision on possible implementation of arrangements and measures that can improve the plant safety and avoid accidents. The authorities are informed about the assessment conducted. The assessment would also provide a basis for concluding on the need for involving other stakeholders, like neighbors to the plant.

The main deliveries of the risk assessment would be a conclusion to the executive board summarizing the findings related to the objectives. No specific risk or decision criteria are established.

You will lead the assessment. A project group is established consisting of in-house and external experts. The group covers personnel with competencies within operations of the plant, fire and explosions, structural analysis and human and organizational factors, as well as risk analysis. The work is to be carried out over a period of two months, and about 500 hours are budgeted.

10.3 TREATMENT OF UNCERTAINTY IN RISK ANALYSIS

If one challenge is to be highlighted in risk analysis, it is uncertainty. It is an extremely important concept in risk analysis, but it is difficult to conceptualize, assess, characterize, communicate and handle. This section will address some practical issues linked to all of these tasks and difficulties, for short referred to as the treatment of uncertainties in risk analysis. Two main recommendations are provided:

- Be careful specifying what the quantities of interest are, and focus on understanding and communicating the uncertainties about these quantities
- In the risk communication, seek balance between confidence and humbleness

We end the section with a list of misinformation sources in relation to risk analysis, which to a large extent are about uncertainties.

10.3.1 Be careful specifying what the quantities of interest are, and focus on understanding and communicating the uncertainties about these quantities

It is commonly expressed that a risk assessment and its results are inherently uncertain. What is meant by that is not clear. Using the risk nomenclature of Chapters 2 and 3, it is the future consequences C of the activity that are uncertain. The risk assessment is a tool we use to understand, describe and communicate the uncertainties about C. This tool does not introduce uncertainties; the uncertainties about C are already present when conducting the assessment. If the risk assessment adds uncertainty, there is something fundamentally wrong about the risk assessment.

Unfortunately, in practice, we often see that risk analysts complicate risk assessments by introducing probability models and frequentist probabilities that do not have proper real-life interpretations. If X represents the number of deaths as a result of a pandemic in the next 10 years, we all acknowledge that there are uncertainties about X. Some risk analysts may, however, introduce a probability model for X, which we refer to as $F(x) = P_f(X \leq x)$, expressing the frequentist probability that X is less than or equal to x for different values of x. The function F is unknown and hence uncertain. Often a parametric form of F is used, specified with a parameter θ, meaning that the form of F is assumed known whereas θ is unknown and hence uncertain. The parameter θ may, for example, refer to the average value of X if the situation studied could be repeated over and over again infinitely.

In this example, quantities have been introduced which do not exist in the real world, and uncertainties about these quantities have become an issue. The analysts have added uncertainties to the analysis that were not originally present. In this example, it is X that is the quantity of interest, not θ or F. The analysts have created confusion by introducing a probability model and a parameter that cannot be meaningfully interpreted. Communicating F or θ in a way that is understandable is impossible, as these concepts have no practical meaning.

This does not mean that frequentist probabilities and probability models have no place in risk analysis. The point being made is that care has to be shown in identifying the key quantities of interest – fictitious quantities as in the pandemic example should not be introduced. In many cases, we deal with large populations, and we can repeat the situation considered under similar conditions. For example, it makes sense to talk about the frequentist probability p of a person from a huge population having a specific disease, interpreted as the fraction of persons in the population suffering from this disease. The quantity of interest is then p. In the risk framework of this book, p is then an example of the consequences C'.

From a practical point of view, we need not be concerned about the fact that the population is finite, as long as it large. The frequentist probability p is a model concept approximating the real world.

If p is the quantity of interest, we can start assessing uncertainties about p, as it is unknown. If we produce just an estimate of p, or more generally an estimate (prediction) of C', we do an incomplete risk assessment. We always need to address uncertainties in relation to C' and X or p in the previous examples. Uncertainly analysis is at the core of the risk assessment. When hearing people refer to uncertainty analysis as something in addition to the risk assessment, the reference seems to be that risk assessment is restricted to estimates of the frequentist probabilities. Such a framework is inadequate for assessing risk, as it does not address uncertainties. The frequentist probability does not reflect uncertainties but variation in the underlying population considered. The reference to stochastic or aleatory uncertainties when speaking about frequentist probabilities is unfortunate and should be avoided, as these probabilities do not reflect anyone's uncertainties. There is only one type of uncertainty, and that is epistemic uncertainties, stemming from lack of knowledge of the true value of C' (e.g., p or X).

An important category of risk assessment is causality analysis, were we seek to establish a cause-effect relationship, for example, between smoking and cancer. Schematically we can formulate this by establishing a model g linking a cause A' with effects C', such that $C' = g(A')$. A quantity of interest, then, is the function g, as if we know g, we can predict C' from A'. Again, a risk assessment is not only about determining g but also addressing uncertainties about the model error, $g(A') - C$, for different As.

The next question, then, is how to conduct the assessment and characterization of the uncertainties of the quantities of interest. Our recommendation – see Chapter 3 – is to use knowledge-based probability P (precise or imprecise), together with strength of knowledge (SoK) judgments. Formally, we write the uncertainty descriptions as (P,SoK,K), always adding the knowledge K supporting P and SoK. In addition, we recommend a specific treatment of the uncertainties with the aim of identifying potential surprises relative to K.

To be able to assess the strength of the knowledge and communicate its basis, it is essential to clarify the knowledge type. One aspect relates to which inputs and knowledge are empirically 'objective' and which inputs are based on subjective expert elicitation, which inputs and knowledge are generated or founded by testing, modeling, argumentation and reasoning and which aspects are treated with assumptions.

Weak knowledge could lead to calls for some specific management strategies, for example, more detailed studies and research, and weight on robustness and resilience types of measures. Specific focus should always be made on 'other events and scenarios' to think outside the standards normally used for identifying threats and risks. In security contexts, this is of special importance, as attackers may deliberately design attacks that are not on the defender's list. To identify potential surprises, it is essential to check for unknown knowns, for example, using an independent review of the analysis, as well as the possibility that some events have been disregarded because of very low probabilities, in particular when these probabilities are based on critical assumptions. It should always be asked if relevant signals and warnings concerning the existing knowledge basis have been examined.

Assumptions strongly influence the risk assessment results. If it is assumed that a component will not fail, uncertainties about the functioning of this component are ignored. The actual knowledge of the analysts could be different and stronger than reflected by the K used in the assessment because of these assumptions. Identifying the assumptions is consequently critical for the understanding and communication of the results.

It should be highlighted in the communication of the risk assessment results, each assumption represents a risk; there is a potential for additional negative consequences as a result of a deviation between the assumption made and the actual or true state. For each significant assumption, it is thus relevant to question if this risk has been considered. For the proper understanding of the risk assessment findings, clarifications on this point is essential. We recommend using sensitivity analysis to study the effect of changes in the assumptions and combine these with judgments of uncertainties. In this way, the importance of the assumptions can be evaluated. More explicit approaches for handling this type of risk also exist, one based on the concept of 'assumption deviation risk'. The idea is to consider potential deviations from the initial assumption and the potential effects on the quantities of interest and then assess the uncertainties about these deviations and effects (using qualitative probability and related SoK judgments), leading to an overall judgment of the risk associated with the assumption; see also Section 11.1.2 and Aven (2013).

Scope boundary issues of the assessment, such as a specification of what types of hazards to include in the study, can also be considered as assumptions but should be presented distinct from the other assumptions.

It is common to speak about data uncertainties, but this term is misleading. Data are not uncertain. There can be variation in the data, and the data can be of varying quality, for example, when it comes to the relevancy for the situation considered. The more relevant the data, the stronger the knowledge base available for assessing the risk. Data are also referred to as a source for uncertainty, but again, it can be questioned what that means. Rather we would speak about the data as an element of the knowledge K and the quality and amount of data as aspects to consider when making judgments about the strength of the knowledge supporting the risk characterization (C',P).

The variation in data samples is treated in traditional statistical analysis. A confidence interval is commonly referred to as an uncertainty interval, but it does not reflect any person's uncertainty or lack of knowledge. The interval is interpreted using variation on samples. If the interval is $[Z_1,Z_2]$ and the level of confidence 90%, the interval expresses that if we consider an infinite number of similar samples, the fraction of these samples for which the interval $[Z_1,Z_2]$ contains the true value of the parameter θ considered would be 0.90. If we make a 90% credibility interval $[a,b]$ for the parameter θ, we do, however, talk about uncertainties, as the interval $[a,b]$ contains the true value of θ with a 90% probability, where probability is knowledge based (subjective).

The data can be derived from measurements or expert judgments. For expert judgments, special cases arise when different experts provide conflicting judgments. However, for any risk assessment, the expert judgments represent input to the analysis. It is a fundamental principle of risk assessment as used in this book that it is the risk analysts who are ultimately responsible for the assessment, and as such, the analysts are obligated to make the final call on uncertainty characterizations. 'Experts' have advanced knowledge in rather narrow subjects and disciplines and are unlikely to devote the time necessary (even with training) to become as familiar with the unique demands of the risk assessment as the analysts. Of course, changing the experts' uncertainty descriptions should not be done if such a possibility is not a part of an agreed-upon procedure for elicitation between the analysts and the experts. The case of expert disagreement can be reflected in the strength of knowledge judgments and may be a point that is especially highlighted in the risk characterization and communication of the results of the assessments.

The idea that risk assessments produce uncertain results is commonly rooted in the idea that these results are in the form of estimates P_f^* of frequentist probabilities P_f. From such a perspective, it is relevant to speak about uncertainties of the estimates relative to true underlying value P_f, as well as sources for the uncertainties such as little relevant data and ignorance of some contributors to P_f^*. In the general risk framework in this book, the situation corresponds to the risk assessment producing estimates or predictions C^* of C (or C'). Suppose, for example, that a set of event tree models have been developed to produce C^*, but these models have missed an event A because the analysis group lacks knowledge on the specific topic. Failure to include relevant events can be viewed as an uncertainty source for C^*. Inspired by the aspects to consider when making judgments about the strength of knowledge (refer to Section 3.2), we can point to uncertainty sources of the following types:

- Few data, non-relevant data
- Erroneous assumptions
- Poor understanding of the phenomena studied, knowledge gaps
- Inaccurate models
- Lack of agreement between experts

These factors are, however, not to be considered uncertainty sources for the risk characterization (C',P,SoK,K), as this characterization has already assessed and described the uncertainties. They are, however, aspects related to the knowledge K and its strength SoK supporting the probability judgments made, as discussed in Section 3.2. The risk characterization may also have specifically addressed some of the uncertainty sources, for example, by performing an assumption deviation risk assessment or explicitly showing how the differences by experts affect the output results.

Model (output) uncertainty, which is uncertainty about the deviation between the model output and the actual quantity being modeled (refer to Problem 3.4) – is often discussed in relation to risk assessment (Aven and Zio 2013). In the framework presented in this book, it is addressed as an aspect of K and SoK that supports the probability judgments. In practice as well as in the scientific literature, the issue of quantifying the model uncertainty has been frequently discussed and procedures developed for how to conduct the quantification. In general, we do not, however, recommend quantification of model uncertainty when using models in a risk assessment. The reasoning is as follows:

Risk is (C,U), and the risk assessment produces a risk characterization (C',Q,K) based on one or more models. For the sake of argument, let us simplify and say it is only one, g, linking some input quantities X to C'. Knowledge-based probabilities P are used to express the uncertainties about X and C'. The difference between g(X) and C' is the 'error' introduced by the model g, and model (output) uncertainty refers to uncertainty about this error. This error and uncertainty obviously need to be addressed, as the uncertainty assessments are conditional on the use of this model. But how should we best do this – should we try to quantify the model uncertainties?

It would depend on what the purpose of the analysis is and the stage of the assessment process. Of course, when observations of C' are available, we would compare the assessments of C' which are conditional on the use of the model g, with these observations. The result of such a comparison provides a basis for improving the model and accepting it for use. Unfortunately, in many cases when conducting risk assessment, such observations are not available, and the justification of the model will be based on reasoning and understanding the phenomena considered. Anyway, at a certain stage, we accept the model and apply it for supporting the relevant decision-making concerning, for example, choice of arrangements or measures. The risk characterization is (C' = g(X),Q,K), where K and the SoK judgments address the accuracy of the model. In theory, we can assign knowledge-based probabilities on the model error, but the information value of these probabilities could be small in many situations because of a weak knowledge base. A qualitative judgment as a part of the SoK would normally be sufficient in a practical context.

This discussion shows how important it is to be clear what the uncertainties are about and what theoretical risk framework will be used. It is a misconception that precision on the foundation of risk is not of importance for the practical execution of a risk assessment. To be able to understand and communicate results from risk assessment, it is necessary to have clarity on concepts and structures in relation to risk, uncertainties and knowledge. If risk analysts are to be able to explain to risk management decision-makers what model uncertainty means and its implications for decision-making, such clarity is required.

Sensitivity analysis is commonly used to show how sensitive a risk metric or risk characterization is with respect to changes in conditions and assumptions. If the risk results depend strongly on an assumption, it is informative to show the results for different assumptions, in particular when there are large uncertainties about issues being reflected by the assumption. A risk assessment may, for example, show how an enterprise risk depends on the oil price. Different scenarios may be selected to simplify the assessment, a best estimate, a high value and a low value of the oil price. The risk analysts may discuss the likelihood of the various scenarios, but when the uncertainties are very large, it is also common to present the results for the specific scenarios, leaving it to the decision-makers to evaluate the assessment for the different possibilities. Yet we may consider this a type of risk assessment, even though the assessment is conditional on a set of assumptions.

In a scenario analysis (refer to Section 5.1), a set of possible future scenarios (events and consequences) of an activity is studied. It is common to consider an optimistic, a pessimistic and a 'probable' scenario, spanning the range of potential future developments. Criteria for the selection of these scenarios are not trivial to establish. It is often stated that the scenarios should be plausible (refer to Problem 5.15), which then relate to terms like being reasonable, believable and credible. We prefer using the likelihood concept, as it has a precise interpretation, in combination with knowledge judgments. We address in a risk assessment extremely unlikely events but also events not known today, with of course no intent to quantify the likelihood for these events. Optimistic and pessimistic scenarios can be specified that typically would be very unlikely but still may have some justification given the knowledge available. It could, for example, be a signal observed pointing to a development in a specific direction or a preliminary hypothesis that has not yet been investigated.

10.3.2 In the risk communication, seek balance between confidence and humbleness

As discussed in relation to climate change risk in Section 8.1.5, science balances 'confidence' (for example, stating that climate change is mainly man made) and 'humbleness' (reflecting that there are uncertainties). Risk analysis and risk science are critical for finding this balance, clarifying the meaning of the 'humbleness' part and pointing to the strengths as well as the limitations of the 'confidence' part. Uncertainty treatment and in particular uncertainty communication are important in this regard. The risk assessment has given us increased understanding of the risks involved. Communicating this message is the confidence part, stating, for example, that a drug or vaccine is safe and that there is no need to be worried about flying on a commercial plane. The humbleness part relates to explaining what safe means and what the uncertainties are about. The key is to explain the risk concept and what it means that the risk is low or high.

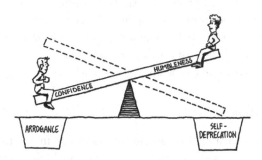

Too often, we see such explanations based on historical data alone. They may work for traveling by plane but not to convince people that nuclear industry or other activities with the potential for extreme outcomes are safe. It is a common misconception that risk can be explained by referencing historical data or risk characterizations based on expected value metrics. As highlighted in this book, for an activity to be judged safe, it is necessary that the analysts' probabilities for undesirable outcomes be small and the supporting knowledge strong. What these undesirable outcomes can be is also an issue. In addition, the question about trust is relevant, as these assessments are conducted by someone. Can we trust that risk assessments are based on the best knowledge available and the analysts do not have any type of agenda? As discussed in Section 6.2.1, we need a skeptical trust – not a blind one – which critically reviews all judgments made, both on the confidence part and the humbleness. Many issues in life are complex, and the way we frame the problem, formulate hypotheses and select data and information can strongly influence the results of the assessments. The underlying values of analysts and scientists may make them unaware of their own biases. The perspective on risk adopted in this book acknowledges the importance of the uncertainties in risk communication and handling by stressing the humbleness dimension, as well as critical trust.

Here is a checklist of factors that decision-makers, policy makers and others (e.g., lay people) should address when seeking to be informed by science on a risk issue (related to, for example, medicine and health, technology, finance or security):

Framing: What is the framing of the risk issue? Is it biased in any way? Is it possible to identify an underlying agenda of the analysis, for example, a political one? Are the choices of null hypotheses justified? Is the selection of data and information sufficiently comprehensive and balanced? Is the overall analysis approach justified? Does the analysis group have the necessary competence and resources to conduct the analysis?

Analysis message (confidence): Is there a clear message from the analysis, for example, how risk can most effectively be reduced? Is the message 'purely scientific', or does it also reflect some stakeholders' values? What are these values? For example, if a conclusion is made that an activity is safe, it needs to be based on someone's judgments of acceptable risk.

Limitations of the analysis (humbleness): Are key assumptions identified? Is a monitoring program of key assumptions in place? Have potential deviations from these assumptions been addressed? Have assumption deviation risks been assessed? Has the strength of knowledge supporting the analysis message been assessed? Have potential surprises relative to the analysts' knowledge been addressed?

10.3.3 Misinformation

This section presents a list of misinformation sources in relation to risk analysis, summarizing issues addressed in earlier chapters and highlighting some additional ones considered important in different types of risk assessment, communication and management contexts. The list can be viewed as a checklist for issues to consider when evaluating and discussing the quality of a risk analysis. The list is by no means all inclusive.

- *Reference to obsolete risk science knowledge.* If risk is described and communicated on the basis of such knowledge, for example, that societal risk in general can be adequately characterized by expected values, decision-makers and other stakeholders would be misinformed.

- *Value-laden use of terms.* For example, using the word 'surge' to describe some quantitative level suggests to the audience that the described quantitative level is unacceptably high, discouraging the audience from making their own assessments.
- *Using vague terms.* An example is using the word 'likely' without a reference to the 0–1 scale, thus allowing for completely different interpretations of what the term means.
- *Overconfidence*, that is, having confidence that is stronger than justified. An example is a judgment about the performance of a system when the system limitations and uncertainties are ignored or suppressed.
- *Model limitations and assumptions not communicated.* All models are simplified representations of the real world and are often based on strong assumptions. Misinformation occurs when the uncertainties and the assumptions are not highlighted.
- *Selection of data supporting a specific stand.* For many issues, there are numerous data sources, and to support a specific view or perspective, only some selected sources are addressed.
- *Use of non-relevant data.* For example, when studying a new, untested technology, data related to this technology may not be available, so experts may choose to use data from a related, more established technology.
- *Use of other types of biased data or biased data collection schemes.* For example, confirmation bias may encourage experts to only use information that seems 'correct'. Analytical tools, such as data cleaning, could disregard seemingly anomalous data, possibly removing important data.
- *Expert judgments constrained by social, political or cultural factors.* Experts and non-experts alike may be encouraged to agree or disagree with various ideas, people and entities due to social or cultural expectations. For examples, physicians in a professional society may feel pressured to agree with the society's stance on various issues, such as methods used to treat patients.
- *Authority granted to non-experts.* For example, social media platforms have contributed to the emergence of non-experts forming opinions on complex matters using polarizing language/images, memes, biased news and stances from virtual communities.
- *Restrictions on scientific discourse.* For example, social media platforms acting as arbiters for the truth about scientific issues and practices.
- *Confusion of science and politics.* For example, referring to a policy as science based when in fact being informed by some scientific findings and ignoring others. See discussion in Section 7.2.3.

Finally, we would like to stress the importance of being precise on the meaning of all concepts introduced. A fundamental feature of scientific work is that a concept used is accurately defined and an interpretation provided. If this is not the case, misinformation is likely to occur. Refer also to Problem 7.3. The issue is further discussed in Aven and Thekdi (2021).

10.4 PROBLEMS

Problem 10.1

The ISO 31000 standard on risk management (ISO 2018) defines risk as "the effect on uncertainty on objectives". Likelihood is defined as the chance of something happening, "whether defined, measured or determined objectively or subjectively, quantitatively or qualitatively, and described using general terms or mathematically (such as a probability or a frequency over a given time period)". Do you find these definitions useful? Explain.

Problem 10.2

You consider two alternative models in relation to a risk assessment. You are not sure which model to use. Would you refer this as 'model uncertainty'?

Problem 10.3

In relation to risk and uncertainty assessments, the following questions are fundamental: What is uncertain? Who is uncertain? How are the uncertainties represented or expressed?

Using the framework in Chapters 2 and 3, answer these questions in general terms.

Problem 10.4

In many cases, more than one analysis approach could be applied. A requirement that can be made is that the choice is logical and justified. Any comment to this?

Problem 10.5

A common quality assurance tool is to explain why the results of a study are as they are. Explain why this is also a useful thing to do for risk assessments.

Problem 10.6

Many risk issues these days are strongly politicized. Do you think risk analysis and risk science can produce neutral information to the benefit to all parties? Should the aim of the risk analysis be to produce such information? Is it acceptable to use risk analysis to support a particular political stand?

Problem 10.7

You are acting as a reviewer for a risk assessment process and find that many of the criteria on the checklist in Section 10.3.2 are not met. How would you then conclude?

Problem 10.8

Is practical risk analysis a multidisciplinary/interdisciplinary activity? Why/why not?

Problem 10.9

Provide some illustrations of misinformation in relation to current risk issues. For example, consider issues related to climate change and mask-wearing during a pandemic.

Problem 10.10

Debate the following topic with a classmate: Is it possible to create an 'objective' risk assessment? Explain why or why not.

Problem 10.11

Make a list of at least five success criteria and five pitfalls for the practical execution and use of risk assessments.

KEY TAKEAWAYS

- The different topics of risk analysis and risk science are closely related. Analysts with competences in all main topics are needed
- Applied risk analysis – tackling practical risk issues – is a multidisciplinary and interdisciplinary activity, supported by risk analysis and risk science
- Standards should be used with care, as they are not necessarily risk science based
- Be careful specifying the quantities of interest, and focus on understanding and communicating the uncertainties about these quantities
- Do not introduce frequentist probabilities and probability models when they cannot be justified
- Data are not uncertain. There can be variation in the data, and the data can be of varying quality
- Assumptions should be highlighted and assessed. They represent risks
- In risk communication, seek balance between confidence and humbleness
- See the checklist of factors that decision-makers, policy-makers and others (e.g., lay people) should address when seeking to be informed by science on a risk issue

11

Cases

Personal, organizational and public

CHAPTER SUMMARY

This chapter discusses the three cases introduced in Chapter 1, covering the individual level (a student's life), an organization (an enterprise addressing supply chain risks) and the societal level (a pandemic). The cases are used to illustrate some basic risk analysis concepts, principles, approaches and methods in relation to completely different situations. In the student example, the focus is on the planning, execution and use of a risk assessment. In the enterprise example, we look closer into the process of integrating performance, risk and vulnerability (resilience) analysis and management. In the social case, we highlight issues linked to science, policies and politics.

LEARNING OBJECTIVES

After studying this chapter, you will be better prepared for conducting real-life risk analysis in different settings:

- by observing how risk are described and risk assessments are conducted
- by observing how performance, risk and vulnerability (resilience) analysis and management can be integrated
- by observing how values and goals form the basis for risk strategies
- by observing how science relates to risk policies

The student case of Section 11.1 is inspired by Aven and Flage (2018), whereas the enterprise case, Section 11.2, is partly based on Thekdi and Aven (2016, 2019) and Aven and Thekdi (2020). The final Section 11.3 on the pandemic is to a large extent built on Aven and Bouder (2020) and Aven and Zio (2021).

11.1 PERSONAL CASE: STUDENT RISK MANAGEMENT

We remember from Section 1.1 that you are to mentor a college student on risk analysis and risk management issues. A number of risk issues were indicated in Section 1.1.

DOI: 10.4324/9781003156864-11

The first thing you and the student would like to do is to structure them, identifying the main criteria and objectives for the student. It turns out that the dialogue between you and student quickly leads to a conclusion that the main values of importance for the student are academic performance and health. All other issues raised are directly influenced by these two main values.

You then conduct a risk assessment in order to improve the understanding of what risks are most important for obtaining top performance and having good health and guide the student in making the best follow-up decisions. In the following, we discuss the planning of this risk assessment and its execution and use.

11.1.1 Planning of the risk assessment

You review the main activities of the planning stage of a risk assessment, including the clarification and specification of the decisions to be supported by the risk assessment; the development of the objectives of the assessment; and establishing the scope of the assessment, clarifying what aspects and features to include and not include in the work.

As noted, the overall main objective of the risk assessment is to support the student's life in the coming years, with a special focus on academic performance and health issues. The criteria (ideal goals) specified for the student are summarized in Table 11.1.

From this basis, the following specific aims of the risk assessment are formulated:

1   identify gaps between these goals/criteria and the current status
2   identify events and factors that could have strong effects on the achievement of the goals/criteria or other issues of importance for the student's academic performance and health
3   Suggest a set of measures that are considered necessary to bridge the gap, meet these goals/criteria and avoid potential negative consequences and surprises

TABLE 11.1 Ideal goals/criteria specified for the student

Actor	Ideal goals/criteria	
Student	High academic performance (A and B grades, thesis grade A)	Good health

4   Assess the risk associated with implementing or not implementing these measures, as well as cost-benefit analysis in a wide sense, highlighting all relevant pros and cons
5   For risk judgments, follow the approach recommended in Chapter 4, which in brief means that the following tasks should be carried out:

- Identify events that could result in goals/criteria being/not being met
- Identify underlying events that could be of importance in this regard
- Assess the probability for these events using a suitable interval probability scale
- Assess the strength of knowledge supporting these judgments. Identify assumptions on which these probability judgments are based. Assess possible deviations from these assumptions. Consider ways of strengthening the knowledge
- Scrutinize the assessments by letting others (for example, friends of the student) check the assessments, in particular for knowledge gaps
- Assess the robustness and resilience if something unlikely/surprising should occur
- Consider measures to improve the robustness and resilience
- Carry out an ALARP process to further reduce risk

6   Derive a priority list of the measures for implementation, based on different potential policies

The assessment is carried out by the student with help from you.

11.1.2 The execution of the risk assessment

Brainstorming sessions (student and you participating) were conducted to carry out the activities associated with aims 1–3: identify gaps, events and factors and measures to bridge the gaps.

A list of gaps was identified and divided into two categories: those judged the most important and others. Table 11.2 summarizes the results for two such gaps, some exams with results C or worse and some days mentally not in a good state.

TABLE 11.2 Examples of identified gaps, important events and factors and measures to bridge the gaps

Identified gaps	Events and factors				Measures		
Some exams with results C or worse	These results occur for the most for math-oriented subjects	Attends about 50% of classes. Not always prepared for the classes	Very active in student organization on campus	Part-time job takes time	Work harder on math-oriented subjects, seek guidance	Attending all classes, always strongly prepared (M1)	Consider dropping part-time job
Some days mentally not in a good state	Stress because of too many things going on	Affected by stress in relationship with others	Concerned about not being able to obtain top results	Concerned about issues in society	Seek professional help	Reduce activity level, plan activities better	More focus on physical training

The next task is to conduct a risk assessment of these measures. We restrict attention to measure M1 (attending all classes, always strongly prepared) and the base case, doing nothing and proceeding as is. We then go systematically through the issues for 5 listed in the guideline presented previously. Tables 11.3 and 11.4 summarize some key points in these assessments for the base case and M1, respectively.

We see from Table 11.3 that the student judges that, by proceeding as is, it is quite likely to get subjects with grades C or worse. This judgment is based on some assumptions: that the health condition continues as is, and there are no major changes in the way courses

TABLE 11.3 Summary of risk judgments 5 for the base case: no specific measures implemented

Events that could result in goals/ criteria not being met	Events/ factors that could be of importance in this regard	Assessed probabilities for these events	Key assumptions made	Assessment of deviation risk	Assessment of the strength of knowledge supporting the probabilities assigned
Subjects with grades C or worse (A_1)	The student's health, how the university and professors organize and conduct teaching	$P(A_1) \geq 0.10$	Current health state, no changes in the way courses run	Low medium	Medium-strong

TABLE 11.4 Summary of risk judgments 5 for measure M1 (attending all classes, always strongly prepared)

Events that could result in goals/ criteria not being met	Events/ factors that could be of importance in this regard	Assessed probabilities for these events	Key assumptions made	Assessment of deviation risk	Assessment of the strength of knowledge supporting the probabilities assigned
Subjects with grades C or worse (A_1)	How the classes are run compared to plans outlined	$P(A_1) \leq 0.10$	The preparations are effective, the professors have outlined clearly what topics the next classes are addressing	Medium-large	Medium

are run. The first assumption is given a low assumption deviation risk score, as the student considers it very likely that this assumption will hold. The other assumption risk is judged as a medium assumption deviation risk. This is based on an overall evaluation of the probability of deviations from the assumption, the implications from these deviations and the strength of knowledge supporting these judgments.

To assess the strength of knowledge supporting the probability assignments, the assumptions' risks are taken into account, as well as the availability and amount of data/information and judgments of the basic understanding of the phenomena and processes being studied, in line with the approach outlined in Section 3.2.

Table 11.4 is analogous to Table 11.3 but relates to the case when measure M1 is implemented, that is, the student will attend all classes and always be strongly prepared for these classes. Given this measure, the student considers it rather unlikely that there will be subjects with grade C or worse. This judgment is conditional on the assumption that the preparations are effective and the professors have outlined clearly what topics the next classes will address. The assumption deviation risk is judged to be medium large, as the student is not sure that all professors would make proper plans and run the classes consistent with these plans. The knowledge strength supporting the judgments is considered medium strong.

Table 11.5 gives further details about the risk judgments 5 for measure M1. The first point covers a scrutiny of the assessments by letting others (friends) check the assessments, in particular for knowledge gaps. One of the interesting outcomes of these judgments was the information that one particular professor nearly always failed to follow his plans in classes – he was enthusiastic and interesting to listen to but was not talking about the announced topics.

With regard to robustness/resilience, a key point is the potential health issue, which could affect the student's performance. Measures to increase robustness/resilience would

TABLE 11.5 Summary of additional risk judgments 5 for measure M1 and event A_1

Events that could result in goals/ criteria not being met	Scrutinize the assessments by letting others check the assessments, in particular for knowledge gaps	Assess the robustness and resilience if something unlikely/ surprising should occur	Consider measures to improve the robustness and resilience	Are there relevant signals and warnings?	Carry out an ALARP process to further reduce risk
Some exams with results C or worse	Revealed special problems with one professor not following plans	If the student's health condition becomes worse, this could strongly influence academic work and results	Physical training Professional help Contact the professor not following plans to see if improvements can be made	The observation that there has been a slight increase in the number of days with mental health challenges	Moving to a flat closer to university to reduce traveling time

be physical training and seeking professional help. The fact that the number of challenging days has increased over the last year can be viewed as a signal of a potential more serious problem. Also, contact should be made with the professor not following plans to see if improvements can be made. Other measures were also considered, including living closer to the university to reduce travel time, but this was not implemented because the costs would be too large.

A ranking of the risk of specific events was made based on the following classifications (refer to Section 3.3.2):

1   Very high risk: Potential for extreme consequences, relatively large associated probability of such consequences and/or significant uncertainty (relatively weak background knowledge)
2   High risk: The potential for extreme consequences, relatively small associated probability of such consequences and moderate or weak background knowledge
3   Moderate risk: Between small and high risk. For example, a potential for moderate consequences, and weak background knowledge
4   Low risk: Not a potential for serious consequences

For the risk related to academic performance and no risk reduction measures implemented, a very high risk (category 1) is assigned, as the failure to get a top score (A and B) is considered serious for the student's career, and the probability is considered large. However, with the planned measures implemented, the risk is reduced to level 2, high risk. Still, with the measures implemented, the student considers the risk high, showing the importance of additional measures and being careful in following up the decisions made.

The ranking of risks does not provide clear guidance on what to do. For that purpose, we also need to address issues of cost and benefits in a broader sense. First, we ask, what is the manageability of the risk? How difficult is it to reduce the risk?

For the A_1 event, measures can be implemented that are strongly believed to reduce the risk and improve the current situation. Dropping the part-time job is judged to reduce the risks considerably, as it allows more time to study, but it would be very costly. The other measures discussed could have a similar effect on performance and risk but basically without costs.

Opportunities are also an issue. Here is an example: If the student attends all classes and is well prepared, professors will notice, and it could lead to a positive relationship, which could benefit the student over time.

11.1.3 Use of the risk assessment

The risk assessment is used to provide insights about the risks and provide decision support. In this case, the main findings from the assessment were:

- The two key issues for the student are obtaining top grades and having good health.
- Without implementing some measures, the risk of failing on the academic performance is considerable.
- Several measures are considered to reduce the risks considerably

It is also important to acknowledge the limitations of the risk assessment, for example, in being able to reveal all relevant risk sources and events. Also, perceptional aspects are considered. The student is worried about his health worsening. Based on this review, the student decides to focus on measures that can contribute to strengthening his health. In addition, measures are implemented to improve the efficiency of the learning processes. The student will attend all classes and be well prepared. The student is somewhat less motivated to work harder on math-oriented subjects and seeks guidance on that but also concludes that this measure needs strong attention if the overall goals are to be met. The motivation issue needs to be confronted.

REFLECTION

A friend of the student who is informed about the risk assessment is skeptical about the quality of the risk assessment, expressing: "this is not science, just subjective judgments". Any comments?

Yes, the assessment is subjective but based on risk assessment knowledge (science) on how to perform the study. The assessment is not adding anything new to risk science but can still be useful for the student in improving the understanding about risk and hence being able to make more informed decisions. See also Appendix B.

11.2 ORGANIZATIONAL CASE: ENTERPRISE RISK MANAGEMENT

We refer to Section 1.2 and the presentation of the case. The executive team of Total Business Management (TBM) has contracted you and your organization to give a response to the following questions:

1 What are the relevant risks for TBM?
2 How should the risks be measured and characterized?
3 What are the relevant risk management issues? What risk strategies and policies are relevant for TBM?

As a first step, you clarify what type of risks to address; see Problem 2.9. The general theory of risk conceptualization and characterizations is still relevant (Chapters 2 and 3), but more specific concepts, principles, approaches and methods are also needed. The main focus in this section is issues related to 3.

A basic concept is enterprise risk, which is defined as risk of an enterprise where the consequences are related to the principal objectives or overall performance judged important for the organization (Aven and Aven 2015; Aven and Thekdi 2020). The setup is the same as in Chapter 2 – we just need to clarify the context. Different types of consequences can be defined, but they should be related to the principal objectives or overall performance of the organization. A common principal objective for an (profit-maximizing) organization is to maximize the value but at the same time to avoid health, safety and environment (HSE) and integrity incidents. The consequences can be formulated as deviations in relation to change in monetary value and occurrences of incidents.

In an enterprise context, it is also common to distinguish between strategic risk, financial risk, operational risk and social risk, where:

- strategic risk is risk where the consequences for the enterprise are influenced by mergers and acquisitions, technology, competition, political conditions, laws and regulations, labor market and so on;
- financial risk is risk where the consequences for the enterprise are influenced by the market (associated with changes in the value of an investment due to movements in market factors: the stock prices, interest rates, foreign exchange rates and commodity prices); credit issues (associated with a debtor's failure to meet its obligations in accordance with agreed-upon terms) and liquidity issues, reflecting lack of access to cash; and the difficulty of selling an asset in a timely manner, that is, quickly enough to prevent a loss (or make the required profit);
- operational risk is risk where the consequences for the enterprise are a result of safety- or security-related issues (accidental events, intentional acts), as well as supply chain disruptions; and
- social risk, covering risk where the consequences for the enterprise relate to, for example, customer satisfaction and reputation.

Your recommendation for TBM is to be clear on the definitions of enterprise risk, project or task risk and personal risk (Aven and Aven 2015). A number of tasks/projects are carried out in the enterprise, but the risks and related deviations are not explicitly linked to the principal objectives of the enterprise. We speak about task or project risk, and related task (project) risk management. The long-term aim of these tasks/projects could be to contribute to meeting the principal objectives, but there is not always a clear link between the task goals and these impacts. In some cases, we may even experience the tasks having a negative influence on the principal objectives of the enterprise.

In addition, we use the concept of personal risk and personal risk management. Managers may, for example, have a goal of increasing their income by 50%, and the satisfaction of this goal is seen as dependent on meeting specific task goals. It could also be that a bonus scheme is directly linked to the achievement of these goals.

Personal risk management is not a formal part of the management of an enterprise, but as it could strongly affect the task or enterprise risk management, it needs to be given due attention. The challenge is to design the enterprise's incentive scheme to ensure consistency between personal risk management and the task/enterprise risk management. In the following, our focus is on enterprise risk and related enterprise risk management. We use the term performance also to refer to the consequences related to meeting the overall principal objectives of the enterprise.

More specifically, you recommend an approach for understanding, assessing, communicating and handling the performance (enterprise) risks based on these three key steps (Aven and Thekdi 2020):

1 Establish a performance risk (enterprise risk) foundation built on basic concepts as defined in this book, highlighting enterprise risk, events (threats, opportunities), risk sources, vulnerabilities, resilience, improvements and antifragility.

2 Define risk-performance objectives and policies; develop guidelines for how to assess the performance and risks.
3 Implement the risk-performance management and review process (build accountability, integrate into organizational processes, acquire resources, communicate, improve, monitor).

The approach is applied to the TBM case in the following, mainly focusing on steps 2 and 3, using the framework established in Thekdi and Aven (2016, 2019) and Aven and Thekdi (2020).

High-level measurement of performance for TBM includes the following aspects:

- Shareholder value
- Sales
- Product quality
- Customer satisfaction
- Safety or security incidents
- Carbon footprint
- Employee satisfaction

The performance risk is defined by these performance measurements and related uncertainties. More specifically, we consider performance events, expectations, exposure assessments, vulnerability and resilience, improvement and competitive edge, as shown in Table 11.6.

TABLE 11.6 Performance-related concepts for the TBM

Performance concept	Examples relevant for TBM
Performance events	• Competition • Changes in consumer shopping • Problems with traditional shipping vendors • Changes in technology landscape (see Section 1.2)
Performance expectations	• Financial goals, expressed as shareholder expectation • Sustainability goals, though goals extending beyond regulatory goals are not binding • Safety goals • Shareholder expectations • Customer expectations
Performance exposure assessments	• Threat/hazard/opportunity analysis • Customer surveys • Cause analysis of past incidents within the company and in the sector • Study effects of building up specific TBM supply chain capabilities (see Section 1.2)
Performance vulnerabilities and resilience	• Perform vulnerability and resilience analysis • Study effects of building up specific TBM supply chain capabilities (see Section 1.2)
Performance improvement and competitive edge	• Invest in a culture of continuous improvement • Investment in active use of the performance-risk framework

A coarse risk assessment is conducted to assess the effects of TBM to build up its own supply chain capabilities, by 1) dedicated shipping channels with TBM's privately owned infrastructure, to move product among its regional warehouses; and 2) some dedicated shipping channels to deliver products directly to customers located near regional warehouses; and 3) capability to outsource delivery to customers by using delivery workers who accept delivery jobs on a case-by-case basis and perform the task using their own vehicles.

The costs of these measures are considerable, but the potential profits are also large. The risks need to be assessed to support the decision-making. Table 11.7 illustrates the basic elements of such a risk assessment for two events: technology changes faster/slower than expected (A_1) and quality issues due to delivery workers not performing as intended (A_2). For event A_1, TBM expects a rather fast and extensive technology development to occur in the coming years, making the company rather vulnerable to scenarios not expected or foreseen. Measures to reduce the vulnerabilities and enhance resilience should therefore be prioritized. For event A_2, the risk is judged as small provided that due attention is placed on quality issues for the delivery workers.

TBM develops a risk program to document activities necessary for implementation of the enterprise risk management. The risk program activities involve oversight and implementation of the risk management, including the hazard/threat/opportunity identification, risk assessment and the risk treatment. A risk program is created to ensure that the risk management process is performed in an effective manner and in line with contemporary risk science principles. The program provides clear and concise steps for systematic planning of risk activities, including clear project milestones, goals and responsible parties. Based on the guidelines of Aven and Thekdi (2020) and the COSO

TABLE 11.7 Crude risk assessment for the TBM implementing measures 1–3 (illustration of analysis for two selected events)

Event	Consequences	Likelihoods	Knowledge strength	Vulnerabilities/ resilience	Risk contribution (potential deviations from an expected profit)
Technology changes faster/ slower than expected	Large	High	Strong	Considerable vulnerabilities	High Measures needed to reduce vulnerabilities and enhance resilience
Quality issues due to delivery workers not performing as intended	Moderate	Small if quality is given due attention	Strong	Quality issues could lead to loss of reputation	Small if risk followed up on carefully

enterprise risk management framework (COSO 2020), TBM develops a risk program containing the following items:

1 Introduction: This should articulate the following:

- Why a risk program is needed for the organization
- The need for widespread adoption of the risk program and coordination among stakeholders
- The scope of the risk program (what types of risks are included, justification for what specifically is not included)
- Leadership levels that have approved and adopted the risk program, including leadership's commitment to ensuring the risk program has the resources and executive support necessary for a successful implementation

TBM: The program aims at ensuring that all important risks are identified, assessed and handled, and an oversight and proper coordination of all risk-performance management activities are provided.

2 Risk policies

- Standardized processes for how to handle risk (see Chapter 8)
- Templates or forms to be used for documenting risk policies and implementation activities. For example, documentation could include output of the steps shown in Section 8.3, standardized communication with parties responsible for implementing risk treatment plans and other documentation needed for audit purposes

TBM: Guidance for how to conduct and use risk assessments (including performance-risk assessments) are documented, as well as how to build robustness and resilience. Incorporate risk management within normal business processes. An overview of strategies and instruments for different types of risk problems are presented, as in Table 8.1.

3 Criteria for evaluation of risks. This should include guidance on the organization's risk appetite, broad priorities for risk assessment (e.g., safety, financial, etc.) and any related strategic initiatives that can influence how risk is treated in the organization.

TBM: The company's attitude to risk is reflected by the following statements:

- TBM is willing to take on risks in pursuit of values and meeting the company's principal objectives
- The risk related to negative consequences will be reduced according to the ALARP principle
- TBM wants to be seen as best in class when it comes to safety and product quality

4   Risk activities. This should contain clear guidelines for the following elements:

- Specification of risk activities, including a process for selecting responsible parties for implementing risk activities, setting deadlines for risk activities and monitoring of implementation
- Evaluation and monitoring of the current risk program, recognizing that the risk program should adapt to current conditions (strategic, economic, etc.)

TBM: An overview is provided for all main risk activities.

5   Related documentation. Any requirements for paperwork and regular meetings should be balanced with any organizational efforts toward minimizing (or increasing) bureaucracy. Overburdensome documentation can lead to a lack of efficiency, while the use of some documentation can improve standardization. Related documentation can include guidelines for:

- Responsible parties for managing the risk program
- Responsible parties for implementing risk treatment activities
- Responsible parties for overseeing coordinated activities among business units
- Guidelines for reporting and transparency of risk activities
- Guidelines for the risk program evaluation process, including a timeline for evaluation (e.g., annually, every five years, etc.). In some cases, it may be more appropriate to recommend a risk program evaluation following major events that can signal major changes in the operating environment. Examples of major events include changes in political leadership and associated initiatives, advancements in technology capabilities, political conflicts, changes in organizational leadership/ownership and so on
- Guidelines for effective governance/oversight of risk activities. This oversight should carefully monitor effectiveness, support internal auditing and promote transparency of risk practices. This oversight should also include continuous improvement and re-evaluation of risk program

TBM: The documentation is restricted to key aspects of the risk management, including the overview of the risks and the risk management activities.

The risk program defined previously requires that the program be supported by all levels of organizational leadership, include adequate coordination of activities and ensure that risk goals are recognized and maintained. The risk program, as part of larger enterprise risk management activities, is a management process that involves stakeholders from all levels of the organization. While it cannot operate as a standalone process, it should instead be embedded into all business processes.

To perform an assessment of the company's risk management, TBM decides to use a taxonomy for enterprise risk management (ERM) maturity (Aven and Thekdi 2020). It is based on three levels of maturity: beginner, intermediate and advanced. While the overall ERM principles are the same for all three maturity levels, increased levels of maturity involve more sophisticated approaches. The assumption is that organizations that are new to risk management practices start at the beginner level. Then, with experience and increased resources, organizations can achieve higher levels of maturity. TBM's ambition is to be at the advanced level.

Table 11.8 presents the taxonomy of ERM maturity, with scores for TBM. The types of maturity are defined by the following types of characteristics: Resources in the organization that are responsible for ERM; Expertise that is gained through training; Culture that involves organizational support for ERM and the creation of a risk-conscious workforce – risk culture is here understood as shared beliefs, norms, values, practices and structures with respect to risk in an organization (Aven and Ylönen 2020); and Practices involving the ERM procedures in place. In the table, each characteristic

TABLE 11.8 Taxonomy of ERM maturity, with TBM assignments (based on Aven and Thekdi 2020)

	Characteristic	Beginner	Intermediate	Advanced	TBM
Resources					
	R.1. Dedicated risk manager		✔	✔	Yes
	R.2 Dedicated risk management business unit (proportional to size/importance of organization)			✔	No
	R.3 Documented risk guidelines and policies, available to all organizational stakeholders			✔	Yes
	R.4 Clear and detailed risk strategies (risk-informed strategies, cautionary/precautionary/robustness/resilience strategies and discursive strategies)	✔	✔	✔	Yes
	R.5 Resources for regular risk management benchmarking and reporting		✔	✔	Yes
Expertise					
	E.1 Some employees trained on risk management practices	✔	✔	✔	Yes
	E.2 All employees trained on risk management practices, with training aligned with each role's function in risk management processes		✔	✔	No
Culture					
	C.1. Agreement among board and other leadership on the organization's risk appetite	✔	✔	✔	Yes
	C.2 Regular assessment and accountability at all levels of the organization to ensure risk policies are properly implemented	✔	✔	✔	Yes

		Characteristic	Beginner	Intermediate	Advanced	TBM
		C.3 Risk perception studies to identify major risk concerns, including social, cultural and psychological factors in risk judgment		✔	✔	No
		C.4 Implementation of open, transparent and timely risk communication procedures	✔	✔	✔	Yes
		C.5 Invite feedback from stakeholders engaged in the risk practices and incorporate in risk policies as needed	✔	✔	✔	Yes
Practices						
		P.1 Meets local and industry-specific regulations	✔	✔	✔	Yes
		P.2 Meets local and industry-specific non-regulatory risk and safety guidelines	✔	✔	✔	Yes
		P.3 Knowledge-dependent prioritization of risk informed by formal tools	✔	✔	✔	Yes
		P.4 Formal procedures for balancing risk concerns, such as cost-benefit methods taking into account risk and uncertainties		✔	✔	Yes
		P.5 Formal procedures for identifying appropriate risk control, risk treatment, risk response strategies that are in agreement with the overall risk appetite of the organization		✔	✔	Yes
		P.6 Active stakeholder involvement in risk management processes		✔	✔	Yes
		P.7 Formal processes for assessing risk for high uncertainty and black swan surprises			✔	No
		P.8 Continuously monitor and audit the ERM process while adapting to changing conditions and stakeholder feedback	✔	✔	✔	Yes

is followed by a checkmark designating whether the presence of that characteristic is deemed to be at the beginner, intermediate or advanced level. Some characteristics may apply to multiple maturity levels.

As we see from Table 11.8, TBM scores at the advanced level for most items. The TBM leadership has considered establishing a dedicated risk business unit but has not yet done so. The reasoning was that it would be too resource demanding to establish such a function, and it could also hamper the proper incorporation of risk management within all the normal business processes of the company. Currently not all employees have been trained on risk management practices, with training aligned with each role's function in risk management processes. The task will, however, be prioritized. TBM is also working on improving its assessment concerning black swan types of risk, for example, as a result of erroneous assumptions, in line with the ideas of Section 3.4.

11.3 SOCIETAL CASE: PANDEMIC

In this case, you are a risk advisor to the leadership for a nation, and the concern is a previously unknown virus that has begun spreading globally, with potential to spread to your nation. You are to give guidance on how risk science, with its concepts, principles, approaches, methods and models, can support the actual assessments, communication and handling of the risks related to the virus. This includes responding to challenges related to, for example

- How should the risks be measured and characterized?
- What are the relevant issues related to risk perception and communication when consulting with political leaders and the public?
- How should the risks be communicated?
- What are the relevant risk management and government issues?
- How should the risks be handled? What risk strategies and policies should be adopted?

The following sections provide some input to the response to these challenges; see also Problems 2.10, 8.16 and 8.17, as well as Sections 5.5, 6.1.3, 7.1.5, 8.1.4 and 8.7.

11.3.1 How risk should be described and communicated

Literature, both scientific and popular, was rapidly emerging on risk issues associated with the virus. In this huge corpus, risk and related concepts, such as probability and vulnerability, are commonly referred to, described and measured, although they are seldom precisely defined or interpreted. You as a risk science expert and advisor do not consider this satisfactory. It could seriously hamper risk communications and risk handling.

Risk science can provide essential help in this regard. You highlight that to understand risk, it is important to clarify the role of uncertainty and knowledge. It is not enough to talk about probabilities and historical data. Think about the risk related to the number of deaths in the coming months as a result of this virus. The way risk is conceptualized and described could be very important for how the authorities will judge the magnitude of the risk, communicate the risk to the public and conclude what to do. With large uncertainties, as in this case, it is clear that accurate predictions cannot be made. Yet such predictions are made, usually based on models of the phenomena studied. However, as the knowledge is weak, the models are based on strong assumptions, which could turn out to be far from reality. Models to express risk thus need to be used with care, to avoid extreme scenarios being given stronger authority than is justified.

You stress that the number of deaths due to the virus in the coming month is not risk. It is an unknown quantity. To talk about risk, we also have to include uncertainty – we do not know today what this number will be. We assess these uncertainties in risk assessments using all relevant knowledge, founded on data, information, models, tests, analysis and argumentation. Probabilities are used to express these uncertainties, but probability is just a tool and has limitations. For example, in the previous case, when considering risk related to deaths, analysts could perform a risk assessment establishing a probability for a maximum number of deaths, given the implementation of a specific policy. However, such a probability should be communicated with care, as the knowledge supporting it is rather weak. It is a fundamental risk science insight today that probabilities should always be accompanied by judgments of the knowledge supporting these probabilities. In line with this, you recommend that the communication of risk be formulated as follows:

> The result of the risk assessment is that the number of deaths is unlikely (less than 5%–10%) to exceed x in the coming month, given the implementation of policy y. This assessment is based on current knowledge on the topic using the best models and data available. There are, however, considerable uncertainties about the underlying phenomena and how the epidemic will develop – many of the assumptions of the models used are subject to large uncertainties. Overall, the knowledge supporting the risk assessment is considered rather weak.

An alternative is to indicate a prediction interval [a,b], expressing that the analysts are confident (90%–95% probability) that the actual number of deaths is in the interval given a specific policy. In addition, knowledge strength judgments are needed.

As another example, consider the message that the authorities would like to convey, that one stays safe when following the guidance on how to protect oneself against getting the virus (washing your hands frequently, maintaining social distancing, etc.). However, understanding what 'safe' here means is not straightforward. You will provide clear explanations. Despite these measures, one would face some risks. The point is that the probability of one getting the virus when following these guidelines is judged to be very low, and this conclusion is supported by strong knowledge/evidence; refer to discussion in Section 2.3.4. You make the point that using such words would strengthen the risk communication.

More explicitly, think about the risk of getting infected by going to the grocery store. A related risk judgment can be summarized as follows. If a large number of people (n) go to the grocery store in the coming days, some (x) will be infected as a result. If the experts consider that the ratio x/n will be very low and have good reasons for doing so, the risk is considered small.

When we go to the grocery store, we are not primarily concerned with how many people are infected in the entire population but whether we will be infected. Then you can make your own assessment. Perhaps a conclusion is drawn that the risk is low on the basis of assessments of probability and knowledge strength as referred to previously, and then suddenly you are affected by a person coughing near you in the grocery. Then the risk changes. The probability is considered slightly higher and the supporting knowledge weaker. Some may claim that these are not proper risk assessments but risk perception. But it is not. The perception of risk takes into account emotions and affects, but what we are talking about here are conscious assessments of whether infection can occur, with assessments of probability and underlying knowledge. Fear and other emotions can greatly affect one's risk perception, but with practice and increased risk science knowledge, we can better distinguish between proper risk assessments and feelings.

11.3.2 The difference between professional risk judgments and risk perception

Many people experienced stress from the risks related to the virus. People's risk perception is strongly influenced by repeatedly hearing about cases and deaths. Media stories, death counts and public campaigns that amplify danger may stir emotions and promote the sense of risk while increasing the attention on negative events. Risk science has developed considerable knowledge about how and why this happens, as discussed in Chapter 6. As we easily retrieve from memory a vast number of 'alarms', the conclusion is that the risk is high. The representativeness is not reflected. The virus hits all the hot buttons: unknown, new and delays in effects, lack of control and catastrophic potential, often summarized by the two dimensions, newness and dread, as discussed in Section 1.3. The result is that the risk is amplified.

You summarize this knowledge, but you underline that lay people's risk perception is not only about feelings. It can also capture conscious judgments of uncertainties. History has shown many examples of this, where highly relevant uncertainties were ignored by the professional risk judgments but included in lay people's risk appraisals (for example, in relation to nuclear risk). It is well documented that traditional, professional perspectives on risk, which are built on probability, historical data and models, have failed to properly reflect important uncertainty aspects of risk, as discussed in Chapter 6. You explain that contemporary risk science provides clarity on these issues by showing the importance of knowledge and lack of knowledge when characterizing risk. The concept of risk includes uncertainties, and people may have good reasons in many cases for questioning issues linked to these uncertainties. If the experts build their risk assessments on a 'narrow' perspective on risk, they may be tempted to downplay such questioning, considering it influenced by affects and not the result of conscious judgments of uncertainties and risk.

You emphasize that it is essential for the quality of the analysis that the risk assessments be placed in a 'broad' risk framework, which gives due attention to all aspects of uncertainties and knowledge. Traditional probabilistic risk assessments are not sufficient for adequately studying risk in the case of large uncertainties. Using such a broad framework makes us better able to understand and value lay people's risk perception.

11.3.3 Science-based decision-making and the precautionary principle

In your guidance to the leadership in the country, you give a lecture on the fundamentals of risk decision-making. You stress that, inevitably, there is always a balance to be made between measures to create values, on the one hand, and measures to protect, on the other. Science does not give us the formula for finding the right balance. It provides relevant knowledge and thus informs the decision-maker, but it does not prescribe what should be done. The authorities in different countries have all labeled their policies 'science-based', indicating that their policies are determined by scientific findings. This notion is misleading, as the policies are at best science informed. As we know, science produces justified claims about the world, but claims are neither facts nor the truth. Policies also reflect values, in particular related to how to give weight to uncertainties. In the case of large uncertainties, there is a considerable leap between insights provided by science and the considerations and policies derived by political institutions.

When faced with high stakes and large (scientific) uncertainties, precautionary measures will and should be given weight, as in the risk handling related to the coronavirus. The point is simply that we will avoid very high losses, and, given the uncertainties about the virus and the potential spread, this leads to extraordinarily cautious policies. All governments seem to have adopted such policies. Some scholars are skeptical about the use of the precautionary principle/approach, indicating that it is not rational. However, the principle, interpreted here as a guiding perspective, is in no way irrational: because of the uncertainties, there is not sufficient knowledge available to guide us scientifically. A key precautionary measure becomes, then, research to strengthen the knowledge base.

It is acknowledged that different situations call for different approaches for assessing and handling risk – 'simple' problems with no or very limited uncertainties require different strategies than those required by problems with high stakes involved and large uncertainties.

You make it clear that the health experts and agencies provide guidance, nothing more. Their recommendations need to be based on scientific findings. There is, however, a need to see beyond these to make proper policy and political decisions.

The country is faced with a situation of considerable uncertainty as to what the consequences will be in terms of the spread of the virus and what effects various policies and measures would have. In such a situation, there is no strategy that can be argued to be objectively and scientifically the best. The health agencies need to have a clear medical professional justification, which can be interpreted as expressing that if one does not have such a justification, the measure cannot be recommended. The way of thinking is the same as used in science where the starting point is that drugs are not approved until it is scientifically proven that they work and are safe. The burden of proof is on those who want to introduce and profit from the drug. It is a sensible general principle but cannot be used in situations where we are facing an acute danger, with potentially very serious consequences and where the uncertainties are large. This is the situation here faced, and the government cannot rely on a traditional scientific approach. The risks and uncertainties are too great and the time aspect critical.

You stress that the government must take these risks and uncertainties into account. It must apply risk science, which involves seeking a balance between various considerations, including the protection of life and health, on the one hand, and maintaining social activities and economic values, on the other. This means in particular to give weight to the precautionary principle. You explain that this is an appropriate principle for the government to use in this situation. The principle is, however, outside the tool box of the health sciences. It is not traditional science but risk science and risk policies. What should one do if one is faced with a threat that is to a large extent unknown, that has the potential for enormous negative impact? Wait until you have gained more comprehensive and stronger knowledge about this threat? No, you cannot do that, because it is urgent to make decisions. One implements measures because one does not know, because the threat is considered so serious. This is what precaution is all about. And it is this way of thinking that you recommend following, at the same time as efforts are made to continuously strengthen the knowledge base so that future decisions can be based on knowledge that is as strong as possible. The application of precaution and traditional science go hand in hand, as we have seen in this case.

You explain that health scientists may also refer to a need for balancing different concerns when evaluating potential measures. Some level of proportionality should be sought. However, this involves weightings of risk and uncertainty and is not of a scientific nature. It is about politics. It is therefore correct that it is the politicians who must draw the final conclusions as to which strategies and measures should be implemented. Health experts and others provide decision support. The strength, relevance and suitability of this support must also be considered, and that cannot be done by the health professionals alone.

11.4 PROBLEMS

Problem 11.1

A traditional qualitative risk assessment typically covers the following points: hazard/threat identification; assessment of the consequences of these hazards/threats; probability judgments; and a summarizing risk metric, typically a risk matrix. What are the main problems with using a risk matrix to characterize risk?

Problem 11.2

See case discussed in Section 11.1. If we focus on the risk sources/hazards/threats and the measures needed to meet these, it may seem unnecessary to describe risk using probabilities; the numbers seem rather arbitrary. Any comments?

Problem 11.3

Would you characterize the concerns many people had in relation to the coronavirus of going to the grocery store as based on perceptual factors like fear and dread?

Problem 11.4

Why is it not necessarily convincing to argue that a specific policy is science based?

Problem 11.5

Can prediction intervals be used in case of rare events?

Problem 11.6

Consider the taxonomy of ERM maturity shown in Table 11.8. For the TBM and pandemic case examples explored in this chapter, explain the following: Is it critical for the organization to achieve advanced maturity during the program implementation? Explain.

KEY TAKEAWAYS

- When conducting a risk assessment, identify the values, goals and criteria of the relevant stakeholders.
- Performance and risk can be elegantly integrated using the (C,U) perspective.
- Prediction intervals can be informative ways of informing about risk. Related likelihood and knowledge strengths are required to accompany these intervals.
- Using the precautionary principle is to apply risk science but not traditional science.
- Agency professionals (like health scientists) are not in a position to prescribe decisions when facing scientific uncertainties.

Appendices

Basic terminology

This appendix summarizes the risk analysis and management terminology used in the book. The appendix includes three sections, with definitions presented in alphabetical order:

I. Terminology of basic concepts
II. Terminology of related concepts, methods, procedures
III. Terminology of risk management actions

The definitions are in line with those given in the SRA (2015) Glossary. This reference contains additional and alternative definitions that are not described in this appendix. Additionally, this reference does not include the concept of managerial review and judgment presented in this appendix.

A.1 TERMINOLOGY OF BASIC CONCEPTS

Ambiguity

The condition of admitting more than one meaning/interpretation.

Complex/complexity

- A system is complex if it is not possible to establish an accurate prediction model of the system based on knowing the specific functions and states of its individual components.
- Complexity: A causal chain with many intervening variables and feedback loops that do not allow the understanding or prediction of the system's behavior on the basis of each component's behavior.

Event, consequences

Event

- The occurrence or change of a particular set of circumstances such as a system failure, an earthquake, an explosion or an outbreak of a pandemic.
- A specified change in the state of the world/affairs.

Consequences: The effects of the activity, with respect to the values defined (such as human life and health, environment and economic assets), covering the totality of states, events, barriers and outcomes. The consequences are often seen in relation to some reference values (planned values, objectives, etc.), and the focus is often on negative, undesirable consequences.

Exposure

Exposure to something:

- Being subject to a risk source/agent (for example, exposure to asbestos).

Harm, damage, adverse consequences, impacts, severity

Harm: Physical or psychological injury or damage.
Damage: Loss of something desirable.

Adverse consequences: Unfavorable consequences.

Impacts: The effects that the consequences have on specified values (such as human life and health, environment and economic assets).
Severity: The magnitude of the damage, harm and so on.

Hazard

A risk source where the potential consequences relate to harm. Hazards could, for example, be associated with energy (e.g., explosion, fire), material (toxic or eco-toxic), biota (pathogens) and information (panic communication).

Knowledge

Two types of knowledge:

Know-how (skill) and know-that of propositional knowledge (justified beliefs).
Knowledge is gained through, for example, scientific methodology and peer review, experience and testing.

Model

A model of an object (e.g., activity, system) is a simplified representation of this object. A probability model is a special type of model, based on frequentist probabilities (often referred to as chances in a Bayesian context).

Opportunity

An element (action, sub-activity, component, system, event, . . .), which alone or in combination with other elements has the potential to give rise to some specified desirable consequences, that is, a risk source where there is a potential for desirable consequences.

Probability

Either a knowledge-based (subjective) measure of uncertainty of an event, conditional on the background knowledge, or a frequentist probability (chance). If a knowledge-based probability is equal to 0.10, it means that the uncertainty (degree of belief) is the same as randomly drawing a specific ball out of an urn. A frequentist probability (chance) of an event A is the fraction of times this event occurs when the situation under consideration can be repeated over and over again infinitely. See SRA (2015) and Section 3.1 for further details.

Resilience

Overall conceptual definition:

Resilience is the ability of the system to sustain or restore its basic functionality following a risk source or an event (even unknown).

Resilience metric/description (example):

- The probability that the system is able to sustain operation when exposed to some types of risk sources or events (which can be more or less accurately defined).

A resilient system is a system for which the resilience is judged high (this is a value judgment).

Risk

See Chapters 2 and 3. In its most general form:

Risk is the potential for undesirable consequences.

Risk is the two-dimensional combination of the consequences C of the activity (with respect to something that humans value) and associated uncertainties about C.

Risk description: A qualitative and/or quantitative picture of the risk, that is, a structured statement of risk usually containing the elements: risk sources, causes, events, consequences and uncertainty representations/measurements. Formally, we write:

Risk description = (C',Q,K), where C' is the specified consequences of the activity considered, Q the measure of uncertainty used and K the background knowledge that C' and Q are based on.

Risk source or risk agent

An element (action, sub-activity, component, system, event, . . .), which alone or in combination with other elements has the potential to give rise to some specified consequences (typically undesirable consequences).

Robustness

The antonym of vulnerability.

Safe, safety

Safe: Without unacceptable risk.
Safety:

- Interpreted in the same way as safe (for example, when saying that safety is achieved).
- The antonym of risk (the safety level is linked to the risk level; a high safety level means a low risk level, and vice versa).

Sometimes limited to risk related to non-intentional events (including accidents and continuous exposures).

Security, secure

Secure: Without unacceptable risk when restricting the concept of risk to intentional acts by intelligent actors.
Security:

- Interpreted in the same way as secure (for example, when saying that security is achieved).
- The antonym of risk when restricting the concept of risk to intentional acts by intelligent actors (the security level is linked to the risk level; a high security level means a low risk level, and vice versa).

Threat

Risk source, commonly used in relation to security applications (but also in relation to other applications, for example, the threat of an earthquake).

Threat in relation to an attack: A stated or inferred intention to initiate an attack with the intention to inflict harm, fear, pain or misery.

Uncertainty

Overall conceptual definitions:

- For a person or a group of persons, not knowing the true value of a quantity or the future consequences of an activity.
- Imperfect or incomplete information/knowledge about a hypothesis, a quantity or the occurrence of an event.

Uncertainty metrics/descriptions (examples):

- A subjective probability.
- The pair (Q,K), where Q is a measure (interpreted in a wide sense) of uncertainty and K the background knowledge that supports Q.

Epistemic uncertainty: As previously for the overall conceptual definition of uncertainty and uncertainty metrics/descriptions (examples).

Aleatory (stochastic) uncertainty: Variation of quantities in a population of units (commonly represented/described by a probability model).

Vulnerability

Overall conceptual definitions:

- The degree to which a system is affected by a risk source or agent.
- The degree to which a system is able to withstand specific loads.
- Vulnerability is risk conditional on the occurrence of a risk source/agent.

Vulnerability metrics/descriptions (examples):

As for risk, but conditional on the risk source or event (load).

- Expected loss given a failure of a single component or multiple components.
- Expected number of fatalities given the occurrence of a specific event.
- Expected system loss under conditions of stress.
- The probability that the system capacity is not able to withstand a specific load (the capacity is less than the load).
- A probability distribution for the loss given the occurrence of a risk source.
- (C',Q,K| risk source or event), that is, a risk description given the occurrence of a risk source; see Chapter 3.

As for risk, the suitability of these metrics/descriptions depends on the situation.

A vulnerable system is a system whose level of vulnerability is judged to be high.

A.2 TERMINOLOGY OF RELATED CONCEPTS, METHODS, PROCEDURES

Model uncertainty

Uncertainty about the model error, that is, about the difference between the model output and the true value being modeled.

Precautionary principle

An ethical principle expressing that if the consequences of an activity could be serious and subject to scientific uncertainties, then precautionary measures should be taken, or the activity should not be carried out.

Risk analysis

Systematic process to comprehend the nature of risk and to express the risk, with the available knowledge.

Risk analysis is often also understood in a broader way, in particular in the Society for Risk Analysis community: risk analysis is defined to include risk assessment, risk characterization, risk communication, risk management, and policy relating to risk, in the context of risks of concern to individuals, to public and private sector organizations and to society at a local, regional, national or global level.

Risk appetite

Amount and type of risk an organization is willing to take on risky activities in pursuit of values or interests.

Risk assessment

Systematic process to comprehend the nature of risk and express and evaluate risk with the available knowledge.

Risk aversion

Disliking or avoiding risk.

Technical definition: Risk aversion means that the decision-maker's certainty equivalent is less than the expected value, where the certainty equivalent is the amount of payoff (e.g., money or utility) that the decision-maker has to receive to be indifferent between the payoff and the actual 'gamble'.

Risk characterization, risk description

A qualitative and/or quantitative picture of the risk, that is, a structured statement of risk usually containing the elements: risk sources, causes, events, consequences, uncertainty representations/measurements (for example, probability distributions for different categories of consequences – casualties, environmental damage, economic loss, etc.) and the knowledge that the judgments are based on. See also the definition of risk description in relation to the definition of the concept of 'risk'.

Risk communication

Exchange or sharing of risk-related data, information and knowledge between and among different target groups (such as regulators, stakeholders, consumers, media, general public).

Risk evaluation

Process of comparing the result of risk analysis (see 'Risk analysis') against risk (and often benefit) criteria to determine the significance and acceptability of the risk.

Risk framing (pre-assessment)

The initial assessment of a risk problem, clarifying the issues and defining the scope of subsequent work.

Risk governance

Risk governance is the application of governance principles to the identification, assessment, management and communication of risk. Governance refers to the actions, processes, traditions and institutions by which authority is exercised and decisions are made and implemented.

Risk governance includes the totality of actors, rules, conventions, processes, and mechanisms concerned with how relevant risk information is collected, analyzed and communicated and management decisions are made.

Risk management

Activities to handle risk such as prevention, mitigation, adaptation or sharing.

It often includes trade-offs between costs and benefits of risk reduction and choice of a level of tolerable risk.

Risk perception

A person's subjective judgment or appraisal of risk.

A.3 TERMINOLOGY OF RISK MANAGEMENT ACTIONS

Managerial review and judgment

Process of summarizing, interpreting and deliberating over the results of risk assessments and other assessments, as well as of other relevant issues (not covered by the assessments) in order to make a decision.

   This definition is not given in the SRA Glossary.

Risk acceptance

An attitude expressing that the risk is judged acceptable by a particular individual or group.

Risk policy

A plan for action on how to manage risk.

Risk prevention

Process of actions to avoid a risk source or to intercept the risk source pathway to the realization of damage, with the effect that none of the targets is affected by the risk source.

Risk reduction

Process of actions to reduce risk.

Risk regulation

Governmental interventions aimed at the protection and management of values subject to risk.

Risk sharing or pooling

Form of risk treatment involving the agreed-upon distribution of risk among other parties.

Risk tolerance

An attitude expressing that the risk is judged tolerable.

Risk trade-offs (risk-risk trade-offs)

The phenomenon that intervention aimed at reducing one risk can increase other risks or shift risk to another population or target.

Risk transfer

Sharing with another party the benefit of gain, or burden of loss, from the risk.
Passing a risk to another party.

Risk treatment

Process of actions to modify risk.

Stakeholder involvement (in risk governance)

The process by which organizations or groups of people who may be affected by a risk-related decision can influence the decisions or their implementation.

Risk science fundamentals

A new science has developed over the last four decades, meeting the need for systematic knowledge generation in relation to understanding, assessing, communicating, managing and governing risk: risk science. The growing understanding and definition of risk science can be found in several recent publications, including work by the Society for Risk Analysis (SRA 2015, 2017a, 2017b) and in particular the document referred to as *Core Subjects of Risk Analysis* (SRA 2017a). This document provides a list of subjects that are considered essential in a study program on risk science. The documents also reflect on the main subjects of risk science: risk fundamentals (for example, on basic concepts like risk and probability), risk assessment, risk perception and communication and risk management and governance. The present book presents and discusses what we consider the essential topics of these subjects.

The following text, which to a large extent is based on Aven and Flage (2020), Thekdi and Aven (2021b) and Aven (2018a, 2020a, 2021), reviews basic knowledge about this science, with a focus on educational programs in the area.

As high-profile issues, such as climate change, health-related epidemics and economic fluctuations have a significant impact on the practice of business, engineering and so on, students across academic disciplines are becoming increasingly aware of risk within their career functions. While students seek educational programs that will prepare them for these career functions, it is imperative for universities to develop curriculums that integrate and balance relevant risk science topics and the more discipline-oriented subjects. As a result, study programs are increasingly being accompanied by a broad range of curriculums that include elements of risk science topics.

In the following, we provide an assessment of topic areas for both risk science and non-risk domain programs that include risk in the curriculum. The referenced programs are based on a collection of university programs that are listed by the Society for Risk Analysis, which was collected using a survey of society members (see Thekdi and Aven 2021a, 2021b for details). The authors have performed a secondary assessment of the required coursework and other degree requirements of these programs in order to understand the as-is status of risk science within university curriculums. We also present an ontology and taxonomy to understand how to classify programs encompassing both risk science and non-risk domain areas, and we provide some perspectives on the status and future of the risk science education. Our focus is on university training that is for credit at accredited universities, distinguishing between undergraduate and graduate (masters and PhD) programs.

We clarify language by noting the following definitions. A *program* refers to a single degree earned for coursework at a university, such as a masters or doctoral degree. *Curriculum* refers to the academic content, courses, subjects, project work and theses that are required for earning a degree within a *program*. A *domain* is referred to as a topic area for a program, such as engineering, business and healthcare. When referring to risk science topics covered within the *curriculum*, we consider both topics covered as dedicated courses and also topics existing within course modules, case studies and other types of assignments.

Section B.1 will provide more detail on the components of risk science. Section B.2 will present an ontology of risk science concepts within university programs, a taxonomy for risk science programs and the use of these concepts as a basis for a review of current risk-related programs. Section B.3 provides a discussion of the as-is status of risk science programs and topics, with an emphasis on how to further develop risk science studies at colleges and universities. A key point raised is the need for study programs in generic risk science and the interactions between generic risk science and applied risk science, as will be further discussed in the following section.

B.1 WHAT IS RISK SCIENCE?

Risk analysis as a field and discipline includes concepts, principles, approaches, methods and models for understanding, assessing, communicating, managing and governing risk, with applications. The foundation of this field and discipline has been subject to continuous discussion since its origin some 40 years ago with the establishment of Society for Risk Analysis and the *Risk Analysis* journal. This field and discipline include all relevant study programs, researchers, journals, scientific conferences, societies and so on. A suggestion for what specific subjects to include is presented by SRA (2017a), as mentioned previously. This SRA document summarizes the key questions that risk analysis can help answer, such as:

- What are people worrying about?
- What is worth worrying about?
- What is not worth worrying about?
- What might go wrong?
- Why and how might it go wrong?
- What are the consequences?
- What happens if we intervene?
- What should we do next, given the resources, risks, uncertainties, constraints and other concerns?
- What are unacceptable risks?
- How can we be prepared in case something happens?
- How can we build robust and resilient systems?
- Who should say what to whom?

The document also refers to different types of risk analysis, including descriptive, causal, predictive, prescriptive and learning.

As for all fields and sciences, there are topics and subjects of risk analysis that can belong in multiple disciplines. An important issue in this regard is the relationship and delineations between different fields and sciences, for example, between risk analysis and statistics. We will come back to this issue later in this section.

This leads us into the discussion about what a science is. The topic has been discussed by philosophers and other scholars for hundreds of years. Recent work by SRA (SRA 2017a) has pointed to the suitability of a knowledge field (discipline) perspective as thoroughly studied by Hansson (2013), in contrast to a method-based approach. According to this perspective, we can define risk science as the practice that provides us with the most reliable statements (i.e., most epistemically warranted statements or most justified beliefs) that can be produced at the time being on the subject matter (scope) covered by the knowledge field/discipline (Hansson and Aven 2014; Aven 2020a). The scope of this field and science covers, as mentioned previously, concepts, principles, approaches, methods and models for understanding, assessing, characterizing, communicating, managing and governing risk, with applications. See Figure B.1. There will always be discussion about what the most justified knowledge is. The present book is based on SRA work and related research.

> *Risk science is the most updated and justified knowledge on risk fundamentals (concepts), risk assessment, risk perception and communication and risk management and governance. This book aims at presenting the core of this knowledge. Risk science is also about the process – the practice – that gives us this knowledge.*

A distinction is made between generic, fundamental risk analysis (B) and applied risk analysis (A); see Figure B.2. Generic risk analysis covers generic concepts, principles,

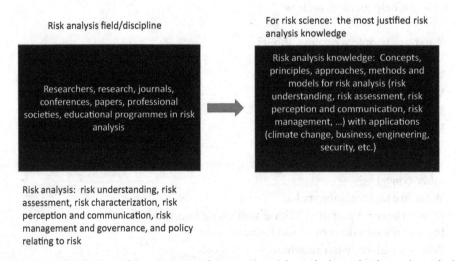

Risk analysis field/discipline

For risk science: the most justified risk analysis knowledge

Researchers, research, journals, conferences, papers, professional societies, educational programmes in risk analysis

Risk analysis knowledge: Concepts, principles, approaches, methods and models for risk analysis (risk understanding, risk assessment, risk perception and communication, risk management, ...) with applications (climate change, business, engineering, security, etc.)

Risk analysis: risk understanding, risk assessment, risk characterization, risk perception and communication, risk management and governance, and policy relating to risk

FIGURE B.1 Illustration of the process of generating risk analysis and science knowledge

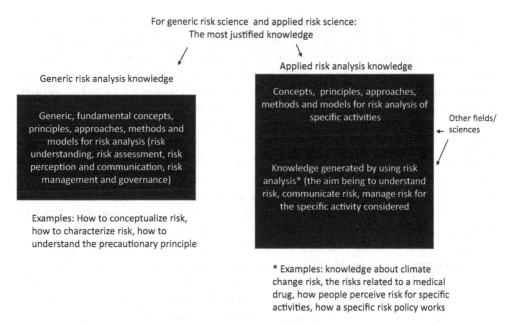

For generic risk science and applied risk science:
The most justified knowledge

Generic risk analysis knowledge

Generic, fundamental concepts,
principles, approaches, methods and
models for risk analysis (risk
understanding, risk assessment, risk
perception and communication, risk
management and governance)

Examples: How to conceptualize risk,
how to characterize risk, how to
understand the precautionary principle

Applied risk analysis knowledge

Concepts, principles, approaches,
methods and models for risk analysis of
specific activities

Other fields/
sciences

Knowledge generated by using risk
analysis* (the aim being to understand
risk, communicate risk, manage risk for
the specific activity considered

* Examples: knowledge about climate
change risk, the risks related to a medical
drug, how people perceive risk for specific
activities, how a specific risk policy works

FIGURE B.2 The difference between generic and applied risk analysis (science)

approaches, methods and models for understanding, assessing, characterizing, communicating, managing and governing risk. Research aiming at improving the way we
should describe uncertainties in risk assessments is an example of generic risk knowledge
generation. The development of the SRA documents referred to previously (SRA 2015,
2017a, 2017b) is another example. If we review risk analysis journals, we observe that
a considerable proportion of the scientific papers belong to the B type of contributions.

Applied risk analysis supports risk knowledge generation and communication in
relation to specific activities and supports the tackling of specific risk problems or
issues. For example, to gain knowledge about the risk related to climate change or the
operation of a concrete process plant, risk assessments offer applied risk analysis. The
process of establishing suitable concepts, principles, approaches, methods and models
for this application is also a part of applied risk analysis.

Based on the concepts of generic and applied risk analysis, a distinction can also
be made between generic and applied risk science, see Figure B.2. There could be many
relevant fields for applied risk analysis, for example, if the problem is climate change
risk, the key sciences are the natural sciences – the analysis is a multidisciplinary or
interdisciplinary process. There are many ways of classifying sciences, but again following Hansson (2013), we can distinguish between five main categories:

1   nature (natural science),
2   ourselves (psychology and medicine),
3   our societies (social sciences)
4   our own physical constructions (technology, engineering)
5   our own mental constructions (linguistics, mathematics, philosophy).

The generic, fundamental risk analysis B is founded on 5. Concepts, principles, approaches, methods and models are studied in order to understand, assess, describe, communicate, manage and govern risk. Generic risk analysis uses different types of applications to illustrate generic knowledge, but it is not dependent on one specific application to be justified. The applied risk analysis A relates to all five categories, 1–5 – it is supported by the generic risk analysis knowledge 5.

A risk assessment of a specific activity, for example, a risk assessment of a process plant, may produce new knowledge about the activity – for example, expressing that the plant is subject to considerable risks and as such providing useful decision support. To varying degrees, depending on the scope and originality of the study, such a risk assessment may also give new insights on concepts, principles, approaches, methods and models for understanding, assessing, characterizing, communicating, managing and governing risk and as such contribute to developments in B.

There is and should be a strong interaction between the applied and generic activities. Developments in B could influence practices in A, and experiences from A can lead to new research and developments in B. The authors of the present book can refer to many examples of this, for example, how industrial work on risk characterizations has led to generic research on that topic, see, for example, Aven and Reniers (2013).

Other related fields and sciences can be defined as previously for risk analysis, for safety, reliability, resilience and uncertainty, replacing 'risk' with one of these concepts. However, adopting a broad understanding of risk as in the SRA (2015) glossary and this book, all these fields (sciences) can be seen as parts of the risk field (science); refer to discussions in Aven (2020a). To justify a distinct science, it needs to be supported by a sufficiently strong basis of study programs and research. Only if risk analysis is interpreted as broad, as presented here, can we find risk science to meet this requirement and to be robust. Highlighting the generic topics of risk analysis (B) is also an argument for ensuring robustness, as it is relevant for different types of applications.

Analogous to risk science, we define statistical science as the practice that provides the most epistemically warranted or justified statements or beliefs that can be produced at the time being on the subject matter (scope) covered by the statistical field, which captures the collection, analysis, presentation and interpretation of data (Aven 2018a, 2020a). As for risk, we can distinguish between applied and generic (fundamental) statistics. Risk science uses statistics but covers many topics not addressed in statistics, for example, on how to use the precautionary principle. Neither statistics nor any other science can replace risk science, as risk science is the only one with a scope as defined here.

> **REFLECTION**
>
> Mathematics is difficult to define but is a science as defined previously, covering the most justified knowledge produced by the math field or discipline. It is commonly referred to 'pure mathematics'. What is that?
>
> *It is generic, fundamental mathematics, relevant for different types of applications (business, engineering, etc.). As for statistics and risk analysis, we can distinguish between a generic and an applied math science.*

Research methods

To obtain knowledge about the risk related to an activity, we perform testing (if possible) and analysis. A probability model is established, and by observations, we are able to accurately estimate the distribution. Most textbooks on risk show how to conduct such analyses. We are applying the 'scientific method' (also referred to as the 'hypothetico-deductive method'). It typically has the following four steps (Wolfs 1996; Wolf 2018): 1) observations and descriptions of a phenomenon; 2) formulation of a hypothesis to explain the phenomenon, for example, using a mathematical relationship; 3) use of the hypothesis to predict the existence of other phenomena or to predict the results of new observations; and 4) performance of experimental tests to verify or falsify the hypothesis. For the risk setting studied here, the hypothesis in 2) is formulated using a probability model, and, in 3), based on the observations, this model is used to make predictions for a new activity. If the data show that the model is inaccurate, model changes are needed, and the analysis process repeats. The common framework for carrying out this method is statistical inference.

The scientific method is a cornerstone in the natural sciences whose scope covers the study of the physical world and its phenomena, such as physics and biology. The method is also used to describe how people perceive risk; refer to Chapter 6. It is essential to make a clear distinction between the science and the scientific method applied to obtain knowledge within that science. Other methods also exist to gain knowledge and scientific knowledge. 'Conceptual research' is an example and plays an important role in risk science. An historical example of such research is the development of the risk concept – with its characterizations – used in risk science, as thoroughly discussed in this book. A substantial volume of research has been conducted to develop the theories, as demonstrated by the many scientific papers published on these topics. 'The scientific method' is not relevant. The conceptual research is about developing suitable concepts, principles, approaches, methods and models, to a large extent based on reasoning. There is always an element of empiricism – we observe how different risk definitions work in practice – but the research is mainly about analysis and argumentation. For the risk concept example, a challenge was to generalize the idea that risk is seen as an expected value or a probability distribution. How should this be done? Alternative ideas were presented and formalized through the introduction of new concepts and theories. Their strengths and weaknesses were discussed and conclusions made.

The research typically covers one or more of the following elements: identification (for example, clarifying which features that should be included in a definition of risk), revision (for example, modifying current definitions to better reflect uncertainties), delineation (for example, focusing on some type of activities), summarization (to see the wood for the trees, for example, allowing for only a finite set of events or outcomes), differentiation (for example, differentiating between different categories of activities), advocating (for example, argumentation to justify or support a specific formulation of risk) and refuting (for example, argumentation aimed at rebutting a given perspective) (MacInnis 2011; Aven 2020a). The quality of conceptual research is evaluated in the same way as other types of research: by references to criteria such as *exposition* (conceptual clarity and internal consistency), *theory building* (e.g., precision and rationale), *innovativeness, potential impact* and *validity* (reflecting the degree to which one is able to conceptualize what one would like to conceptualize) (Yadav 2010; Aven 2020a). It is also common to refer to the following four criteria: originality, solidness, relevancy and usefulness (see e.g., Aven and Heide 2009). Scientific work is also characterized by a set of general norms and standards, such as the four institutional imperatives: universalism, communality, disinterestedness, and organized skepticism (Merton 1973; Hansson and Aven 2014). Research meeting these criteria and published in well-recognized scientific journals is today broadly acknowledged as scientific.

The methods for deriving scientific knowledge are continuously developing. This is a feature that characterizes science. A major strength of science is its capability of self-improvement (Hansson and Aven 2014). A method-based delimitation of science can only have temporary validity, as discussed by Hansson (2013). A famous example of a 'method-founded science' is Karl Popper's falsifiability criterion, according to which "Statements or systems of statements, in order to be ranked as scientific, must be capable of conflicting with possible, or conceivable observations" (Popper 1962, p. 39). The idea is that a statement is scientific if it could, at least in principle, be falsified by some data.

This criterion, as all such criteria, has severe problems, as thoroughly discussed in the literature. Most of them are suitable only for some, not all, of the science disciplines, and all of them make the science of previous centuries unscientific, although it was the best of its day.

The most justified knowledge

As mentioned, risk science is defined as the practice that provides us with the most epistemically warranted or justified statements or beliefs that can be produced at the time being on the subject matter (scope) covered by the risk field. A number of papers have been produced on risk, but what are the most justified concepts, principles, approaches, methods and models of the field?

In some cases, the answer is clear; in others, it is not. Which is the best is contested. Science is characterized by a continuous 'battle' on what these statements and beliefs are – it is about institutions and power. Different schools of thought argue for their beliefs and try to influence and control the field (Bourdieu and Wacquant 1992). It is the same for risk science. For example, for years, we experienced an intense discussion about the suitability of the Bayesian perspective (see, e.g., Easterling 1972; Bernardo and Smith 1994; Lindley 1985, 2000; Aven 2020c); see also Problem B.1. Argumentation was provided for why this perspective was unscientific, the main problem being the use of subjective probabilities. However, the Bayesian advocators rejected this view. Instead, they highlighted the need to use all relevant information and knowledge to adequately support the decision-making. A different perspective on science was required.

For any field and science to develop, it is essential that there be continuous questioning and scrutiny of the current concepts, principles, approaches, methods and models. Critique is a cornerstone of the scientific system. However, at the same time, any field and science needs to clarify what is its core knowledge today. This amounts to concepts, principles, approaches, methods and models. If we look at the present risk field, it is not difficult to identify a number of topics for which there is broad consensus about belonging to the core subjects of this field. Both the traditional statistical approach and the Bayesian approach are used to analyze risk. Through integrative thinking, the scientific 'battle' about Bayesian ideas and methods has led to new insights and a broader set of instruments for understanding and analyzing risk. An integrative process is a form of thinking which reflects a strong

> ability to face constructively the tension of opposing ideas and instead of choosing one at the expense of the other, generate a creative resolution of the tension in the form of a new idea that contains elements of the opposing ideas but is superior to each.
>
> (Martin 2009, p. 15)

In this particular case, there were different perspectives on how to approach and analyze data, which can be considered to create tension. However, integrative thinking makes the analysts see beyond these perspectives – it utilizes the opposing ideas to obtain a new and higher level of understanding.

Despite broad agreement on many areas, it is not difficult to point to issues where there are discussions on what are the most warranted statements and justified beliefs concerning concepts, principles, approaches, methods and models. An example is how to express uncertainties in relation to risk characterizations. For many scholars, probability is the answer, whereas others argue for the need to use alternative approaches. Different 'schools' have been developed with rather limited interactions and communication. However, as for most situations, when this type of disconnection occurs, it is important to seek arenas for dialogue where relevant views and assumptions made are discussed. The result is often new knowledge and improved concepts, approaches, methods and models. An example is the work reported in Aven et al. (2014), in which, over a five-year period, the authors studied different perspectives on how to best treat uncertainties in a risk context. The work did not result in full consensus on the preferred approach in all types of situations, but it clarified what the different perspectives mean, their strengths and weaknesses and how they are to be used in practice. Then it is up to the analyst to choose an approach in a specific case, considering the arguments provided for the alternatives available. Our studies on this topic have shown that many risk researchers are not familiar with the concept of subjective (knowledge-based, judgmental) probability to describe uncertainties. They apply an alternative approach by arguing that, because of uncertainties, probability cannot be quantified. However, a subjective probability can always be quantified. The problem is that the knowledge supporting it can be more or less strong. Addressing this knowledge and its strength, and allowing for imprecise probabilities in addition to precise probabilities, a practical framework for uncertainty descriptions can be made as described in this book (Chapter 3). The risk science develops this type of knowledge, building a knowledge base of the most justified beliefs of the field.

B.2 TAXONOMY OF RISK SCIENCE PROGRAMS

Risk science is both a generic discipline and a topic area component of specific domain areas, such as healthcare, business and policy. In practice, there is a mix of purely risk science curriculums and a wide variety of ways in which universities are incorporating risk science topics into domain area curriculums. The following presentation is based on Thekdi and Aven (2021b).

It can be argued that the boundaries between application areas and risk science are blurry; some concepts and skills can be classified as both within the realm of risk science and application area. This can be seen as a strength of risk science as a discipline because it contains elements that are founded and well accepted in other disciplines. Risk science, like any other discipline should not be siloed but instead find strength in being accepted and embraced by other disciplines. Conversely, these other disciplines can benefit from studying generalizable or generic features of risk science. Developing a common language of risk science provides a systematic way of describing terminologies and practices that exist within various application areas, allowing for learning and sharing among application areas to explore the wide range in which entities interpret and practice risk. This serves as a method for bridging gaps in interpretations and practices with the intent to improve and grow all involved fields.

While acknowledging the different perspectives, there remains a need to classify how various programs approach this issue. This classification must acknowledge that risk knowledge generation in practice and tackling real risk problems are and will remain an interdisciplinary field that is based on a set of core topic areas that are adapted to the various application areas. This section will describe an ontology chart for understanding the relationship between risk topics and application areas and also describe an approach to creating a taxonomy of those programs.

Section B.2.1 proposes an ontology for understanding the relationships among risk programs and topic areas. Section B.2.2 proposes a classification scheme for risk programs. Section B.2.3 provides an overall assessment of risk-science related programs.

B.2.1 Ontology for risk science topics in university programs

Figure B.3 provides an ontology chart for risk science topics within university programs, in line with the terminology introduced in Section B.1. Academic programs can be viewed as being either dedicated risk science programs or programs in non-risk domains. Non-risk domains encompass areas such as healthcare, business and policy. Risk science programs are dedicated to risk science topics: *Concepts, principles, approaches, methods and models for understanding, assessing, characterizing, communicating, managing and governing risk* (SRA 2017a; Hansson and Aven 2014; Aven 2018a, 2020a). For short, we refer to the italic text in the previous sentence as 'concepts etc. for risk analysis'.

The application of risk science concepts can be tailored to fit the individual needs of the various non-risk domains. For example, in a healthcare setting, issues surrounding risk assessment may be more important compared to other risk science topic areas.

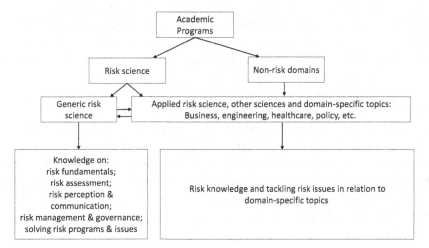

FIGURE B.3 Ontology chart for risk science topics within university programs (based on Thekdi and Aven 2021b)

B.2.2 Classification of single risk programs

To develop a classification system for risk-related programs, it is evident from the discussion previously that we need to distinguish between generic and applied studies. Is the main aim to learn about risk concepts, principles and methods as such, or is it to learn about engineering, medicine or finance? In practice, there could be varying degrees of generality, for example, a program could be generic in health risk analysis but still not generic when considering all types of applications. Second, the study programs could to a varying degree cover the different topics of risk science, such as risk assessment, risk perception and communication and risk management and governance.

On this basis, a taxonomy for classification for both risk science and non-risk domain programs is developed, as shown in Tables B.1 and B.2. While the taxonomy described in these tables is applied to hypothetical university programs, the intent is for the taxonomy to be applied to a variety of individual programs for several purposes. First, this taxonomy would allow for a structured understanding of how risk topics fit into individual programs, to understand questions such as: 1) What risk topics are necessary for particular application areas? 2) What risk topics are specifically not included in the curriculum? Why? 3) If in-depth knowledge of particular risk topics is not in necessary for a program, is there a need for general awareness of these risk topics? 4) Are students able to gain understanding of non-covered risk topics outside of the academic program, for example, during internships or licensing exams? Second, this taxonomy could be used to provide a collective assessment of focus areas of programs existing across universities. This type of analysis could start to develop a status quo for various types of academic programs and help provide guidance for identifying curriculum gaps, potential topic area redundancies or potential opportunities for cross-program collaborations.

Table B.1 represents a program highlighting applied risk assessment and management topics related to engineering, environmental and security issues. This type of program does not require coverage of all broad risk science topics but instead focuses on the most relevant aspects of risk assessment and management that can be applied to engineering and security. Table B.2 represents a program focusing on generic, fundamental risk science topics. This generic program covers all broad risk science topics in detail but does not cover domain knowledge related to application areas. Different types of application examples are used to illustrate the concepts, principles, frameworks, approaches, methods and models discussed, but the main focus is on the risk science knowledge.

Broad categories of application areas are used in Tables B.1 and B.2. Depending on the purpose, there could be a need for refinements. For example, the category business/finance/insurance covers a number of different types of applications, yet there is a common basis, with students typically recruited from the same types of institutions and schools, using rather similar curriculums on basic subjects. The risk science categories follow the scheme developed by SRA in the *Core Subjects of Risk Analysis* document (SRA 2017a) mentioned previously.

B.2.3 An overall review of risk-related programs

This section provides an overall assessment of risk-related programs, using the classification system developed in the previous section as a way of structuring the programs. The assessment is informed by a data collection exercise that involves study programs being offered worldwide on risk science–related topics at our colleges and universities; see Thekdi and Aven (2021b). The review is limited to a sample of English language–taught programs that are listed on the Society for Risk Analysis website.

The review shows that risk-related topics are included in a number of study programs in disciplines like those referred to as applied in Tables B.1 and B.2 (engineering, business, etc.). The level of risk-related training ranges from discussion of risk science topics in specific courses and dedicated risk science courses (for example, on risk assessment), to full and comprehensive risk science programs, for example, a master degree program in risk analysis and management. To illustrate this range, consider studies in risk in an engineering master degree program, as shown in Table B.3. Consider a program in civil engineering: most programs in this field have modules on risk, safety and reliability, covering topics like risk measurements, structural reliability analysis and Bayesian updating. The teaching and training material may

TABLE B.1 Taxonomy for classifying risk science programs. A hypothetical university program indicated highlighting applied risk assessment and management topics related to engineering and security issues (based on Thekdi and Aven 2021b)

		Risk topic areas				
	Overall	Risk fundamentals (concepts)	Risk assessment	Risk perception and communication	Risk management and governance	Solving risk problems and issues
Risk science	Generic					
	Applied ✔		✔		✔	✔
Application areas	Engineering/ technology		✔		✔	✔
	Business/finance/ insurance					
	Medicine/healthcare/ toxicology					
	Environment/ecology/ climate change		✔		✔	✔
	Security		✔		✔	✔
	Policy					
	Law					
	Food					
	Transportation					

TABLE B.2 Taxonomy for classifying risk science programs. A hypothetical university program indicated highlighting generic risk science topics (based on Thekdi and Aven 2021b)

| | Overall | Risk fundamentals (concepts) | Risk topic areas | | | |
			Risk assessment	Risk perception and communication	Risk management and governance	Solving risk problems and issues
Risk science						
	Generic ✔	✔	✔	✔	✔	✔
	Applied					
Application areas						
	Engineering/technology					
	Business/finance/insurance					
	Medicine/healthcare/toxicology					
	Environment/ecology/climate change					
	Security					
	Policy					
	Law					
	Food					
	Transportation					

TABLE B.3 An illustration of the different levels of risk science training in relation to a master engineering program

Applied: Master Degree Engineering			
Risk science topics addressed in courses but not dedicated risk science courses	A dedicated risk science course	A specialization on risk within the engineering program, for example, covering 30 credits on risk topics and the master thesis risk related	A full master program on risk-related topics

be more or less tailor made for the civil engineering students. Commonly, the risk-related subjects are offered as generic courses, relevant for different types of engineering students. We also see programs offered having several courses on risk, forming a specialization in risk science within the engineering program. Finally, there are full engineering master programs in risk-related topics. The title of the program relates to risk or related terms like safety, reliability and resilience. Examples include master programs in, for example, risk assessment, risk management, risk management policy, risk communication and risk – reliability – resilience. Many of these programs are based on degrees in engineering or business (finance), but the review shows that there are a large number of courses on different types of risk science topics for various applications, including those mentioned in Tables B.1 and B.2. There exist, however, very few programs with a scope covering generic risk science, including all its main topics, as outlined in Table B.2.

There is wide variability in master programs related to risk, with durations ranging from one to two years. The shortest typically highlight specific applications of risk science. These programs are typically designed for candidates who have work experience and would like to develop expertise and leadership in risk-related topics. The longest are typically more scientific oriented with a stronger generic risk science component, which prepares students for risk-related PhD programs.

Programs at the bachelor degree level allow students to combine domain-level competence (e.g., engineering, business, etc.) with risk knowledge to become skilled in applied risk science. Most risk programs are, however, on the master level. There are some, but not many, dedicated risk science programs on the PhD level. The large majority of PhD programs with risk-related topics are applied, with domains such as those listed in Tables B.1 and B.2.

Very few universities and colleges refer to risk science in relation to their programs and courses. The various risk-related topics (e.g., risk assessment, risk management, etc.) are often taught within the context of the program-related applications (e.g., engineering, transportation, etc.). The studies are often referred to as multidisciplinary or interdisciplinary. Students can design their curriculum by cutting across multiple disciplines. When risk science is specifically mentioned, it is commonly linked to disaster risk science, health and the environment. However, we also see some examples of more broad risk science programs covering many or all of the topics referred to in Tables B.1 and B.2, but the number of such programs is small. It increases if we allow for programs on related topics like safety science. Here quite a few programs exist, but they are also usually linked to specific applications.

B.2.4 Discussion

To further develop risk science education, it is essential to acknowledge the need for both generic and applied risk science. From a societal perspective, it can be argued that the main driver for further enhancements should be applied risk science, to understand and solve real issues and problems. However, in order to do this, we need to develop the most suitable concepts, principles and methods. This cannot be obtained by limiting the risk science to applications, as such an approach would not sufficiently stimulate learning across applications. The advantage of having a generic risk science is that it builds on the knowledge gained in all types of applications and seeks to further develop this knowledge to the benefits for all application areas. 'Risk' is a generic concept; it is not linked to a specific application. Its understanding and use can therefore be studied without linking it to concrete activities. Insights can be obtained through applications, which can then be included in the total knowledge basis of risk science.

Consequently, there is no clear one-size-fits-all approach for the dissemination of risk science knowledge. Each risk program is different, as the taxonomies in the previous section demonstrated. The heterogeneity of university programs should be seen as a strength because universities are able to adapt the programs and curriculums for their own individual needs. For example, universities may gain their funding from particular industries, public sector organizations, non-profits and so on. Funding opportunities can dictate the availability of domain areas for study, the demand for related academic programs and also the training needed for students seeking employment in the region. The individual needs of various domain areas can also indicate the relative importance of various risk science topics. For example, policy programs might see a higher need to cover topics such as risk management and governance. These same programs might see less relative importance for in-depth coverage of risk assessment topics, as those types of activities may be more guided by analysts and other experts. Refer to discussion in Section 10.1.

The need for both a generic and applied science is acknowledged in many other areas. Think about applied and generic and fundamental mathematics or statistics. Thousands of professors worldwide are researching and teaching generic mathematics and statistics, and many study programs are mainly generic. The generic part is acknowledged for its importance for the applications but also that it has a value in itself. For example, the Abel (2020) prize is established to award outstanding scientific work in the field of mathematics within the scope of generic mathematics.

The development of statistics as a discipline and the widespread adoption of statistical theories and approaches is not known to be instantaneous. Consider the fact that Bayes' theorem was first published in 1763, and now the fundamental concepts continue to rapidly gain popularity and acceptance 250 years later (Bayes 1763; Murphy 2012). In fact, the growth of statistics can be attributed to several factors: 1) the growth in computational power and availability of data; 2) the need for statistical modeling for needs within various disciplines; 3) the growth of machine learning and related data science fields; and 4) comfort with and acceptance of statistical thinking, such as with aspects of variability, sampling and statistical significance. This comfort with statistical thinking may be rooted in general familiarity with statistics, potentially beginning from exposure in an undergraduate curriculum and also exposure to these topics in highly visible applications. One can even think of knowledge trends as being generational, such that features of academic contributions and the state of the art become more accepted over time and benefit from greater exposure with junior academics (students) who progress through graduate education, practice and research endeavors.

Currently, risk science does not have the broad acceptability that exists for mathematics and statistics. There are few generic risk science programs offered today worldwide and also a very small number of professors of generic risk science or generic risk science topics, if we compare with mathematics or statistics. Also, many research funding schemes do not include risk as a distinct category (Hansson and Aven 2014). On the other hand, risk science can refer to many high-quality journals and conferences, and different types of societies are actively engaged in enhancing this science.

Similarly to statistics, the risk science discipline can grow and adapt to current and future needs. There are signals indicating that the demand will grow in the fashion of statistics. These signals include the incorporation of risk principles within accreditation standards (see Thekdi and Aven 2021b), as well as the widespread visibility and concern over risk-related issues, such as climate change, cybersecurity events, pandemics, natural disasters and so on. As seen with statistics, the adoption of risk science will continue to grow as students from these academic programs are exposed to risk concepts and practice in industry and conduct their own academic research.

B.3 PROBLEMS

Problem B.1

Consider the work by Kaplan and Garrick (1981). Is it generic (fundamental) risk science or applied risk science?

Problem B.2

One of the oldest tensions in statistics – as well as in risk science – is the controversy between frequentists and Bayesians. The tension reflects two different philosophies and approaches to probabilistic thinking and statistical inference. Review what this controversy is about. Has the debate between these two groups faded out, or can we still talk about a serious tension or controversy?

Problem B.3

Another tension is the controversy between advocates of probability to assess uncertainties in risk assessment and advocates of non-probabilistic approaches. Review what this controversy is about. Has the debate between these two groups faded out, or can we still talk about a serious tension or controversy?

Problem B.4

As a third tension, we refer to scientific controversy between risk science and a specific perspective on resilience. Risk and resilience are both terms with a long history, but the respective sciences are not more than some decades old. These concepts and sciences are linked, but how they are related and should be related are strongly debated; see, for example, Linkov et al. (2014), Aven (2019c), Linkov and Trump (2019). Whereas risk according to risk science (Kaplan and Garrick 1981; SRA 2015; Aven 2020a) typically covers events, their consequences and uncertainty/probability, the previous references on resilience limit risk up to the point in time of the event. For example, from Linkov et al. (2014) we read: "Risk management helps the system prepare and plan for adverse events, whereas resilience management goes further by integrating the temporal capacity of a system to absorb and recover from adverse events, and then adapt". Discuss to what degree this tension is about terminology or more fundamental differences in thinking and practical approaches for how to conduct analyses and manage relevant situations.

Problem B.5

Review a risk-related study program at a university, such as yours. Apply the ontology and taxonomy introduced in Section B.2.

Outline of solutions to problems

Problem 2.1

A person, Tom, walks underneath a boulder that may or may not dislodge from a ledge and hit him.

a   Risk is understood as the potential for the boulder to dislodge from the ledge and hit Tom, with the result that he is injured and killed. Risk can be formalized as (A,C,U), where A is the event that the boulder dislodges from the ledge, C the consequences for Tom if the boulder dislodges from the ledge and U the associated uncertainties (will A occur, and what will be the effects C?).

b   Vulnerability is the potential for Tom to be injured or killed given that the boulder dislodges from the ledge or alternatively given that the boulder dislodges from the ledge and hits Tom. Formalized vulnerability takes the form (C,U|A), where A is one of these two events and C the consequences for Tom.

c   The boulder could dislodge from natural reasons (safety) or because of people intentionally pushing the boulder so that it dislodges from the ledge (security).

d   The boulder can be referred to as a risk source.

e   We consider all of these risk statements consistent with the definition of risk provided in Section 2.1.

Problem 2.2

For Sara, who invests $1 million in a project, risk is the potential for some losses. Risk is conceptualized as (C,U), where C is the investment losses (or win) and U the associated uncertainties. The consequences from the investment are considered for a period of, say, one week.

Problem 2.3

The human body is commonly considered resilient "in its ability to persevere through infections or trauma. Even through severe disease, critical life functions are sustained and the body recovers, often adapting by developing immunity to further attacks of the same type" (Linkov et al. 2014). An example of a critical life function is breathing. However, the human body can also be considered vulnerable – history has shown that, if medical advances like penicillin had not been made, the consequences of some bacterial infections would have been devastating. This may suggest a characterization of the human body as quite resilient but not highly resilient. See discussions in Section 5.6.

Problem 2.4

That the food is safe means that the risks related to eating the food are acceptable or tolerable. See also Section 3.2.2.

Problem 2.5

In security contexts, it is common to refer to the triplet: value, threat and vulnerability (Amundrud et al. 2017). This perspective is included in the Chapter 2 framework. The values are identified, and the consequences C of the events A (including risk sources) relate to these values. The threats T are defined as such events A, and the uncertainty U associated with the occurrence of the threats is addressed. Given the occurrence of a threat, we look into the consequences, together with the associated uncertainties, which are referred to as the vulnerability.

Problem 2.6

As the concept of sustainability relates to meeting the needs of the present without compromising the ability of future generations to meet their needs, we can define a reference value r given by the objective 'meeting the needs of the present without compromising the ability of future generations to meet their needs'. Risk is then related to deviations from this goal and related uncertainties. Another perspective is to consider sustainability the same as sustainable (when, for example, saying sustainability is achieved), meaning an acceptable or tolerable risk. We see the similarity between sustainability and safety. We can consider the term *sustainability* a special case of safety when the reference value is 'meeting the needs of the present without compromising the ability of future generations to meet their needs'.

Problem 2.7

Risk is understood in many ways in the scientific and popular literature. Many of these definitions are considered ways of measuring or describing risk and are discussed in Chapter 2. We often refer to these definitions as risk metrics. These metrics are to varying degrees able to reflect all aspects of risk. If we try to define risk using one specific metric, it would not be possible to agree, as no measurement would be suitable for all types of situations. However, if we focus on the ideas that the concept of risk is to convey, we can obtain broad consensus, as shown by the SRA Glossary (SRA 2015).

Here are some other definitions of risk that have been suggested (Aven 2012a).

RISK IS AN EVENT

A definition of risk often referred to is: "Risk is a situation or event where something of human value (including humans themselves) is at stake and where the outcome is uncertain" (Rosa 1998, 2003). Comparing this definition to the (A,C,U) risk concept, risk is here understood as the event A. A classical example is lung cancer being the risk for a person, which is in this book is considered a risk source or event (hazard). If, on the other hand, we refer to the 'situation' in this definition, smoking can be considered a risk, which in our setup can be viewed as the activity leading to risk. The definition is thoroughly discussed by Aven et al. (2011) and Aven (2014). To us, it is more natural to refer to risk as something 'produced' by an activity than risk being the activity in itself. Also, if risk refers to the event (for example, lung cancer), conceptual difficulties arise. We cannot conclude, for example, about the risk being high or low or compare different options with respect to risk. For example, it does not make sense to say that the event is large.

RISK IS UNCERTAINTY AND OBJECTIVE UNCERTAINTY

In economic applications, and in particular investment analysis, it is common to refer to risk as uncertainty. The uncertainty is typically seen in relation to a reference value, for instance, a historical average value for similar investments. Risk captures a potential deviation relative to this level. Without such a reference level, this way of understanding risk does not work. Uncertainty seen in isolation from the consequences and the severity of the consequences is not meaningful as a general definition of risk. Only if the potential outcomes are large/severe in some respect do large uncertainties need attention.

A closely related definition of risk is that of Knight (1921), which is commonly referenced by economists. The basic idea is that 'risk' is present when an objective probability distribution can be obtained. Otherwise, we face 'uncertainty'. This means that risk becomes a very narrow concept and not suitable for any of the examples discussed in Chapter 2. It is a definition conflicting with daily language. See discussion in Aven (2010, p. 75).

RISK IS THE SAME AS RISK PERCEPTION (REFER TO PROBLEM 6.2)

Risk research also refers to risk as the same as risk perception (Jasanoff 1999, critical comments in Rosa 1998; Douglas and Wildavsky 1982; Beck 1992). However, such a perspective is not in line with contemporary risk science, which distinguishes between the risk concept and how it is described. In contrast to risk perception, a professional judgment of risk does not reflect perceptional aspects like fear and dread, nor how one likes or dislikes features of C and U. Risk perception also commonly includes judgments of the acceptability of risk, which is not a part of a professional measurement or description of risk. See Chapter 6.

Problem 2.8

The student faces different types of risks, which can all be understood using the (C,U) and (A,C,U) formulations. Here are some examples:

- Academic performance: Risk is the potential for undesirable grades or not qualifying for prestigious graduate programs or jobs. The consequences C is defined as the grades earned and U the associated uncertainties, or alternatively as meeting or not the goals of qualifying for prestigious graduate programs or jobs.
- Financial security: Risk is the potential for financial problems. The consequences C is defined as the available economic funds and income, and U the associated uncertainties.
- Physical, mental and emotional wellness: Risk is the potential for issues linked to wellness. The consequences C is defined as a degree of wellness and U the associated uncertainties.
- Sleep: Risk is the potential for sleeping problems. The consequences C is defined as the number of hours sleeping per night and U the associated uncertainties.

As a concrete example, think about the risk related to academic performance for a student called Tom. He is active in sports, in particular soccer. Then we may introduce A as the event 'injury' in the (A,C,U) formulations, with C as the performance given that this event occurs. The focus is on performance risk in relation to sports activities and potential related injuries. Other examples of events A for this example could be 'financial insecurity' or 'sleeping problems'.

Problem 2.9

There are a number of risks related to this example. The key one for TBM is the potential for economic losses. The consequences C is defined as the profit and U the associated uncertainties. Many of the issues addressed in Section 1.2 can be considered risk sources (also opportunities), for example, the gig economy, self-driving vehicles, sustainability initiatives, electronic materials, subscriptions, rising cost of packaging for delivery and fluctuating shopping habits.

Problem 2.10

Examples of risk is in the virus example of Section 1.3 are:

- Public health: Risk is the potential for undesirable health consequences of a virus outbreak. The consequences C is defined by reference to loss of lives and well-being, with associated uncertainties. See Section 2.2.3.

- Individual health risk: Risk is the potential for undesirable health consequences of a virus outbreak for the individual. The consequences C is defined by reference to loss of life and well-being, with associated uncertainties.
- Economic risks: Risk is the potential for undesirable economic consequences of a virus outbreak in society. The consequences C is defined by reference to the economic loss, with associated uncertainties. See Section 2.2.3.
- Individual economic risks: Risk is the potential for undesirable economic consequences of a virus outbreak for the individual. The consequences C is defined by reference to the economic loss for the individual, with associated uncertainties.

Problem 2.11

She is vulnerable in the sense that her activity (with its stressors) often leads to injuries, and the fact that she is often injured can lead to fewer sponsors and reduced economic support. The vulnerability when injured is not only about returning quickly to a normal state (resilience) but possibly also about other consequences (here economic).

Problem 2.12

The activity considered is the recovery phase. We define a reference value r as successful recovery suitably defined also reflecting time, for example, an objective of obtaining a normal state within six months. The risk is then related to failures according to this objective and reference. Risk can be written formally as (D,U), as introduced in Section 2.1, with D expressing deviations from this reference and U the associated uncertainties.

Problem 3.1

p = the fraction of time the event 'C = 1' occurs if we could repeat the situation considered over and over again (i.e., throw the die over and over again) infinitely under the same conditions.

We assign P(p = 1/6) and P(p = 1/3) (where the sum is 1), for example, 0.5 for each. These probabilities are conditional on the assessor's knowledge K.

Problem 3.2

Two interpretations:

- The person estimates $P_f(A)$ to be 3/10
- The person's knowledge-based probability is 3/10.

Problem 3.3

Here f(y) is the fraction of times that Y = y if the experiment can be repeated over and over again infinitely. $E_f[Y]$ is the average value of Y if the experiment can be repeated over and over again infinitely. $E_f[Y]$ estimate: $0 \cdot 0.80 + 1 \cdot 0.10 + 2 \cdot 0.05 + 3 \cdot 0.03 + 4 \cdot 0.02 = 0.37$.

Problem 3.4

Poisson model $f(x) = P_f(X = x) = \lambda^x e^{-\lambda}/x!$, $x = 0, 1, 2, \ldots$, where $\lambda = E_f[Y]$ is the parameter of the model.

The variation in events per day is described by $f(x)$: $f(0)$ is the fraction of days with 0 events, $f(1)$ is the fraction of days with 1 events and so on.

$D = f(x) - g(x)$, which is the difference between the model (frequentist probability) and the actual observations. Hence, it makes sense to refer to D as a model error. Since g is unknown, D is unknown, and we are facing uncertainties about D. It can be referred to as model uncertainty.

Problem 3.5

$P(\lambda = 1) = 0.5$, $P(\lambda = 2) = 0.5$; P is a subjective probability.

Problem 3.6

Meaning of $P(N = 1) = 0.10$, for example: The uncertainty (degree of belief) is the same as drawing a particular ball out of an urn that contains ten balls (under the standard conditions).

$E[N] = 0.37$, the centre of gravity of the uncertainty distribution. It is not uncertain but depends on the background knowledge K.

Problem 3.7

Interval set: $\{0,1,2\}$. The assessor is 95% certain that N is in this interval – where certain is with respect to the relevant subjective/knowledge-based probability.

Problem 3.8

$f(x)$ is the fraction of times that $X = x$ if the experiment can be repeated over and over again infinitely. A uniform distribution over $[0,1]$. Hence $P(p \leq p') = p'$, for p' in the interval $[0,1]$.

Problem 3.9

$FAR = (10^{-4}/10^3)10^8 = 10$. First we calculate the expected number of fatalities per hour exposed, which is $10^{-4}/10^3$. Then we multiply by 10^8, as FAR is the expected number of fatalities per 100 million exposed hours.

IR: probability that a specific person is killed in a year = $PLL/10 = 1/10,000$.

Problem 3.10

$E[C'|A']$ could be a poor representation of the actual consequences, such as

- The centre of gravity of the distribution could deviate strongly from the actual value.

- The probabilities and expected values could be based on more or less strong background knowledge, but this is not reflected in the placing in the risk matrix.

If the event A' is a pandemic, there is a specter of potential consequences if this event occurs, which is not reflected by $E[C'|A']$. Unconditionally, the specter is even bigger, as it also covers the non-occurrence of the pandemic.

Problem 3.11

The strength of knowledge supporting the probabilities could be completely different. This aspect should also be seen as an aspect of risk.

Problem 3.12

Typically, the consequences C' is specified as a result of a choice of the analysts to focus on particular quantities, such as the number of serious injuries or loss of lives. Then (C',P) can be seen as a probability distribution with related metrics based on this distribution. Hence, the background knowledge is the same for both P and (C',P). However, there could be situations where there is a difference. Suppose the analysts seek to describe all relevant hazardous events related to an activity subject to large uncertainties. A key focus is on the events A'. Hence, it could be important to distinguish between the knowledge supporting P(A') for defined events A' and the knowledge supporting the identification of events A'. We may, for example, have a case where the knowledge supporting the latter is rather weak, whereas the knowledge supporting the former is rather strong for the events A' considered.

Problem 3.13

In order to understand what p means, we need to introduce a mental construction. We have to contemplate an infinite population of similar situations to the one considered: p is the fraction of situations for which the event occurs. In this case, this population relates to similar car trip from Paris to Rome. This is a thought construction, a model of the real world. To define the population, we need to specify a set of frame conditions, for example, what type of cars we are considering, what time periods we are considering and what type of drivers. Is the population composed of all types of cars or only those similar to Tim's car. And what is similar? Tim may have a very good car, drive cautiously and not use drugs and alcohol. Should the population reflect that or any car-driving situation? Clearly many ways of defining the reference population can be made, all with some type of justification.

If the population becomes more specific, we reduce the data available. Accurate estimates of p cannot be made. If the population is made less restrictive, more data would be available, but the relevancy would be reduced. Thus, a balance has to be found. A population of reasonable similar situations is considered, and estimates can be derived that express some type of average values. Yet it provides some insights about the risk of taking the trip.

Problem 3.14

Risk is in general described by (C',Q,K). If probability P is used for Q, the expected disutility −E[u(C')] can be seen as a metric combining C' and P. Hence, the problem discussed in Section 3.2 about using probability-based metric also applies here. The strength of knowledge supporting the metric is not included and not the knowledge K, including key assumptions.

The problems with using the expected value are serious. The expected disutility metric seeks to meet these. The introduction of the utility function u means that the preferences of the decision-maker (and other stakeholders) are reflected in the risk characterization. The result is a mixture of professional judgments of uncertainties about C' and the decision-makers' preferences concerning different values of C'. It is difficult to specify the utility function, and its assignment will easily introduce an element of arbitrariness. Decision-makers would also be reluctant to specify the utility function, as it would reduce their flexibility for weighting different concerns. In practice, decision-makers would like to have a full display of the magnitudes and sources of uncertainties rather than integrating these with their own priorities (see discussions in Paté-Cornell 1996; Aven 2020b).

See Chapter 9. Although difficult to assign, these utilities and the theory behind the use of expected utilities have a strong rationale.

Problem 3.15

If an unknown known A occurs, it is not covered by the events A' specified in the risk characterization (A',C',Q,K).

Problem 3.16

Taleb defines a black swan in this way: First, it is an outlier, as it lies outside the realm of regular expectations, because nothing in the past can convincingly point to its possibility. Second, it carries an extreme impact. Third, in spite of its outlier status, human nature makes us concoct explanations for its occurrence after the fact, making it explainable and predictable.

The first two points convey basically the same ideas: a surprising event with large, extreme consequences/impact, which is the definition used in this book. The third point is a typically a feature of such events but is not included in the definition of the term in this book.

Problem 3.17

For perfect storms, we understand the relevant phenomena. The experts can calculate the probabilities of such events and the associated risks with a high degree of precision (using frequentist probabilities). They can make accurate predictions of what will happen, stating that in 1 in 100 such situations the waves will be like this, and in one in 1000 such cases, the waves will become so big and so on. From that perspective, it can be argued that the occurrence of the extreme storm did not come as a surprise: if we conduct the 'experiment' over and over again, we know that this event will occur. From that perspective, it is not a black swan. Weather experts could accurately predict the event. There is no surprise aspect.

However, it can also be argued differently. The crew of the fishing boat was convinced based on their knowledge that the waves would not be a problem. Their knowledge-based probability for a capsizing event to occur was very low, with strong support based on many years of experience. For them, the extreme weather – the amplifying interactions between a confluence of factors – came as a surprise, and the event can be seen as a black swan for them.

We see how important it is to relate the surprise and black swan concepts to whose knowledge it relates.

Problem 3.18

The occurrence of a virus-type pandemic such as COVID-19 cannot be seen as a black swan – its occurrence was rather likely. Health organizations have warned about this type of pandemic for many years. This does not mean that the severity of the pandemic did not come as a surprise for many.

Problem 3.19

Here is a suggested approach (Aven 2017b, 2020a; Aven and Flage 2018):
The knowledge K is judged as weak if one or more of the following conditions are true:

w1   The assumptions made represent strong simplifications.
w2   Data/information are/is non-existent or highly unreliable/irrelevant.
w3   There is strong disagreement among experts.
w4   The phenomena involved are poorly understood; models are non-existent or known/believed to give poor predictions.
w5   The knowledge K has not been examined (for example, with respect to unknown knowns).

If, on the other hand, all (whenever they are relevant) of the following conditions are met, the knowledge is considered strong:

s1   The assumptions made are seen as very reasonable.
s2   Large amounts of reliable and relevant data/information are available.
s3   There is broad agreement among experts.
s4   The phenomena involved are well understood; the models used are known to give predictions with the required accuracy.
s5   The knowledge K has been thoroughly examined.

Cases in between are classified as medium strength of knowledge. To obtain a wider strong knowledge category, the requirement that all of the criteria s1–s5 need to be fulfilled (whenever they are relevant) could, for example, be replaced by a criterion expressing that at least one (or two, three or four) of the criteria s1–s5 need to be fulfilled while, at the same time, none of the criteria w1–w5 are fulfilled.

A simplified version of these criteria can be obtained by applying the same score for strong but assigning the medium and weak scores when a suitable number of conditions are not met, for example, a medium score if one or two of the conditions s1–s5 are not met and a weak score otherwise, that is, when three, four or five of the conditions are not met.

Problem 3.20

No, knowledge is justified beliefs – claims about the truth. Think as an example about a person who is to estimate the frequentist probability of a specific die showing 2 in a trial. The person studies the die and argues that, because of symmetry, the probability is 1/6. The statement is a justified belief, but it could be wrong. The truth – the actual frequentist probability – is not known. The true outcome of future events cannot be known with certainty; nevertheless, we can have knowledge about this future.

Problem 3.21

No, we need also to take into account the duration of the trip. If a plane trip takes one hour and the driving the same distance takes eight hours, the death probability is four times higher when driving compared to flying for this trip.

Problem 3.22

Vaccines basically always lead to some side effects, but it is difficult to know the type and magnitude of these.

Problem 3.24

See Aven (2019b).

Problem 3.26

In this alternative approach, we define a set of consequence categories $C_1, C_2, \ldots C_m$, reflecting, for example, different loss of lives categories (e.g., 0, 1–2, 3–5, 6–10, 11–20, 21–100, >100). Then probabilities are established for events leading to these consequences, for example, the probability that events occur having at least 100 fatalities. Categories of probabilities are commonly used as in Figures 3.1 and 3.2. The attractiveness of this approach compared to the traditional one is that the consequences are specified; they are not expected values. Strength of knowledge judgments should be added, as always.

Problem 4.1

A report of accident data as such is not a risk assessment. Such data describe historical facts but only when we address the future the concept of risk applies. To assess what will happen, we can decide to use the data, and they will provide input to the risk assessment.

Problem 4.2

Meeting regulatory requirements is of course important, but if this is the case, the full potential of these assessments is not reached. The main reason for conducting a risk assessment is to improve the understanding of risk to support decision-making on choice of arrangements and measures. Such a culture can clearly hamper the proper use of risk assessment, leading to inefficient use of resources and possibly also compromising safety.

Problem 4.3

Possible events: head-on crash, rear-end crash, lane changing accidents, bus fire

Corresponding causes: Turn around in tunnel due to long queue, smoke, exhaust and so on; wrong-way entry; slow moving vehicles; breakdown (speed, light conditions, road marking); technical failure (collision, ignition)

Corresponding consequences: 1–3 fatalities and serious injuries; less serious injuries; serious injuries; many fatalities

Corresponding likelihood judgments (one year): 1–10%, several times; 10–50%, less than 0.1%

Corresponding strength of knowledge judgments: strong, strong, strong, medium strong

Corresponding risk judgments: high, low, moderate, moderate-high

Corresponding measures to consider: signals for wrong-way driving, detection of slow-moving vehicles, detection of slow-moving vehicles, fire-extinguishing equipment, ventilation

Problem 4.4

FMEA form for the components V1 and LSHH, addressing function, failure mode, effect on other units, effect on the system, failure probability, strength of knowledge and failure effect ranking:

V1:

- Functions: Stop the fluid supply when the fluid level is high. The value is normally open (function)
- Failure modes: Does not close on signal. Close when not intended. Significant leakage
- Effect on other units: None

- Effects on system: The fluid level may increase abnormally. The fluid supply stops. The fluid supply stops (corresponding to the three failure modes)
- Probability: 2% of total number of demands. Once in 10 years. Once in 10 years (corresponding to the three failure modes)
- Strength of knowledge: Strong. Strong. Medium (corresponding to the three failure modes)
- Failure effect ranking: Low-moderate for all three failure modes

LSHH

- Functions: Switch that sends stop signal to V2 and open signal to V3 if the fluid level is abnormally high
- Failure modes: Does not send signal when the fluid level is abnormally high. Sends signal when the fluid level is not abnormally high (corresponding to the two failure modes)
- Effect on other units: V2 does not close, and V3 does open. V2 closes when not intended, and V3 opens when not intended (corresponding to the two failure modes)
- Effects on system: The tank is overfilled. The tank is drained (corresponding to the two failure modes)
- Probability: 1% of total number of demands. Once every second year (corresponding to the two failure modes)
- Strength of knowledge: Strong. Strong (corresponding to the two failure modes)
- Failure effect ranking: high, moderate (corresponding to the two failure modes)

Problem 4.5

The FMEA method was developed in the 1950s and was one of the first systematic methods used to study failures in technical systems. It is also commonly referred to as failure modes, effects and criticality analysis (FMECA). In this book, we use FMEA when the analysis also includes a criticality ranking.

An FMEA gives a systematic overview of component failures in a system. Typical technical failures are highlighted, but human failures can also be included. In an FMEA, failures are considered in isolation from others, which means that the method is not suitable for identifying how combinations of component failures can lead to system failure. The storage tank example of Problem 4.4 demonstrates this, as to avoid overfilling of the buffer storage, it is sufficient that valve 1 close or that valve 2 close or valve 3 open. The analysis is, however, a good starting point for a fault tree analysis, as discussed in Problem 4.6.

A main challenge in using FMEA is that it could be resource demanding and include a lot of non-interesting results. Following the method, all components are analyzed and documented, as well as the failures that have minor effects. This problem can to a large extent be reduced by proper component definitions. For the storage tank system, the focus is on the components V1, V2, V3, LSH and LSHH and not subcomponents of these units. For larger systems, it is essential to define subsystems (system functions) on a rather high level to avoid analysis of non-critical elements. An initial FMEA may then address failures of these subsystems, and more detailed FMEA studies can be conducted for specific subsystems.

Problem 4.6

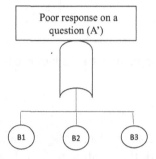

A fault tree that comprises AND and OR gates can be transformed to a so-called reliability block diagram, which is a logical diagram which shows the functional ability of a system where each component in the system is illustrated by a rectangle. The corresponding reliability block diagram for the tank example takes the form shown in Figure C2.

We see that for the system to work (not fail), V1 and LSH need to work, or alternatively LSHH and either V2 or V3. We refer to the subsystem 14 as a series system of the components 1 and 4, and 23 as a parallel system of the components 2 and 3.

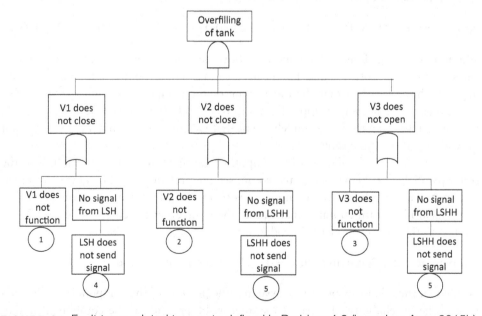

FIGURE C1 Fault trees related to events defined in Problem 4.6 (based on Aven 2015b)

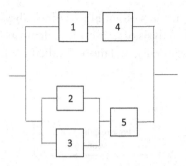

FIGURE C2 Reliability block diagram for the tank example (based on Aven 2015b)

Problem 4.7

The minimal cuts sets are 15, 45, 123 and 234. If, for example, V1 and LSHH both fail, the tank will be overfilled. If we reduce this set, it will not ensure system failure. The minimal cut sets are important for understanding the reliability (or unreliability) of the system. If a minimal cut set fails, the system fails. And, reversed, if the system fails, a minimal cut fails. We will see in the coming problem how the minimal cuts play an important role when quantifying the probability of system failure, that is, the probability that the top event occurs.

Problem 4.8

Simplified and using that system failures occurs if at least one of the minimal cut sets fail, it follows from probability calculus that

$$P(\text{top event}) \approx 0.02 \cdot 0.01 + 0.01 \cdot 0.01 + 0.02 \cdot 0.02 \cdot 0.02 + 0.01 \cdot 0.02 \cdot 0.02 = 0.03\%,$$

as a minimal cut set fails if all components of the set fail. For example, for the minimal cut set 15 to fail, both V1 and LSHH need to fail. The probability for that is $0.02 \cdot 0.01$. We have assumed that the functioning of the components is independent. The approximation is in fact an upper limit and is a good approximation, as the formula is based on ignoring the here-unlikely event that two or more minimal cut sets are not functioning at the same time.

The exact value is obtained by computing the reliability and unreliability of the series-parallel subsystems 14, 23 and so on; see the previous block diagram. We introduce p_i as the probability that component is functioning and q_i as the probability that component i is not functioning, that is, that the basic event i occurs, where i = 1,2, ... 5 refers to the five components introduced previously. We then obtain

$$P(\text{top event}) = 1 - P(\text{system failure}) = 1 - (1 - p_1 p_4)[1 - p_5(1 - q_2 q_3)],$$

noting that the reliability of a series system is the product of the component reliabilities p_i, and the unreliability of a parallel system is the product of the component unreliabilities q_i. The exact formula produces in this example the same number, 0.03%, when using this level of precision.

Hence, if the system is exposed to 25 situations in a year, the probability of overfilling during one year is approximately $25 \cdot 0.03\% = 0.75\%$.

Which component is most important? It depends on what we mean by importance. If we think about the improvement potential, component 5 (LSHH) is the most critical one, as if this one becomes perfect, 100% reliable, it leads to the highest improvement in the system reliability, from 0.03% to 0.0012%. For the other components, the difference is considerably less.

Another way of measuring importance is to look at how small changes in the component reliabilities affect the system reliability, or how small changes in p_i and q_i influence the probability of the top event. The method represents a classical sensitivity analysis. It is implemented in practice as the partial derivative of the system (un)reliability with respect to the component reliability p_i (unreliability q_i). Again, the computations will show that component five (LSHH) is the most important, together with V1 and LSH. This is intuitive, as if we can improve LSHH, we can make the system very reliable given that LSHH is in series with V2–V3, which is a reliable parallel unit. A component importance measure based on the partial derivative as here outlined is referred to as Birnbaum's measure. See Section 5.1 for further discussion of this type of sensitivity and importance analysis.

Problem 4.9

The probability $P(Y \geq 1|A)$ expresses the probability of at least one fatality given a leakage. It is equal to

$$10^{-3} \cdot 0.90 \cdot 0.5 + 10^{-3} \cdot 0.1 \cdot 0.7 = 0.52 \cdot 10^{-3} \approx 5 \cdot 10^{-4}.$$

The first term is based on the scenario A occurring, B occurring, not C and one fatality, whereas the second term reflects A, B, C occurring and at least one fatality.

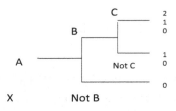

FIGURE C3 Event tree for problem defined in Problem 4.9

Problem 4.10

Examples of branches: V1 does not close, V2 does not close, V3 does not open. Then fault trees are developed for each branch event.

Problem 4.11

We remember the definition of causality referred to: We say that for B to cause A, at a minimum, B must precede A, the two must covary (vary together) and no competing explanation can better explain the correlation between A and B.

The causality concept is commonly used in risk assessment, and in many contexts, the meaning is rather straightforward, for example, when used in fault tree analysis. The tank fault tree shows that, for example, failures of V1 and LSHH cause system failure. The fault tree analysis can be seen as a causality analysis. The conclusion is based on a model of the system studied.

For the smoking example, the conclusions are more difficult. There is a shown correlation, but causality is more difficult to establish. However, by thorough analysis, we will be more and more confident that there are no other factors that could better explain the correlation between A and B, and that is the best we can do using the statistical approach.

Root cause is a difficult concept, as discussed, for example, by Hollnagel (2004). For example, what does it mean that 'poor quality of the maintenance work' (referred to as R) is the root cause of a type of accident? There may be no direct link from R to the accident. A number of other conditions and events need to be in place to ensure the occurrence of the accident. Rather we should simply refer to R as an influencing factor or a risk source.

Problem 4.12

We introduce X as the state of the patient, which is 1 or 0 depending on whether the test gives positive or negative response. Furthermore, we let θ be the true health state of the patient, defined as 2 if the patient is seriously ill, 1 in case the patient is moderately ill and 0 if the person is not ill at all.

From the information provided we have the following *prior probabilities* for θ:

$P(\theta = 2) = 0.02$, $P(\theta = 1) = 0.10$, and $P(\theta = 0) = 0.88$.

Moreover, we have information expressing that

$P(X = 1 \mid \theta = 2) = 0.90$, $P(X = 1 \mid \theta = 1) = 0.60$, and $P(X = 1 \mid \theta = 0) = 0.10$.

We seek to compute $P(\theta = 2 | X = 1)$. To do this, we first use the rule of total probability to compute

$P(X = 1) = P(X = 1 \mid \theta = 2) P(\theta = 2) + P(X = 1 \mid \theta = 1) P(\theta = 1) + P(X = 1 \mid \theta = 0) P(\theta = 0)$
$= 0.90 \cdot 0.02 + 0.60 \cdot 0.10 + 0.10 \cdot 0.88 = 0.166.$

Then using Bayes' formula, we obtain the *posterior probability*

$P(\theta = 2 | X = 1) = P(X = 1 | \theta = 2) P(\theta = 2)/P(X = 1) = 0.90 \cdot 0.02/0.166 = 0.11.$

Hence, the positive response of the test has increased the probability of being seriously ill from 2% to 11%.

Problem 4.13

If $A_1, A_2, \ldots, A_n$ are the failure events, and system failures occur if at least one of these events occurs, the system failure probability is approximately equal to

$$P(A_1) + P(A_2) + \ldots + P(A_n)$$

for small probabilities. The exact probability when independent events is equal to

$$1 - \prod (1 - P(A_i))$$

where the product is over all components i. Thus, it is not enough to just control just the maximum $P(A_i)$, as the number of terms could be very large.

Problem 4.14

The forward approach makes it possible to perform a comprehensive analysis covering all types of effects C' given the occurrence of A'. This is the strength of the approach, but also it is a challenge, as the analysis can be rather extensive, covering many types of unimportant effects. The risk description could be more complete using a forward approach, but there is a risk that the assessment could become so detailed that we are not able to see what really matters and what does not. The backwards approach, on the other hand, ensures that focus is on the defined, interesting consequences, and it can be less resource demanding. However, the backwards approach requires considerable experience and competence in order for the analysts not to overlook potential relevant scenarios.

Problem 4.15

First consider situations characterized by a lot of relevant data, as in experimental testing cases, where the aim is to estimate frequentist probabilities. Then reliability and validity can be achieved, as consistency of the estimation is possible (reliability), and we can accurately measure what we set out to measure, the frequentist probabilities. For other situations, the concepts are not so straightforward to interpret. Think about the PRA conducted for the Apollo project. Here the uncertainties were large, and the reliability and validity criteria would not be met if the same standards were used as in situations with large amounts of relevant data.

  If the purpose of the risk assessment is to characterize risk, we need to think differently when interpreting these two concepts. Risk characterizations cover $P(A|K)$ and K, and these would not be the same if the assessment were conducted by several teams. The knowledge K would be different, and so would the probability judgments $P(A|K)$, as these depend on K. Reliability in this situation is more about obtaining consistency when, for example, rerunning the computational methods and procedures. It is also important to document the argumentation used for the probability assignments made.

  In general, the validity concept relates to the degree to which the risk assessment describes (characterizes) the concepts that the assessment is attempting to describe, or, rephrased, the degree to which one is able to assess what one sets out to assess. Alternatively, validity may just indicate that the assessment is solid, meeting some relevant criteria (for example, that all concepts are defined and the work is in compliance with all rules used and assumptions made).

Chapter 3 provides guidance for ensuring validity in these senses. Clearly, if risk is described, for example, by an expected value only, the risk assessment would not in general be valid, as the risk is poorly described. Risk assessments that describe risk by such a metric are not valid, as they do not very well describe what they set out to characterize: namely risk. See Aven (2020a) for further discussion of this topic.

Problem 5.2

No, we will argue that different situations call for different measures. What is best depends on what you would like to reflect by the measure. If we are concerned about small changes in a component's reliability, we could use the Birnbaum measure, whereas if we are more interested in identifying the maximum potential of improving a component's reliability, we are led to the improvement potential.

Nøkland and Aven (2013) present an guideline for what measure to use in a safety context distinguishing between the following categorization of decision situations:

a   Design, that is, improvement of design,
b   Operation, that is, operational improvements
c   Maintenance, that is, testing

For the situation a, the guideline expresses that importance measures should reflect large or maximum changes in component performance, which means a measure of the category improvement potential, that is, the component which has the largest change in the reliability or risk performance measure used is most important. The common importance measure risk reduction worth (RRW) also belongs to this category of measures. It is defined by the ratio between the risk metric considered for the base case and a situation where the component is perfect.

For situation b, the guideline expresses that the importance measures should reflect small changes in component performance, which means that a measure based on the partial derivative should be used. Birnbaum's measure is an example of such a measure in a reliability context. For this situation, we are concerned about operational measures, and these will normally give rather small effects on the overall system performance measures; hence, the derivative could provide a useful indicator. Reconfiguration and major design changes are not normally considered relevant options in this operational phase.

For situation c, the guideline states that importance measures should reflect the effect of reduced component performance, which reflects the difference in the risk metric by assuming the component is not working. For example, if components are critical for safe operation of the system, test and maintenance activities should not be performed on a system in operation but instead be performed during planned shut-down periods. The common importance measure risk achievement worth (RAW) is an alternative measure, which expresses the ratio between the risk metric considered for the base case and a situation where the component is not working.

It should be noted that for typical real-life cases, the improvement potential is equivalent to the well-known Fussell-Vesely measure; see Aven and Jensen (2013, p. 28).

These measures do not, however, reflect how uncertainties on the component level affect uncertainties on the system level. A common measure that can be used for all the situations a–c for capturing this aspect is the correlation measure; see Problem 5.5.

Problem 5.3

We may, for example, introduce a more general distribution, such as a Weibull distribution with parameters α and β, and by varying these parameters, we may study the sensitivity of the system reliability with respect to the model choice.

Problem 5.4

Instead of comparing $E[v_i(p_i)]$, we use $Var[u_i(p_i)]$, as the difference is a constant, $Var[h]$, according to the formula $Var[h] = E[v_i(p_i)] + Var[u_i(p_i)]$, where $u_i(p_i) = E[h| p_i]$. To compute these quantities for the reliability example, we first note that

$$E[h| p_1] = E\ [1 -(1 - p_1)(1 - p_2) |\ p_1] = 1 -(1 - p_1)(1 - E[p_2]) = 0.90 + 0.10\ p_1$$

and

$$E[h| p_2] = E\ [1 -(1 - p_1)(1 - p_2) |\ p_2] = 1 - (1 - E[p_1])(1 - p_2) = 0.60 + 0.40\ p_2.$$

It follows that

$$Var[u_1(p_1)] = Var[0.90 + 0.10\ p_1] = 0.1^2\ Var[p_1] = 0.010 \cdot 0.005 = 0.00005$$

and

$$Var[u_2(p_2)] = Var[0.60 + 0.40\ p_2] = 0.4^2\ Var[p_2] = 0.16 \cdot 0.00125 = 0.00020.$$

Thus, the uncertainty distribution of unit 1 has a stronger influence on the variation of the output result than the uncertainty distribution of unit 2. Note that $Var[u_i(p_i)]$ is equal to 'constant $- E[v_i(p_i)]$'.

Problem 5.5

The correlation coefficient is a normalized version of the covariance. It takes values in [–1,1]. Here +1 indicates perfect positive correlation and –1 perfect negative correlation. The measure expresses the strength of the linear relationship between p_i and h. The numerical analysis gives

$$\rho(p_1,h) = Cov(p_1,h)/(SD[p_1]\ SD[h]) = 0.0005/(0.071 \cdot 0.016) = 0.44,$$

$$\rho(p_2,h) = Cov(p_2,h)/(SD[p_2]\ SD[h]) = 0.0004/(0.035 \cdot 0.016) = 0.71,$$

noting that

$Cov(p_1, h) = E[p_1 h] - E[p_1] E[h] = E[(p_1)^2 (1 - p_2) + p_1 p_2] - Ep_1 Eh =$
$= 0.365 \cdot 0.1 + 0.6 \cdot 0.9 - 0.6 \cdot 0.96 = 0.0005,$

$Cov(p_2, h) = E[p_2 h] - E[p_2] E[h] = E[(p_2)^2 (1 - p_1) + p_1 p_2] - Ep_2 Eh =$
$= 0.811 \cdot 0.4 + 0.6 \cdot 0.9 - 0.9 \cdot 0.96 = 0.0004,$

$Var[p_1] = E[(p_1)^2] - (Ep_1)^2 = 0.365 - 0.360 = 0.005,$

$Var[p_2] = E[(p_2)^2] - (Ep_2)^2 = 0.81125 - 0.8100 = 0.00125.$

Thus, the calculations show that the correlation coefficient is largest for unit 2. This was expected, as the unit with the highest reliability dominates the reliability of the system.

Problem 5.6

For our reliability example, such an analysis gives the following intervals for the system reliability:

Unit 1: [0.947, 0.973]
Unit 2: [0.934, 0.986].

To establish these intervals, we first define the 5% and 95% quantiles:

Unit 1: 0.47, 0.73
Unit 2: 0.835, 0.965.

Then we obtain $E[h \mid p_1 = 0.47] = 0.47 + 0.53 \, Ep_2 = 0.947$. The other results follow by similar calculations.

    These intervals are depicted in Figure C.4. We see that unit 2 has the widest intervals, which indicates that variation in the reliability for this unit has the largest effect on the system reliability.

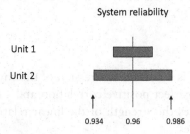

FIGURE C.4 Tornado chart for the system reliability example, expressing the expected system reliability using the 5% and 95% quantiles of the uncertainty distribution of p_i (based on Aven 2010)

Problem 5.7

Yes, but to compute the relevant metrics, they need to be known or estimated.

Problem 5.8

For example, let the unconditional frequentist probabilities for external factor and engine defect be denoted p_1 and p_2, respectively. These are considered unknown. Triangular distributions are then specified expressing the analyst's uncertainties about these frequentist probabilities. Drawing numbers from these distributions allows the analyst to compute related values for the desired probability of the engine being defective, given that the machine is observed not working, and hence a distribution for this probability.

Problem 5.9

The long-term unavailability of each component – that is, the fraction of time the unit is down – will be $q = E[D]/(E[T]+E[D]) \approx E[D]/E[T]$; hence, the sought-after fraction will approximately be the sum of $E[D]/E[T]$ for the three units, as any failure of these three units will lead to a performance level below 100.

Problem 5.10

The benefits would be limited, as the systems and processes of the facility are well understood and the PRA/QRA would consequently not give important decision support. However, it may be useful to perform studies to identify potential deviations from the knowledge and standards established to avoid potential surprises.

Problem 5.11

See, for example, Salignac et al. (2019).

Problem 5.12

We have provided reflections on this in Section 5.5; see also Aven and Zio (2021).

Problem 5.14

There are many standards for doing human health risk assessments. Basic elements included are:

1   Identification of the hazards, which includes clarification of who is exposed, identification of risk sources of concern (chemicals, radiation, biological, etc.) and their origins (contamination, discharges from a factory, etc.) and clarification of how exposure occurs (air, soil, inhalation, ingestion, etc.).
2   Dose-response assessment, establishing what the health problems are at different exposure levels, see Sections 2.2.4 and 4.2.2.

3   Estimation of how many people (the person or population) are exposed to the risk source.
4   Risk characterization combining the previous steps.

Problem 5.15

A hazard and operability (HAZOP) study is a qualitative risk analysis method that is used to identify weaknesses and hazards in a processing facility; it is normally used in the planning phase (design). The HAZOP method was originally developed for chemical processing facilities, but it can also be used for other facilities and systems. A HAZOP study is a systematic analysis of how deviation from the design specifications in a system can arise and an analysis of the risk related to these deviations. Based on a set of guidewords, events and scenarios that may result in a hazard or an operational problem are identified. The following guidewords are commonly used in practice: NO/NOT, MORE OF/LESS OF, AS WELL AS, PART OF, REVERSE and OTHER THAN. The guidewords are linked to process conditions, activities, materials, time and place. For example, when analyzing a pipe from one unit to another in a process plant, we define the deviation 'no throughput' based on the guideword NO/NOT and the deviation 'higher pressure than the design pressure' based on the guideword MORE OF. Then causes and consequences of the deviation are studied. This is done by asking questions. For example, for the first mentioned deviation in the pipe example, the questions would be:

- What must happen to ensure the occurrence of the deviation 'no throughput' (cause)?
- Is such an event possible (relevance/probability)?
- What are the consequences of no throughput (consequence)?

For further details, see, for example, Meyer and Reniers (2013).

Problem 5.16

Some analysts find this term attractive, as likelihood is difficult to assign, and the plausibility concept captures a kind of mixture of likelihood and reasonability justification. In general, we recommend using likelihood (probability) – often imprecise – together with strength of knowledge judgments, as there is a strong theory with meaningful interpretations supporting these terms.

Problem 5.17

Answers will vary, but may contain the following elements:

- C.1. Time dependency should be strongly emphasized for most systems, as conditions or values change over time.

- C.2. One or few components' failures cause failure of the system: This type of property may not exist in all situations but could be a distinguishing factor in the space flight example.
- Q.1. Sensitivity: This property may be of special relevance for systems such as cybersecurity, where conditions (in the form of scenarios) change rapidly.
- Q2. Importance/criticality may have a strong emphasis for systems such as space flight and safety in automobiles, where certain aspects of system design and use have a major impact on overall vulnerability.
- K.1. Degree to which phenomena involved are understood: This property should be emphasized for systems involving significant changes in knowledge over time, such as space flight and cybersecurity.
- K.2. Accuracy of models: It can be argued that modeling accuracy is a critical issue for most systems, but the modeling itself is most critical for engineered systems, such as space flight and automobile safety.
- K.3. Assumptions: This property would be most important for systems involving significant uncertainties, such as with cybersecurity, in which the system is controlled by many actors with very little oversight.
- K.4. Data and information: This property would be most relevant for systems that rely heavily on modeling using data that is difficult or costly to acquire, such as space flight.

Problem 5.18

Answers will vary. Generally, answers should suggest that GK involves available or generally accepted knowledge about the system (or similar systems), users and operating environment. SK involves knowledge that is specific and detailed for the studied system. Systems such as cybersecurity may involve strong GK but low SK because the sub-systems are managed and controlled by independent actors. Systems such as safety in automobiles may have both strong GK and SK, because these are engineered systems that rely on laws of physics and mechanics and have been extensively tested. However, even in this case, SK may be weak when using the automobiles in untested or atypical situations.

Problem 5.19

The increased focus on these methods is, as noted by Aven and Flage (2020), driven by developments in information and communication technology, including sensor and storage technology, coupled with developments in artificial intelligence and statistics, including data mining, machine learning and statistical learning theory (e.g., Hastie et al. 2017; Vapnik 2013).

Supervised learning: based on observed input data x and output y, a model f is established linking x and y. The method is similar to what is conducted in statistics, although no presumed form of f is needed. For example, we may have data x on the health condition of systems together with observations of the total system performance y. Using methods like regression analysis and curve fittings, a function (model) f is established that can be used to predict the performance of a new similar system.

Simplified, we can say that we teach the computer how things are (link x and y) and then let it use its found knowledge on a new situation.

Unsupervised learning: This type of learning refers to methods where there is no target variable y specified – the goal is pattern recognition (as opposed to prediction) to identify useful information about the structure and dependency mechanisms in the data. For example, unsupervised learning techniques could be used in the previous performance example to identify 'hot spots' (e.g., clusters of system components experiencing highest failure rates) without information about the system performance y. In this way, new and important quantities of interests can be identified, as well as dependencies.

Problem 6.2

The statements referred to are based on a different perspective on what risk is compared to current ideas and concepts and as presented in this book. These statements do not make sense following what we consider current risk science knowledge; refer to Figure 6.4. We make a clear distinction between risk as a concept, how it is described on the basis of a professional risk assessment and how it is perceived. The risk concept and the professional description of risk do not reflect the feelings of the assessor, as is the case for risk perception, nor judgments about how one likes or dislikes the risk or features of the risk. Sometimes risk perception also covers judgments of the acceptability of risk, which makes the difference between risk perception and professional risk characterizations even more apparent. Risk as a concept and a professional description of it do not cover the decision-maker's judgments about risk acceptance or tolerance. See also Section 6.3 and Problem 2.7.

Problem 6.3

See Aven (2015a, 2018b). We should be careful concluding that people overrated the risk, as there is no objective, correct risk to compare with. When Kahneman drove away more quickly than usual when the light changed, it was mainly based on System 1 thinking. He should not be chagrined by his cautious behavior, as risk could be judged relatively high (potential for severe consequences and considerable uncertainties).

Problem 6.4

Climate change risk:

High perceived risk for the combined communitarianism and egalitarianism and low risk for the combined hierarchy and individualism.

- Individualistic: Fear of threats or hazards that can obstruct individual freedom. Individualists are typically skeptical of experts.
- Egalitarian: Fear of development that can increase inequalities among people.
- Hierarchical: Fear of aspects that leads to lack of control, such as demonstrations and crime. Hierarchists trust expert knowledge to a large extent.
- Fatalistic: Indifferent attitude to risk, what is feared is to a large extent determined by others.

Problem 6.5

Amplification denotes the process of intensifying and turning up the volume of certain signals during the transmission of information, whereas attenuation refers to the weakening or tuning down the volume of signals. In relation to risk, amplification may be generated by an accident or report showing increased numbers of injuries. Attenuation, on the other hand, may be illustrated by the opposite tendency, when, for instance, the communication of a risk assessment demonstrating no harmful effects of exposure to a certain chemical or a report showing decline in number of injuries results in a lowered risk judgment and heightened perception of safety, relaxed regulations and increased trust in risk-managing institutions.

In short, risk amplification refers to increasing the judged risk, whereas risk attenuation refers to reducing the judged risk. Hence risk amplification and risk attenuation always have to be seen in relation to a reference level, for example, the current level.

Problem 6.6

See Figure C.5.

Problem 6.7

The hypotheses are as in the text,

$$H_o : \mu_{Graduate} = \mu_{High\ School}$$

$$H_a : \mu_{Graduate} \neq \mu_{High\ School}$$

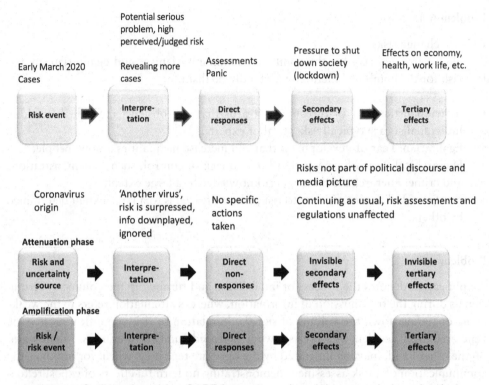

FIGURE C.5 Outline of applying SARF (**FIGURE 6.5**) and the extended version (**FIGURE 6.6**) on the coronavirus case

as well as the *t*-test:

$$t = \frac{\bar{x}_{Graduate} - \bar{x}_{High\ School} - 0}{\sqrt{\dfrac{s^2_{Graduate}}{n_{Graduate}} + \dfrac{s^2_{High\ School}}{n_{High\ School}}}} = \frac{1.95 - 1.93 - 0}{\sqrt{\dfrac{0.84^2}{22} + \dfrac{0.70^2}{54}}} = 0.14$$

Then we need to compare the *t*-value described previously to the critical value found in a *t*-distribution. The critical value is found using a *t*-distribution with degrees of freedom computed using s^2 (the variance) and n (the sample size) as:

$$\upsilon = \frac{\left(\dfrac{s^2_{Graduate}}{n_{Graduate}} + \dfrac{s^2_{High\ School}}{n_{High\ School}}\right)^2}{\dfrac{\left(\dfrac{s^2_{Graduate}}{n_{Graduate}}\right)^2}{n_{Graduate} - 1} + \dfrac{\left(\dfrac{s^2_{High\ School}}{n_{High\ School}}\right)^2}{n_{High\ School} - 1}} = \frac{\left(\dfrac{0.84^2}{22} + \dfrac{0.7^2}{54}\right)^2}{\dfrac{\left(\dfrac{0.84^2}{22}\right)^2}{22 - 1} + \dfrac{\left(\dfrac{0.7^2}{54}\right)^2}{54 - 1}} = 33.25 \approx 33$$

We reject the null hypothesis that the two means are equal when using a 0.05 level of significance if:

$$t > t_{1-\alpha/2,v} = t_{1-0.05/2,33} = 2.03$$

Because the test statistic 0.14 is not greater than the critical value of 2.03, the null hypothesis is not rejected. Therefore, we believe there is no statistical difference in average risk score between the two groups. In other words, because we assumed the null hypothesis was true, the data do not allow us to conclude that there is a difference in risk score between the two education groups.

Problem 6.8

Students may comment on the following aspects:

Which demographic groups are in the highest need of emergency preparedness investments? Why?:

- Populations who show the lowest level of concern, such as those who rate the probability of wildfires and cyber attacks as being relatively low, may have the highest need. For example, answers to questions 2.1.2 (the probability of a wildfire) showed relatively low probabilities associated with the 18–21 age population and the population with some high school education.
- Populations with low strength of knowledge associated with the risks may have the highest need. For example, answers to question 2.3.3 (the strength of knowledge associated with the probability of a cyber attack) showed a relatively low strength of knowledge associated with the 22–39 age population and the population with some high school education.

What types of investments are needed to prepare the highest-need groups?

- Educational materials, such as pamphlets and social media posts
- Community engagement/events to bring attention to emergency preparedness, such as outreach at existing community events or dedicated emergency preparedness events
- Outreach to children in public schools, calling upon children to discuss emergency preparedness with their families
- Partnership with news sources, such as newspapers or TV news, to promote emergency preparedness discussions
- Legislation to support greater access to emergency preparedness supplies, such as tax-free holidays

What additional data should be collected and analyzed?

- Revising the current survey to include information on as-is emergency preparedness. This could include surveys directed at various demographic groups to evaluate steps currently being taken to prepare for various types of emergencies

- Revising the current survey to include information on other factors, such as income, children in households and access to communication

Problem 6.11

Heuristic definition: "A judgment is said to be mediated by a heuristic when the individual assesses a specified target attribute of a judgment object by substituting a related heuristic attribute that comes more readily to mind" (Kahneman 2003). Following this definition, the 'anchoring and adjustment heuristic' referred to by Tversky and Kahneman (1974) would no longer be a heuristic. Instead we can speak about anchoring effects and related biases.

Problem 7.1

The numbers express knowledge-based probabilities as defined in Chapter 3. An interpretation can be provided. In addition to the numbers, the knowledge basis for the numbers as well as a judgment of the strength of knowledge (SoK) supporting the probabilities should be given. The probabilities and the SoK judgments are acknowledged as judgments made by the analysts conducted the study. The risk results are not an objective characterization of risk but an assessment reflecting the analyst's knowledge and judgments.

Problem 7.2

See Section 7.1.4 and Aven (2018c, 2019b).

Problem 7.3

The issue in relation to this discussion is not "exceptional accuracy of the meaning of probability" but having a meaningful interpretation at all. In the case of Section 7.1.1, the management was using the risk analyses in the risk-informed way, but they encountered problems, as they did not understand what the numbers meant and consequently how they should be used in the decision-making context. They were not asking for "exceptional accuracy of the meaning of probability" but simply a clear and easily understandable explanation of what the results expressed.

It would indeed be sensational if a scientific field were happy using a tool where the fundamental concepts adopted are not clearly defined. Both the risk theories and methods, and the applications, need proper conceptual frameworks in which all concepts have meaningful interpretations. If not, the communication would suffer (Aven 2018c, 2020a).

Problem 7.4

Extensive statistical testing showing that the risk related to side effects is very small, meaning that the frequentist probability of such an event is very small with high confidence.

Problem 7.5

Basic principles: Perform risk assessment and characterize risk based on state-of-the-art risk science. Highlight openness and traceability. Involve representatives from the neighbors to take part in the risk assessment process. Build knowledge about risk, risk assessment and management. Form groups where concerns and issues can be addressed.

Problem 7.7

There is a balance to be made. Risk assessment results need to be appreciated for what they are, reflecting the knowledge of the analysts. There is a risk for potential surprises and occurrence of events not known. The issue needs to be addressed, for example, by performing judgments of the strength of the knowledge supporting the results obtained and checking for knowledge gaps. On the other hand, too much focus on this risk issue could easily lead to an unbalanced risk characterization, not highlighting the main message of the risk assessment.

Problem 7.10

Some people may argue that this should not be a problem, as long as the editor is able to distinguish when an editor and when a politician. The editor is highly competent on these topics and, like all of us, this person should be allowed to have views on important issues in society and influence political decisions.

On the other hand, there are good reasons to be concerned. An editor can strongly influence which issues to amplify and which to attenuate or even ignore. Being politically active allows for speculations about the journal having an agenda different than science. If the journal is seen as a vehicle for a political stand, the consequences could be dramatic for the journal's reputation.

Problem 7.11

From a risk science point of view, it cannot be seen as successful, as risk is not adequately described; see discussion in Section 3.3.2. The risk matrices can seriously misguide decision-makers.

Problem 8.1

It simplifies the risk judgments and characterizations, as it means fewer metrics and diagrams. However, it means that a value judgment has been made comparing, for example, multiple deaths with specific environmental damage or economic loss. This value judgment needs to be acknowledged by the decision-makers. Depending on the situation considered, it may be important to highlight the risks related to different types of consequences as a basis for stakeholder deliberations and political processes.

Problem 8.2

This is an unfortunate practice, as risk is poorly described by probability and (expected) consequences, and there are other concerns and issues than risk to consider when making a decision about acceptance or unacceptance and the need for risk-reducing measures. The management review and judgment are ignored. See discussion in Section 8.2.

Problem 8.3

An F-N curve is an informative risk metric but does not capture knowledge strength judgments. See arguments in relation to the previous problem.

Problem 8.4

The ALARA principle states that risk should be reduced to a level that is as low as reasonably achievable, whereas ALARP means that risk should be reduced to a level that is as low as reasonably practicable. There may be theoretical nuances between these two, but in practice, they are used interchangeably.

Problem 8.5

When the company is to specify the criteria, it would like to avoid 'unnecessary constraints' in the optimization of arrangements and measures.

Problem 8.6

A basic principle found in industry is the internal control principle, which states that the operator has the full responsibility for identifying the hazards and ensuring that they are controlled. It can, however, be argued that risk acceptance criteria formulated by the industry would not in general serve the interests of society as a whole. The main reason is that an operator's activity will usually cause negative externalities to society (an externality is an economically significant effect, due to the activities of an agent/firm, that does not influence the agent's/firm's production but influences other agents' decisions). The increased losses for society imply that society wants to adopt stricter risk acceptance criteria than those an operator finds optimal in its private optimization problem (Abrahamsen and Aven 2012).

Problem 8.7

Hedging is a risk management strategy in a finance context where one investment or trade is used to reduce the risk of another. For example, if you buy fire insurance, you are hedging yourself against fires. Another example: You have bought some stocks. You think the price will go up but want to reduce risk against the loss if the price drops. Then you can hedge that risk with a so-called put option, meaning that, for a small fee, you can buy the right to sell the stock at the same price. Thus, if the price falls, you can exercise your put and make back the money you just invested minus the fee.

Problem 8.8

Following Aven (2019a): The five criteria could be rephrased as:

1 There is a potential for extreme positive C values.
2 There is a potential for highly positive C values if the activity is cautiously executed – C is very sensitive to how the activity is run.
3 There is a potential for highly positive C values if a rather likely, specific type of event occurs
4 Weak knowledge about the consequences C of a specific type of activity, for example, about the effect of the use of a specific drug. There is a potential for highly positive C values.
5 There is a potential for highly positive C values, and these are subject to scientific uncertainties.

An example of criterion 3 could be that an opportunity arises as a result of a rival company going bankrupt. Criterion 5 can be interpreted as an 'anti-precautionary principle': If the consequences of an activity could be highly positive and subject to scientific uncertainties, then the activity should be carried out and supporting (stimulating) measures should be taken.

The anti-cautionary principle could reflect a high level of risk appetite in the sense of a strong willingness to take on a risky activity in pursuit of values, the riskiness depending of course on the potential for negative losses. Interpreting the risk appetite concept in this way (in line with SRA 2015), it is a broader concept than the anti-cautionary principle, as it also relates to situations with 'objective probabilities' and negligible uncertainties.

In contrast to the cautionary principle, which is established to protect values, the 'anti-cautionary' principle is designed to generate highly positive values. The 'anti-cautionary' principle represents an extreme form for development principle and, like all such risk management principles, needs to be balanced against other principles, including the cautionary principle. We can, for example, think of a situation where there is a potential for extreme consequences, both positive and negative, and the uncertainties are considerable. Then both of these principles are applicable, and they need to be balanced. Think about the Ebola outbreak in Africa. No cure existed, and many people died or suffered. A vaccine was developed with good effect, but according to international regulation, any new product has to pass a number of tests, usually requiring months of time. The (pre)cautionary principle that applies in medicine of this type will abandon use until all tests are carried out appropriately, since limited knowledge was available on any side effects of the new medicine. However, discussion arose whether the benefits (the wanted, desirable consequence C) were so promising that the (pre)cautionary principle should not be used. Giving weight to the anti-(pre)cautionary principle as defined previously, at the expense of the (pre)cautionary principle, would had led to the conclusion that the vaccine should have been implemented. See also Problem 8.9.

Problem 8.9

Following Aven and Renn (2018), there were major scientific uncertainties about the consequences of the swine flu. No reliable prediction model was available at the time. Risk assessment could have been performed, but, because of the uncertainties, the assessments provided only poor knowledge about the consequences and the fraction of people that would be affected. Yet the authorities needed to act to avoid serious damage. In most European nations, the authorities applied the precautionary principle, which states that, in the face of scientific uncertainties about the consequences of an activity, protective measures should be taken to reduce risks.

But what does the precautionary principle mean from the perspective of each individual who is confronted with the choice of getting vaccinated or not? Each person will be exposed to the side effects of the vaccination, again associated with uncertainties. The decision not to undertake vaccination can be interpreted as an application of the precautionary principle on the individual level. Many people did in fact select this option and avoided vaccination. From a scientific perspective, the odds of suffering from negative side effects caused by the vaccine were judged as significantly lower than the odds of contracting the disease. However, both judgments were associated with a high level of uncertainty, so that unanimous proof in the form of a clear-cut risk assessment was not available.

We are therefore left with a dilemma: the general rule of precaution can lead to different conclusions, depending on the choice of the default option and whose perspective we take. If we regard vaccination as the default option, we should make sure that almost everyone is vaccinated in order to be on the safe side when there is a danger that the flu might spread throughout a population. If, however, non-vaccination is the default option, we would opt for abstaining from any vaccination campaign, since there may be negative side effects associated with the vaccination. Both judgments can be justified in reference to the precautionary principle. The example demonstrates that the application of the precautionary principle cannot be seen in isolation from judgments of risk, uncertainties and other concerns. From an individual perspective, non-vaccination may be seen as the natural default option, and then the application of the precautionary principle needs to be balanced against the risk related to contracting the disease. From the societal point of view, the natural option is the opposite, and the application of the precautionary principle has to be balanced against the risk of getting serious side effects.

Problem 8.10

Ensure that a serious virus is not spreading outside a specific area or ensuring that a fire is not spreading to other areas.

Problem 8.11

Yes, we think so. We note that seizing opportunities is highlighted, which is not always given the attention it should be in risk management.

Problem 8.12

No, as expected values and the decision-maker's certainty equivalent are not easily determined or meaningful in the situations considered.

The certainty equivalent is the amount of payoff (e.g., money or utility) that the decision-maker has to receive to be indifferent between that payoff and the actual 'gamble'. Risk aversion means that the decision-maker's certainty equivalent is less than the expected value; a risk-seeking attitude means that the decision-maker's certainty equivalent is higher than the expected value.

Problem 8.13

The main motivation for applying the de minimis principle in risk management is that it helps us focus on the important risk issues and in this way use our scarce resources in an effective way. In principle, all risks should be paid attention to, and proper risk management does this. A key activity of risk management is to assess risk and implement measures to meet the risks. Such measures may include resilience-based strategies, which are designed to confront even unknown types of hazards and threats. However, in practice, there will always be constraints for risk management. An enterprise, municipality or society in general may face numerous risks and need some practical guidance on what is important and what is not. It cannot give all risks the same attention. A useful mean in this regard is then to be able to conclude, at a specific stage of the risk management process, that some risks are judged so small that they for all practical reasons can be ignored. When designing a technical system, it will be able to tackle some risks but not others. That is the reality of all design processes.

The critical questions then become: What criteria or processes should we apply to determine what is important and what is not? What is the basis for concluding that a risk can be ignored? We refer to discussion in Aven and Seif (2021). A key message is that to ignore a risk, the probability should be sufficiently low, with strong knowledge supporting this judgment.

Problem 8.14

Based on Aven (2010, p. 217). Nowadays it is known that asbestos is the main cause of mesothelioma, a disease with a very long incubation time, which, once it manifests, is normally fatal within one year. The deaths are in the millions. Mining for asbestos began more than 100 years ago – at that time, science was not aware of the dangers of asbestos. Early warnings were many, for example:

- 1898 UK Factory Inspector Lucy Deane warns of harmful and 'evil' effects of asbestos dust
- 1906 French factory report of 50 deaths in female asbestos textile workers and recommendation for controls
- 1911 "Reasonable grounds" for suspicion, from experiments on rats, that asbestos dust is harmful

- 1911 and 1917 UK Factory Department finds insufficient evidence to justify further actions
- 1930 UK 'Merewether Report' finds 66% of long-term workers in Rochdale factory with asbestosis
- 1931 UK asbestos regulations specify dust control in manufacturing only and compensation for asbestosis, but this is poorly implemented
- 1935–1949 Lung cancer cases reported in asbestos manufacturing workers
- 1955 Research by Richard Doll (UK) establishes high lung cancer risk in Rochdale asbestos workers
- 1959–1964 Mesothelioma cancer identified in workers, neighborhood 'bystanders' and the public in South Africa, the United Kingdom and the United States, among others

In 1998–1999, the EU and France banned all forms of asbestos, and in 2000–2001, the World Trade Organization upheld EU/French bans against Canadian appeal.

In the case of asbestos, a lack of full scientific proof of harm contributed to the long delay before action was taken and risk reduction regulation was put in place. The early warnings of 1898–1906 were not followed up by any kind of precautionary action to reduce exposure to asbestos, nor by long-term medical and dust exposure surveys of workers that would have been possible at the time and would have helped strengthen the case for tighter controls on dust levels. A Dutch study has estimated that a ban in 1965, when the mesothelioma hypothesis was plausible but unproven, instead of in 1993, when the hazard of asbestos was widely acknowledged, would have saved the country some 34,000 victims and €19 billion in building costs (clean-up) and compensation costs. This is in a context of 52,600 victims and €30 billion in costs projected by the Dutch Ministry of Health over the period 1969–2030. Today, a substantial legacy of health and contamination costs has been left for both mining and user countries, while asbestos use continues, now largely in developing countries.

Problem 8.15

The failure can be traced back to 1) the origin of the virus, 2) the lack of early control, and 3) the mitigation measures adopted, given its spread.

The first type of failure is difficult to analyze at this stage, but thorough investigations and efforts are needed to see whether there are some fundamental problems in our societies, creating this type of threat. Openness and international cooperation are required to meet this challenge, but it could be difficult to obtain necessary changes in activities that are grounded in traditional and cultural practices. Nonetheless, this issue is a major concern. Is it possible to avoid such a virus occurring? Nothing would be better than the solution to this question.

Second, it was a failure that the signals and warnings were not taken seriously, which led to a delay in actuating measures for the control of the spread. Health organizations have warned governments and politicians for years that a serious pandemic is likely to occur. National and global risk assessments all point to pandemics as a major risk. Yet the signals and warnings were to a large extent ignored. As such, the coronavirus pandemic is not to be seen as a black swan; its occurrence is not a surprising event relative to our knowledge. Maybe we could classify the event as a known unknown, indicating that there was knowledge that a serious pandemic could occur quite soon, but when it would occur was unknown to us. Surely there is a need to improve the way the world deals with these types of alarms and warnings. International cooperation is needed to develop relevant tools and institutions to look into this.

Third, governments struggle to justify investments in preparedness measures. There are many threats that we must be protected from and using a lot of public money for something that might happen requires a strong economy and strict prioritizations. However, for pandemic risks, the evidence has been strong and clear, but still the coronavirus spread has been enormous, whereas its mitigation has been weak in many respects. The problems that we are facing have demonstrated the extreme degree of vulnerability that our societies face today because of globalization. Necessary equipment is not available locally. All countries now depend on deliveries from specific producers often far away. Following the coronavirus case, fundamental pillars for how we have organized the world will be questioned. More decentralized systems will challenge the globalist perspective (Aven and Zio 2021).

Problem 8.16

Failures within the healthcare system, supply chains and social systems were prevalent. However, one can provide many examples of how the disruption created aspects of resilience, such as:

- Universities became better accustomed to online education. Online education can be used to contain and manage risk related to other outbreaks (whether pandemics or seasonal), such as influenza.
- The healthcare system learned how to more effectively: 1) develop aggressive vaccine development and distribution, 2) develop timely testing infrastructure and 3) pivot healthcare operations to treat differing distributions of illnesses and the like.
- Supply chains were able to pivot operations to: 1) brick-and-mortar transition to larger online operations and 2) accelerate production and distribution of pandemic response items.

The disruption demonstrated that organizations can quickly pivot operations to address changing conditions, thereby improving resilience to a wide variety of risk events, not just the coronavirus pandemic.

Problem 9.1

The expected rewards for the two projects are 45 + 100 + 50 = 195, and 60 + 60 + 100 = 220, respectively, which means that the expected cost per expected reward unit is 10/195 = 0.05 and 20/220 = 0.09. Thus, project a) has a cost-effectiveness ratio about half that of project b).

The indices provide decision support, not a prescription for what to do. Other factors also need to be taken into account, for example, the strength of the knowledge supporting the probability judgments. The probability distribution may influence the decision-making; for example, the decision-maker may find project b) attractive because of its higher probability of the high reward 1,000.

Problem 9.2

VSL: how much the decision-maker is willing to pay for reducing the expected number of lost lives by 1.

In a CBA, we compute the expected value related to loss of lives, so if we, for example, expect 0.2 lost lives, a value contribution of $0.2 \cdot 30 = 6$ is computed, which is plugged into the E[NPV] formula.

Problem 9.3

ICAF: expected cost per expected lives saved. VSL: max value the decision-maker is willing to pay for reducing the expected number of fatalities by 1.

If ICAF = 10 < VSL = 20, it is saying that the expected cost per saving an expected life is less than what one is willing to pay; hence, it may provide a justification for the implementation of a safety measure. Conversely, if ICAF > VSL, the cost of the measure is too high to be justified.

Problem 9.4

The ICAF for a safety measure calculates the expected cost per expected number of saved lives. A high ICAF means that the measure is not justified: it costs too much compared to the benefits gained, whereas if the ICAF is low, the costs are relatively low and the measure can be justified. The VSL provides a reference point for what is too high.

The problem is that the approach is based on expected values that do not capture all relevant aspects of benefits and costs. The potential for extreme outcomes is not reflected, nor the strength of the knowledge supporting the judgments made.

Problem 9.5

The answer is that the casino owners should not allow this game to be played, whatever amount a player is willing to stake, as the expected value of the casino payoff for one player is an infinitely large amount of dollars. To see that the expected value exceeds every conceivable large value, note first that the probability of getting heads in the first toss is $\frac{1}{2}$, the probability of getting heads for the first time at the second toss equals $\frac{1}{2} \cdot \frac{1}{2}$ and so on, and the probability of getting heads for the first time at the nth toss equals $(\frac{1}{2})^n$. The expected value of the game for one player is thus

$$\frac{1}{2} \cdot 2 + (\frac{1}{2})^2 \cdot 4 + \ldots + (\frac{1}{2})^n \cdot 2^n + \ldots$$

In this infinite series, a figure of $1 is added to the sum each time, and consequently the sum exceeds any fixed number.

In Bernoulli's days, a heated discussion grew up around this problem. Some indicated that the probabilistic analysis was wrong. But no, the mathematics is correct. However, the model is not a good description of the real world. The model tacitly assumes that the casino is always in a position to pay out, whatever happens, even in the case that heads shows up for the first time after, say, 30 tosses where the payoff is larger than $1,000 million. So, a more realistic model is developed if the casino can only pay out up to a limited amount. Then the expected payoff is limited and can be calculated. See Tijms (2007).

Problem 9.6

There were many stakeholders (company, neighbors, authorities), and one would expect differences in value judgments related to the environment, accident risk etc. Hence, it would be difficult to agree on the different weights to be used in a cost-benefit analysis. This problem could, however, be partly solved by using the CBA as a tool for improved understanding of the differences in view and priorities. There were, however, considerable uncertainties and risks, and these would not be adequately reflected by the CBA.

The utility-based approach is difficult to conduct. The company would not be willing to use time and resources to establish preferences and utility values over consequences with attributes related to costs, lives, long-term exposures, environmental damage and so on. The process is difficult to conduct and would depend on the stakeholders and, with the considerable uncertainties present, not really provide the decisive information.

Problem 9.7

The IRR is defined at the rate of return r such that NPV = 0.

Problem 9.8

An obvious approach would be to assess uncertainties about the NPV value for a fixed discount rate (typically a risk-free discount rate). By specifying probability distributions for the cash flows X_j, a probability distribution for the NPV can be derived, for example, using Monte Carlo simulation. The discount rate would then commonly be specified as a risk-free rate (corresponding to, for example, bank deposits). In addition, a strength of knowledge judgment should be made.

When dealing with uncertainty in project cash flows, we distinguish between systematic and non-systematic risk to the investor, who is commonly assumed to be a shareholder in possession of a well-diversified portfolio of securities. In projects, the term systematic risk (or non-diversifiable risk) refers to uncertainty in factors affecting the cash flow that are also related to other activities in the market like prices of energy and raw material, political situations and so on. Uncertainty in cash flow factors solely impacting the specific project, like operational delays, accidental events or dependency on critical personnel, is denoted non-systematic risk. It will not affect other investments made by the investor. Since the impact of non-systematic risk to the value of the investor's portfolio can be more or less eliminated by diversification, the systematic risk is the main focus in studies of project profitability, for example, NPV analysis.

Example to illustrate the idea of diversification (Aven 2010):

Suppose that you are considering investments in two stocks, i) and ii), with proportion w invested in stock i) and 1-w in stock ii). Let Y_1 and Y_2 be the returns for the two stocks and Y the total return, that is, $Y = wY_1 + (1 - w)Y_2$. It follows that the expected return equals

$$E[Y] = wE[Y_1] + (1 - w)E[Y_2]$$

and the variance is given by

$$Var[Y] = w^2 \, Var[Y_1] + (1 - w)^2 \, Var[Y_2] + 2w(1 - w)Cov(Y_1, Y_2), \qquad (C9.1)$$

where $Cov(Y_1, Y_2)$ is the covariance between Y_1 and Y_2. A positive covariance means that the returns move together, whereas a negative covariance means that they vary inversely (small returns in i) typically correspond to large returns in ii), and vice versa). Independence between Y_1 and Y_2 means a zero covariance.

To simplify, let us assume that the expected values for the stock returns both equal 1; that is, $E[Y_1] = E[Y_2] = 1$, and the variance of the stock returns both equal 0.04; that is, $Var[Y_1] = Var[Y_2] = 0.04$. Furthermore, assume that you consider two alternatives only,

1   Invest in stock i) only; that is, w = 1.
2   Invest in both stocks with equal shares; that is, w = ½.

It follows that E[Y] = 1 for both strategies, and the variance for Y when adopting strategy 1 equals 0.04. In the case of strategy 2, formula (C9.1) gives

$$Var[Y] = 0.01 + 0.01 + Cov(Y_1, Y_2)/2 = 0.02 + 0.02\rho = 0.02(1 + \rho), \qquad (C9.2)$$

where ρ is the correlation coefficient between Y_1 and Y_2, defined by

$\rho = Cov(Y_1, Y_2)/(SD[Y_1] \, SD[Y_2])$, where SD refers to the standard deviation.

The correlation coefficient is a normalized version of the covariance. It takes values in [−1,1]. Here +1 indicates perfect positive correlation and −1 perfect negative correlation.

We see from formula (C9.2) that if the correlation coefficient ρ is zero, the return variance for strategy 2 is just the half of the variance in the case of strategy 1. The uncertainties expressed by the variance are consequently smaller for strategy 2 than 1. By mixing the two stocks, we have reduced the uncertainties (expressed by the variance). Normally that would lead to the conclusion that strategy 2 should be preferred to strategy 1.

The diversification does not need to stop at two stocks. If we diversify into a number of stocks (N), the volatility (variance) would continue to fall. However, we cannot reduce the variance altogether, since all stocks are affected by common macroeconomic factors (such as political events, war etc.). The uncertainties that remain after extensive diversification are called systematic risk. In contrast, the risk that can be eliminated by diversification is called unsystematic risk. We illustrate the difference between these two risk categories by extending the example to N stocks. See the following box.

Variance for a portfolio of N stocks (Aven 2010)

Consider an investment in N different stocks, the same amount for each stock. Let Y_i be the return from stock i. Then the return of the total portfolio equals

$$Y = \frac{1}{N}\sum_{i=1}^{N} Y_i$$

The expected return EY equals:

$$EY = E\frac{1}{N}\sum_{i=1}^{N} Y_i = \frac{1}{N}\sum_{i=1}^{N} EY_i,$$

and the variance equals

$$\text{Var } Y = Var(\frac{1}{N}\sum_{i=1}^{N} Y_i) = \sum_{i=1}^{N}\left(\frac{1}{N}\right)^2 Var(Y_i) + \sum_{i=1}^{N}\sum_{j\neq i,j=1}^{N}\left(\frac{1}{N}\right)^2 Cov(Y_i,Y_j)$$

$$= \frac{1}{N}\overline{VAR} + \left(1 - \frac{1}{N}\right)\overline{COV} \qquad (C9.3)$$

where $\overline{VAR} = \frac{1}{N}\sum_{i=1}^{N} Var(Y_i)$ and $\overline{COV} = \frac{1}{N^2 - N}\sum_{i=1}^{N}\sum_{j\neq i,j=1}^{N} Cov(Y_i,Y_j)$.

Assuming that the variance $VarY_i$ is bounded by a constant c, the average $\overline{VAR}$ is bounded by c, and hence the first term of (C9.3) becomes negligible for large N. For large N, VarY approaches $\overline{COV}$. The terms of formula (C9.3) reflect the unsystematic risk and the systematic risk, respectively. When the number of stocks is large, we see from (C9.3) that the variance for the portfolio is approximately equal to the average covariance. The unsystematic risk is negligible when N is sufficiently large.

The reference for the variance is the expected return. Another risk measure, the beta index (β), is based on a comparison with the market return. If the beta is close to 1, the stock's or fund's performance closely matches the market return. A beta greater than 1 indicates greater volatility than the overall market, whereas a beta less than 1 indicates less volatility than the market.

If, for example, a fund has a beta of 1.10 in relation to a specific market index, the fund has moved 10% more than this index. Hence, if the market index increased by 20%, the fund would be expected to increase 22%. The index beta is formally defined by

$$\beta = \frac{Cov(Y,M)}{Var(M)},$$

where Y is the fund return and M the market return.

The CAPM model referred to in Section 9.3 is used to determine the discount rate. This model produces a rationale for a discount rate given by the following formula:

$$r + \beta(E[M] - r), \tag{C9.4}$$

where r is the risk-free rate. Equation (C9.4) shows that CAPM determines the rate as the sum of the risk-free rate of return and β multiplied by the so-called 'risk premium' of the market $E[M] - r$. We may interpret the β parameter as the number of 'systematic risk units'. Thus, the risk contribution is expressed by the risk premium of the market multiplied by the number of units of systematic risk.

Problem 9.9

Arrow proves that in all cases where preferences are ranked, it is impossible to formulate a social ordering without violating one of the following conditions:

1 Non-Dictatorship: The wishes of multiple voters should be taken into consideration.
2 Pareto Efficiency: Unanimous individual preferences must be respected: If every voter prefers candidate A over candidate B, candidate A should win.
3 Independence of Irrelevant Alternatives: If a choice is removed, then the others' order should not change: If candidate A ranks ahead of candidate B, candidate A should still be ahead of candidate B, even if a third candidate, candidate C, is removed from participation.
4) Unrestricted Domain: Voting must account for all individual preferences.
5) Social Ordering: Each individual should be able to order the choices in any way and indicate ties.

Here is an illustration of the result: Voters are asked to rank their preference of candidates A, B and C:

- 45 votes A > B > C (45 people prefer A over B and prefer B over C)
- 40 votes B > C > A (40 people prefer B over C and prefer C over A)
- 30 votes C > A > B (30 people prefer C over A and prefer A over B)

Candidate A has the most votes, so she/he would be the winner. However, if B were not running, C would be the winner, as more people prefer C over A. (A would have 45 votes and C would have 70).

Problem 9.10

Returning to the setup of Section 9.2, a cost-effectiveness index, $E[C]/E[B]$, was computed, where C denotes the cost of the measure and B the benefits of the measure, for example, expressed through the number of saved lives. Following the ideas outlined for NPV in the previous problem, a probability distribution for C/B is derived, including a strength of knowledge judgment.

See Aven and Flage (2009) for a discussion of alternative approaches.

Problem 9.11

ENPV = − expected costs + VSL · 0.01 = −2 + VSL · 0.01, where VSL is the value of a statistical life (how much the decision-maker is willing to pay to reduce the expected number of fatalities by 1). We assume no discounting, as we would like to give the same weight to life today and in the future.

Problem 9.12

The use of criteria of the form I and III are discussed in Section 8.3 and Problem 8.13. With a typical 'weak limit' for the intolerable risk and 'strong limit' for acceptable (negligible) risk, most risks will be in the ALARP region. Here risk reduction processes are implemented, but, as discussed in Section 9.5.1, implementing ALARP in practice is not simple.

Problem 9.13

Yes, cost-effectiveness and cost-benefit analyses would be useful in this context, as expected values could be accurate predictions of the actual numbers involved. However, if there are some new and unique elements related to the treatment leading to uncertainties, care needs to be shown when using the cost-effectiveness and cost-benefit analyses, as the expected values could lead to poor predictions.

Problem 9.14

The expected utility approach is a normative theory expressing how people should act, and it has a strong rationale. The descriptive research showing that people do not follow the theory is interesting but does not reduce the normative message provided by the theory.

A key point is that people care more about the losses than same-sized gains, which is not reflected in the expected utility theory. Another idea violating the expected utility approach is that people place extra value on certain outcomes, so going from 90% to 100% means much more than going from 40% to 50% (Fischhoff and Kadvany 2011). According to prospect theory, there is a reference point to which people relate their gains and losses, which could be pointing to where they are now, where they expect to be or where someone else is.

The key idea of the rank-dependent utility theory is to allow a different weighing of the probabilities than in the expected utility approach, reflecting that value of an outcome depends on the probability of realizing that outcome and the ranking of the outcome relative to the other possible outcomes.

For further reading, see, for example, Tversky and Kahneman (1992), Diecidue and Wakker (2001), Machina (2008) and Fischhoff and Kadvany (2011).

Problem 9.15

1. Value of a statistical life refers to maximum value the decision-maker is willing to pay to reduce the expected number of fatalities by 1. This interpretation is in reference to risk, conceptually including some version of an expected value function that considers both consequences and likelihoods, which can be interpreted for a single person or broadly across some population group. While VSL is a willingness to pay to reduce risk, the value of a single human life is infinite, as described in Section 9.3.
2. Both terms place monetary value on life but with recognition of key differences described subsequently.
3. VSL is in reference to a measure of risk, in reference to some future event that can occur with some consequence and likelihoods. In contrast, the value of a single human life is not in reference to risk; there is no notion of some risk event, with any consequence or likelihood assumed.

Problem 10.1

This risk definition we consider not understandable, and the likelihood definition is not really a definition, as chance is not explained, and concept and measurement are mixed; see detailed discussion in Aven and Ylönen (2018).

Problem 10.2

It has been referred to as model uncertainty in the literature, but we do not use this terminology in this book. Rather we consider the issue as being about choosing a model. The analysts may conclude to use one of the models, or both, showing results conditional on the use of each of the models. A combined model could also be considered.

Problem 10.3

C and C' are uncertain, the analysts are uncertain and the uncertainties are represented or expressed by (C',Q,K).

Problem 10.4

It is common to refer to established practices and standards, which makes sense, as these are derived on the basis of experience. However, the scientific quality of the standards can be of a varying level, and the reference should always be risk science knowledge and recommendations. In some situations, the use of alternative approaches can provide additional insights, but it will be more resource demanding.

Problem 10.5

The analysts should be able to explain the key contributors and drivers for the results obtained. If the results are not as expected, explanations should be given. If that is difficult, it may indicate that something is wrong in the calculations.

Problem 10.6

Risk analysis (science) is about concepts, principles, approaches, methods and models for understanding, assessing, characterizing, communicating and handling risk, and is as such politically neutral. However, risk analysis (science) *can be used* for different purposes, also political. Many risk assessments are, for example, applied to demonstrate that a particular issue represents a serious societal problem by addressing some specific topics and not including others. Professional risk analysts should, however, be strict on clarifying the premises of the analyses and the key assumptions made, what is science and what is value judgments and politics.

Problem 10.7

Feedback should be produced pointing to the weaknesses of the risk assessment process and encouraging improvements to be made. The criteria can be met in different ways; at a minimum, they should be addressed.

Problem 10.8

Yes, practical risk analysis is commonly a multidisciplinary/interdisciplinary activity, as the assessment and handling of risk issues typically require involvement from many disciplines. A risk analysis of climate change risk may involve natural scientists, economists, sociologists, statisticians, risk scientists and so on.

Problem 10.9

Answers may include:

- News articles that use language that promotes the reader to take a certain 'side' on an issue
- Selection of scientific articles supporting one perspective only
- Ignoring issues of uncertainties
- Social media memes, photos or videos that discuss biased views of the issue
- Charts/graphs that are misleading
- Social media posts that have been flagged for reasons of integrity or other content-related issues
- Social media platforms censoring particular viewpoints
- Information from lobbyist organizations
- Lobbying from businesses encouraging a particular stance on the issue

Problem 10.10

Students who advocate for objectivity may cite:

- Data can come from precise measurements
- The risk assessment characterizes risk on the basis of the available knowledge using risk science principles and methods
- The risk assessment methods can be viewed as objective in the sense that they use systematic and transparent means to understand and describe risk

Students who advocate for lack of objectivity may cite:

- Data used may be relevant to varying degrees
- Models used are based on assumptions that can be contested
- Biases and cognitive limitations exist
- Risk assessments are based on judgments (for example, when measuring or describing uncertainties), which depend on the assessors
- The knowledge supporting the risk judgments are someone's beliefs – they are not universal truths, independent of the assessors

Problem 10.11

Based on the discussion in this book, we would point to the following success criteria and pitfalls:

Success criteria

- Strong foundational basis, with precision and understandable interpretations of all concepts
- Clearly defined objectives of the risk assessment, highlighting the assessment as a tool for understanding and characterizing the risk
- Add considerations of knowledge strength to probabilistic judgments
- Address potential surprises and the unforeseen, using different approaches, including evaluations of the assumptions made
- In the results, balance confidence and humbleness, as discussed in Section 10.3.2.
- Strengthen the knowledge basis of the assessment to the degree possible

Pitfalls

- Vague concepts
- Vague objectives of the risk assessment
- Pure quantitative focus, ignoring many aspects of risk and uncertainties
- Too much weight on 'confidence' or too much weight on 'humbleness'
- The supporting knowledge basis is very weak

Problem 11.1

The problems with using such metrics in describing risk are well known; see, for example, Cox (2008), Flage and Røed (2012) and Aven and Cox (2016) and Section 3.3.2. Some of the main issues are:

- Often the events plotted are not well defined.
- The consequences of the events are in many cases not well reflected by a single point in the matrix but by several with different probabilities. If we restrict attention to one point, we will often think of this value as being the 'expected value', which is the centre of gravity of the probability distribution for the appropriate consequences. In most cases, this value is not very informative in showing the consequence dimension.
- The meaning of the concept of probability is often not explained. Is probability a tool for expressing the analyst's degree of belief or uncertainty, or is it used to reflect variation?
- Two events can have the same location in the risk matrix, but the knowledge supporting these judgments could be completely different: in one case, the knowledge is very strong; in another case, it is very weak. This is not shown in the matrix, and the risk description could be misleading. It is static without the ability to reflect changes in the knowledge.
- Colors are often used in the matrix, indicating that the risk is unacceptable, acceptable, should be reduced and so on. Such a scheme should be avoided, as mechanical conclusions on the basis of likelihood and consequences could be rather arbitrary, not taking into account important aspects of the decision problem, such as the knowledge dimension.

If risk matrices are to be used to summarize the risk description, they need to be supplemented with strength of knowledge judgments.

Summarized, the key problems of using risk matrices are that the full spectrum of consequences is not revealed, nor the strength of the knowledge supporting the probabilities. See alternative approaches presented in Section 3.3.2 and Problem 3.26.

It is also common to use risk matrices on the basis of scores (for example, 1, 2, 3, 4 and 5) for the dimensions consequences and probability and then produce a risk number given by the product. As discussed in Section 3.1.1, expected values are in most cases a poor risk index, but using scores like 1, 2, 3, 4 or 5, the implications are even worse, as the scores do not adequately reflect the real magnitude of the dimensions consequences and probability; see example in Aven (2015b, pp. 144–145).

For the analysis in Section 11.1, we have not considered risk matrices at all; rather, we have sought to highlight a broader information and knowledge basis, capturing aspects such as assumption deviation risk, strength of knowledge and robustness/resilience.

Problem 11.2

To this, it can be argued that we need to prioritize between different risks and measures, and the issue is how to be best informed. We may conclude that some risks need to be reduced and given top priority, but such a conclusion is in most cases better supported if a judgment about the probability can be made. An interval scale commonly balances the need for accuracy without being arbitrary. In the example, we did not use pre-defined probability scales, but this is common, for example: unlikely (≤ 0.05), less likely $(0.05 - 0.20)$, likely $(0.20 - 0.50)$ and very likely (> 0.50).

Problem 11.3

At an early stage of the coronavirus pandemic, the uncertainties were large, and there were reasons to be concerned. These could be based on conscious judgments of risks. With more knowledge, the risk of being infected at a grocery store given implementation of the standard routines is considered low. If still concerned, such factors may be the explanation, not conscious judgments about the risks.

Problem 11.4

Science does not prescribe what is the best policy, but it can give important input in the form of knowledge. However, knowledge is not fact or the truth (reality) but a statement about the truth (reality), and this statement can be more or less supported, and even wrong, and it is almost always contested. Science is not one voice but rather a 'battle' about what the truth is. The media often acts to hide this battle or amplify it, depending what its view is on the matter.

Problem 11.5

It is then more difficult to use and establish easily understandable explanations. However, often it is possible to construct a thought-constructed population of, say, 1,000 similar activities and refer to predictions for the total number.

Problem 11.6

Answers may include discussion of 1) the goal should always be to achieve as high of a maturity level of possible, 2) advanced maturity may not be feasible given the size of the organization and available resources, 3) A beginner level is preferable to no program at all, so too ambitious of a short-term goal may derail efforts to form a basic risk program and 4) given that the pandemic example relates to the health and safety of communities, it is even more critical to achieve advanced maturity.

Problem B.1

It is generic, as it is applicable for all types of applications. It relates to how to quantify risk.

Problem B.2

We refer to discussion in Aven and Ylönen (2021).

The differences relate to a number of technical issues, but the key point is the following: is the perspective of your analysis that there is an underlying 'true' probability (commonly referred to as a frequentist probability) reflecting an inherent property of the activity considered and the aim of your analysis is to accurately estimate this probability, or is probability a way of expressing uncertainty about unknown quantities? In the former case, variation in populations is the fundamental notion providing interpretations of the probability and related metrics, such as variance and quantiles, as well as of concepts like confidence intervals and significance levels in hypothesis testing. This variation forms 'objective' but unknown quantities for science to accurately determine using modeling and data samples from the relevant population. In the latter case, the uncertainty of the assessor is the fundamental notion. It is subjective and based on the knowledge of the assessor. Using Bayes' formula, a systematic approach exists for updating of the probability when new data and information become available.

To illustrate the difference between these two perspectives, think about a fair coin being thrown. Before the throw, both a frequentist and a Bayesian would express a probability of 50% for the coin showing heads, but what happens when the coin has been thrown but the outcome is not revealed? Thus it is either heads or tails but unknown to you. A Bayesian would state that the probability is still 50%, as it just reflects the person's uncertainty about the coin showing heads. No new information has been added. A frequentist would argue differently. Given the throw, the probability is 1 or 0 – 1 if the coin is showing heads up and 0 otherwise. For the frequentists, however, the interest is not post-observation measurements but the underlying true fraction in the population showing heads when mentally constructing an infinite population of such throws and how to estimate that fraction.

The frequentist perspective is supported by a view on science aiming at developing objective knowledge, contrasting the Bayesian approach, which is personal, although it can be argued that the process of processing data using Bayes' formula to a large extent is objective, independent of the analyst. The Bayesian approach allows the analysis to take into account all type of evidence, not only in form of hard data. For example, in a law context where a defendant has been charged with an offence, subjective probability judgments and Bayesian updating can provide useful decision support, but a frequentist approach would not work very well, as the case is unique and the evidence is not only about hard frequency data.

The aim here is not to present all the arguments in favor of each perspective and review the history of the debate between frequentists and Bayesians (interested readers should consult, for example, Efron 2005; Lillestøl 2014). The discussion through the 1950s and the following two three decades was strong and also harsh regarding what the right perspective was. Frequentists argued that Bayesians were unscientific, whereas Bayesians argued that the frequentist approach was not able to provide adequate decision support in many cases. The debate was to a large extent general but also specifically addressing risk issues and risk science; see, for example, Singpurwalla (1988) and Apostolakis (1990).

The tension between these two camps today is more or less non-existent. The debate faded out. We have seen both defensive and active responses to this tension. The active is our focus here, and it involves different strategies (confronting, adjusting and transcending strategies; see Aven and Ylönen 2021). The debate clarified the issues and led to better understanding of the different perspectives, what their foundation was and what purpose it served. It became clear that both perspectives were useful, meeting different situations and aims. Overriding frameworks were developed bridging the gap between the two perspectives. Bayesians also needed probability models for their inference – models of the variation in populations and repeated experiments. In risk analysis, it was referred to the frequency-of-probability approach, where frequency referred to the underlying frequentist probabilities and probability to the subjective probability expressing the analysts' uncertainty about the frequentist probabilities (Kaplan and Garrick 1981; Paté-Cornell 1996; Aven 2012b). Depending on the situation analyzed, different approaches may be useful. It is a pragmatic, instrumental perspective highlighting that different situations call for different perspectives and tools. Research and development are needed within each perspective and tool to prepare them for applications whenever relevant. Today analysts may prefer using a frequentist approach in some situations and a Bayesian one in others.

Such an overriding perspective is also the common risk science perspective today. If one studies publications in top journals of this science, both frequentist and Bayesian perspectives are used. When conceptualizing and describing risk, both perspectives are used. To explain in more detail, consider the risk framework presented in Chapters 2 and 3, in which risk as a concept covers two main features, the consequences C of an activity and associated uncertainties (U), for short denoted (C,U). In its most general form, risk is characterized by (C',Q,K), where C' is some specified consequences, Q a description of the uncertainties U and K the supporting knowledge. A Bayesian analysis can provide input to the risk characterization when Q equals a subjective probability P. If a probability model is introduced – with an unknown parameter θ, the model is a part of K, and C' includes θ. Also, frequentist analysis can be included in this risk framework. The frequentist probabilities are unknown quantities and hence can be viewed as elements of C'. The uncertainties about these frequentist probabilities P_f are not explicitly described but indirectly using variance and confidence intervals, as these express how data sample observations would vary relative to the true underlying frequentist probabilities. Only data variation is covered, not analyst uncertainties because of, for example, data used not being relevant for the situation considered. Risk is in this framework defined by the pair (P_f,U), not P_f, as is common in other frameworks supporting frequentist probabilities. The issue is, for example, whether risk is the fraction in a population having a disease or the combination of this fraction and uncertainties about it. We will leave that philosophical debate here; what matters for the practical risk analysis is the risk characterization, which will be similar. See also Aven (2011a).

Problem B.3

Probability is the common tool used to describe the epistemic uncertainties about unknown quantities. Several authors argue that the problems with using probability in practice lies with the measurement procedures and not with the probability concept itself (Bernardo and Smith 1994; Lindley 2006; O'Hagan and Oakley 2004). Following this line of thought, probability is retained as the sole representation of uncertainty, and the focus is on improving the measurement procedures – no alternative to probability is needed, as "probability is perfect" (Lindley 2006; O'Hagan and Oakley 2004). Supporters of this thesis may acknowledge that there is a problem of imprecision in probability assignments, but this is considered a problem related to the elicitation of probabilities and not a problem of the probability concept (Aven et al. 2014).

Other scholars reject this idea that probability is perfect. The main problem pointed to is that the transformation from knowledge to probability is subjective, adding information that was not available at the point of the assignment. Alternative approaches have been suggested to make this transformation more objective, including probability bound analysis, possibility theory and evidence theory (Dubois 2010; Flage et al. 2014; Aven et al. 2014).

The point here is not to discuss all the details of this controversy but reflect on how the tension has evolved over time. Again we see that overriding solutions and frameworks have been developed (see e.g., Aven et al. 2014). During the last decade, the risk science community has clarified what the different perspectives mean, their strengths and weaknesses and how they are to be used in practical situations. A key point addressed is the importance of distinguishing between the objectivity of the transformation from knowledge to probability and the strength of this knowledge. For risk assessment, objectivity in the transformation may be attractive, but if the supporting knowledge is weak, the assessment would not provide much insights. It is shown that depending on the aim of the analysis and the decision-making context, different approaches can be justified, for example, using both precise and imprecise probabilities. For applications, simplicity and easily understandable ideas are also important criteria, and some of the non-probabilistic approaches are less attractive than probability in this regard.

Fuzzy probabilities are another type of non-probabilistic tool introduced to meet the challenges of subjective probabilities. Fuzzy probability is used to reflect ambiguous statements (like "few failures"). Although commonly used, there is an ongoing controversy between advocators of these probabilities and opponents who argue that these probabilities should not be used, as no meaningful interpretation can be provided. Authors such as Bedford and Cooke (2001) reject such probabilities, as do the present authors. The point being made is that, for any concept to be used for uncertainty characterization, it needs to be defined in a precise way. We can always include data and information that are vague and imprecise in the analysis as a part of the background knowledge for the assignments being made. However, the quantity of interest needs to be well defined, having some true underlying values that can be specified in a meaningful way. There does not seem to be a basis for integrating fuzzy probabilities and 'standard' probabilities (precise and imprecise).

We again refer to discussion in Aven and Ylönen (2021).

Problem B.4

We will argue that the controversy is to a large extent rooted in the definition and understanding of risk. Referring to the quote from Linkov et al. (2014), the critical question then becomes, what is the event referring to? Think about the classical bow tie (see Figure 4.4) used in risk assessments, with an event in the middle, causes to the left and consequences to the right of the event. Risk assessments cover all aspects of the bow tie: events, causes and consequences, as does risk management. However, if the Linkov et al. (2014) perspective applied, risk is only concerned with the left part of the bow tie. Risk science cannot accept such a terminology, as it conflicts the very basic ideas of the risk concept, risk assessment and management. An alternative understanding of the resilience stand would be to interpret the events (the adverse events referred to by Linkov et al. 2014) as accidents with their immediate consequences included. That would provide some rationale, as traditional risk assessments typically stop their analysis at the occurrence of the accident and do not include the recovery phase following the accident. However, adopting more recent risk frameworks, such as the (C,U) perspective adopted in this book, there is no restriction on allowing risk assessment and management also to cover the recovery phase; it is just a matter of defining the consequences C in an appropriate way.

Active response to the tension has been given by clarifying what the issues are about – as previously – also including discussing the issues at conferences. As the previous analysis shows, it seems that the debate is very much about terminology and not fundamental differences in how to handle risk and resilience. Both parties consider the need for both risk and resilience. Yet it seems difficult to solve the issue. It is about what 'school' and terms to highlight, which could be important in relation to funding of research, societal relevance and scientific recognition. Currently resilience is a hot topic in many contexts, for example, in relation to climate change, with many openings for projects and research. Then there is motivation for extending the scope of this field. Risk science welcomes these opportunities and acknowledges the importance of resilience science, but it needs to be properly integrated with current risk knowledge to the benefit of the issues raised, such as climate change. Then we need to join forces, using the relevant terms and marketing this knowledge in the best possible way, taking into account what the current 'selling points' are.

References

Abel (2020) www.abelprize.no/. Accessed November 12, 2020.

Abrahamsen, E.B. and Aven, T. (2012) Why risk acceptance criteria need to be defined by the authorities and not the industry. *Reliability Engineering and System Safety*, 105, 47–50.

Abrahamsen, E.B., Aven, T., Flage, R., Engen, O.A., Røed, W. and Wiencke, H.S. (2020) *Use of Risk Acceptance Criteria* (In Norwegian). Stavanger: Report Proactima.

Abrahamsen, E.B., Moharamzadeh, A., Abrahamsen, H.B., Asche, F., Heide, B. and Milazzo, M.F. (2018). Are too many safety measures crowding each other out? *Reliability Engineering and System Safety*, 174, 108–113.

AIHC (1989) *American Industrial Health Council, US Environmental Protection Agency, Department of Health and Human Services, and Society for Risk Analysis*. Presentation of Risk Assessments of Carcinogens: Report of an Ad Hoc Study Group on Risk Assessment Presentation. Washington, DC: American Industrial Health Council.

AIHC (1992) *American Industrial Health Council, Center for Risk Management (Resources for the Future), and US Environmental Protection Agency. Improving Risk Characterization: Summary of a Workshop Held in Washington, DC on 26 and 27 September 1991*. Washington, DC: American Industrial Health Council.

Ale, B.J.M. (2016) Risk analysis and big data. *Safety and Reliability*, 36(3), 153–165.

Ale, B.J.M., Hartford, D.N.D. and Slater, D. (2015) ALARP and CBA all in the same game. *Safety Science*, 76, 90–100.

Amundrud, Ø., Aven, T. and Flage, R. (2017) How the definition of security risk can be made compatible with safety definitions. *Journal of Risk and Reliability*, 231(3), 286–294.

Andersen, I.E. and Jaeger, B. (1999) Scenario workshops and consensus conferences: Towards more democratic decision-making. *Science and Public Policy*, 26, 331–340.

Anderson, D.R., Sweeney, D.J., Jeffrey, T.A., Camm, J.D. and Cochran, J.J. (2020) *Statistics for Business & Economics*. 14th ed. New York: Cengage Learning.

Apostolakis, G.E. (1990) The concept of probability in safety assessments of technological systems. *Science*, 250, 1359–1364.

Apostolakis, G.E. (2004) How useful is quantitative risk assessment? *Risk Analysis*, 24, 515–520.

Apostolakis, G.E. (2006) *PRA/QRA: An Historical Perspective*. Probabilistic/Quantitative Risk Assessment Workshop, Taiwan, November 29–30.

ARMSG (2019) *Risk Analysis Quality Test*. Applied Risk Management Specialty Group. www.sra.org/wp-content/uploads/2020/08/SRA-Risk-Analysis-Quality-Test-R6.pdf. Accessed November 12, 2020.

Arrow, K.J. (1951) *Social Choice and Individual Values*. New York: Wiley.

Árvai, J. (2014) The end of risk communication as we know it. *Journal of Risk Research*, 17(10), 1245–1249.

Aven, E. and Aven, T. (2015) On the need for rethinking current practice which highlights goal achievement risk in an enterprise context. *Risk Analysis*, 35(9), 1706–1716.

Aven, T. (2007) On the ethical justification for the use of risk acceptance criteria. *Risk Analysis*, 27(2), 303–312.

Aven, T. (2010) *Misconceptions of Risk*. Chichester: Wiley.

Aven, T. (2011a) *Quantitative Risk Assessment: The Scientific Platform*. Cambridge: Cambridge University Press.

Aven, T. (2011b) Selective critique of risk assessments with recommendations for improving methodology and practice. *Reliability Engineering and System Safety*, 96, 509–514.

Aven, T. (2011c) On different types of uncertainties in the context of the precautionary principle. *Risk Analysis*, 31(10), 1515–1525. With discussion 1538–1542.

Aven, T. (2012a) The risk concept – historical and recent development trends. *Reliability Engineering and System Safety*, 99, 33–44.

Aven, T. (2012b) *Foundations of Risk Analysis*. 2nd ed. Chichester: Wiley.

Aven, T. (2013) Practical implications of the new risk perspectives. *Reliability Engineering and System Safety*, 115, 136–145.

Aven, T. (2014) *Risk, Surprises and Black Swans*. New York: Routledge.

Aven, T. (2015a) On the allegations that small risks are treated out of proportion to their importance. *Reliability Engineering and System Safety*, 140, 116–121.

Aven, T. (2015b) *Risk Analysis*. 2nd ed. Chichester: Wiley.

Aven, T. (2015c) Implications of black swans to the foundations and practice of risk assessment and management. *Reliability Engineering and System Safety*, 134, 83–91.

Aven, T. (2015d) The concept of antifragility and its implications for the practice of risk analysis. *Risk Analysis*, 35(3), 476–483.

Aven, T. (2016) Risk assessment and risk management: Review of recent advances on their foundation. *European Journal of Operational Research*, 25, 1–13.

Aven, T. (2017a) How some types of risk assessments can support resilience analysis and management. *Reliability Engineering and System Safety*, 167, 536–543.

Aven, T. (2017b) Improving risk characterisations in practical situations by highlighting knowledge aspects, with applications to risk matrices. *Reliability Engineering and System Safety*, 167, 42–48.

Aven, T. (2017c) On some foundational issues related to cost-benefit and risk. *International Journal of Business Continuity and Risk Management*, 7(3), 182–191.

Aven, T. (2018a) An emerging new risk analysis science: Foundations and implications. *Risk Analysis*, 38(5), 876–888.

Aven, T. (2018b) How the integration of system 1–system 2 thinking and recent risk perspectives can improve risk assessment and management. *Reliability Engineering and System Safety*, 20, 237–244.

Aven, T. (2018c) Perspectives on the nexus between good risk communication and high scientific risk analysis quality. *Reliability Engineering and System Safety*, 178, 290–296.

Aven, T. (2019a) The cautionary principle in risk management: Foundation and practical use. *Reliability Engineering and System Safety*, 191, 106585.

Aven, T. (2019b) Climate change risk – what is it and how should it be expressed? *Journal of Risk Research*, 23(11), 1387–1403.

Aven, T. (2019c) The call for a shift from risk to resilience: What does it mean? *Risk Analysis*, 39(6), 1196–1203.

Aven, T. (2020a) *The Science of Risk Analysis*. New York: Routledge.

Aven, T. (2020b) Three influential risk foundation papers from the 80s and 90s: Are they still state-of-the-art? *Reliability Engineering and System Safety*, 193, 106680.

Aven, T. (2020c) Bayesian analysis: Critical issues related to its scope and boundaries in a risk context. *Reliability Engineering and System Safety*, 204, 107209.

Aven, T. (2020d) Risk science contributions – three illustrating examples. *Risk Analysis*, 40(10), 1889–1899.

Aven, T. (2020e) How to determine the largest global and national risks: Review and discussion. *Reliability Engineering and System Safety*, 199, 106905.

Aven, T. (2021) The reliability science: its foundation and link to risk science and other sciences. *Reliability Engineering and System Safety*. doi: 10.1016/j.ress.2021.107863.

Aven, T., Baraldi, P., Flage, R. and Zio, E. (2014) *Uncertainty in Risk Assessment*. Chichester: Wiley.

Aven, T. and Bouder, F. (2020) The COVID-19 pandemic: How can risk science help? *Journal of Risk Research*, 23(7–8), 849–854.

Aven, T. and Cox, T. (2016) National and global risk studies: How can the field of risk analysis contribute? *Risk Analysis*, 36(2), 186–190.

Aven, T. and Flage, R. (2009) Use of decision criteria based on expected values to support decision-making in a production assurance and safety setting. *Reliability Engineering and System Safety*, 94, 1491–1498.

Aven, T. and Flage, R. (2018) Risk assessment with broad uncertainty and knowledge characterisations: An illustrating case study. In T. Aven and E. Zio (eds.), *Knowledge in Risk Assessments*. New York: Wiley, pp. 3–26.

Aven, T. and Flage, R. (2020) Foundational challenges for advancing the field and discipline of risk analysis. *Risk Analysis*, 40(S1), 2128–2136.

Aven, T. and Heide, B. (2009) Reliability and validity of risk analysis. *Reliability Engineering and System Safety*, 94, 1862–1868.

Aven, T. and Jensen, U. (2013) *Stochastic Models in Reliability*. 2nd ed. New York: Springer Verlag.

Aven, T. and Kristensen, V. (2019) How the distinction between general knowledge and specific knowledge can improve the foundation and practice of risk assessment and risk-informed decision-making. *Reliability Engineering and System Safety*, 191, 106553.

Aven, T. and Krohn, B.S. (2014) A new perspective on how to understand, assess and manage risk and the unforeseen. *Reliability Engineering and System Safety*, 121, 1–10.

Aven, T. and Nøkland, T.E. (2010) On the use of uncertainty importance measures in reliability and risk analysis. *Reliability Engineering and System Safety*, 95, 127–133.

Aven, T. and Reniers, G. (2013) How to define and interpret a probability in a risk and safety setting: Discussion paper, with general introduction by associate editor, Genserik Reniers. *Safety Science*, 51, 223–231.

Aven, T. and Renn, O. (2009) On risk defined as an event where the outcome is uncertain. *Journal of Risk Research*, 12, 1–11.

Aven, T. and Renn, O. (2010) *Risk Management and Risk Governance*. New York: Springer Verlag.

Aven, T. and Renn, O. (2018) Improving government policy on risk: Eight key principles. *Reliability Engineering and System Safety*, 176, 230–241.

Aven, T. and Renn, O. (2019) Some foundational issues related to risk governance and different types of risks. *Journal of Risk Research*, 23(9), 1121–1134.

Aven, T., Renn, O. and Rosa, E. (2011) On the ontological status of the concept of risk. *Safety Science*, 49, 1074–1079.

Aven, T. and Seif, A. (2021) On the foundation and use of the de minimis principle in a risk analysis context. *Journal of Risk and Reliability*. doi: 10.1177/1748006X211028401.

Aven, T. and Thekdi, S. (2020) *Enterprise Risk Management: Advances on Its Foundation and Practice*. New York: Routledge.

Aven, T. and Thekdi, S. (2021) On how to characterize and confront misinformation in a risk context. Forthcoming.

Aven, T. and Vinnem, J.E. (2007) *Risk Management*. New York: Springer Verlag.

Aven, T. and Ylönen, M. (2018) A risk interpretation of sociotechnical safety perspectives. *Reliability Engineering and System Safety*, 175, 13–18.

Aven, T. and Ylönen, M. (2019) The strong power of standards in the safety and risk fields: A threat to proper developments of these fields? *Reliability Engineering and System Safety*, 189, 279–286.

Aven, T. and Ylönen, M. (2020) How the risk science can help us establish a good safety culture. *Journal of Risk Research*, 19.

Aven, T. and Ylönen, M. (2021) Tensions within risk science: what are they about and how should we deal with them? Forthcoming.

Aven, T. and Zio, E. (2011) Some considerations on the treatment of uncertainties in risk assessment for practical decision-making. *Reliability Engineering and System Safety*, 96, 64–74.

Aven, T. and Zio, E. (2013) Model output uncertainty in risk assessment. *International Journal of Performability Engineering IJPE*, 9(5), 475–486.

Aven, T. and Zio, E. (2021) Globalization and global risk: How risk analysis needs to be enhanced to be effective in confronting current threats. *Reliability Engineering and System Safety*, 205, 107270.

Balog-Way, D., McComas, K. and Besley, J. (2020) The evolving field of risk communication. *Risk Analysis*, 40, 2240–2261.

Baram, M. (1984) The right to know and the duty to disclose hazard information. *Public Health and the Law*, 74(4), 385–390.

Barber, B. (1983) *The Logic and Limits of Trust*. New Brunswick, NJ: Rutgers University Press.

Bassarak, C., Pfister, H.R. and Böhm, G. (2017) Dispute and morality in the perception of societal risks: Extending the psychometric model. *Journal of Risk Research*, 20(3), 299–325.

Bayes, T. (1763) LII: An essay towards solving a problem in the doctrine of chances. By the late Rev. Mr. Bayes, FRS communicated by Mr. Price, in a letter to John Canton, AMFR S. *Philosophical Transactions of the Royal Society of London*, 53, 370–418.

Beck, U. (1992) *Risk Society: Towards a New Modernity*. London: Sage.

Bedford, T. and Cooke, R. (2001) *Probabilistic Risk Analysis*. Cambridge: Cambridge University Press.

Bell, T.E. and Esch, K. (2018) *The Challenger Disaster: A Case of Subjective Engineering*. https://spectrum.ieee.org/tech-history/heroic-failures/the-space-shuttle-a-case-of-subjective-engineering. Accessed November 18, 2020.

Bergman, B. and Klefsjö, B. (2003) *Quality*. 2nd ed. Lund, Sweden: Studentlitteratur.

Bernardo, J. and Smith, A. (1994) *Bayesian Theory*. Chichester: Wiley.

Bernstein, P.L. (1996) *Against the Gods: The Remarkable Story of Risk*. New York: Wiley.

Besley, J.C. (2010) Public engagement and the impact of fairness perceptions on decision favorability and acceptance. *Science Communication*, 32, 256–280.

Bier, V.M. (2001a) On the state of the art: Risk communication to the public. *Reliability Engineering and System Safety*, 71, 139–150.

Bier, V.M. (2001b) On the state of the art: Risk communication to decision-makers. *Reliability Engineering and System Safety*, 71, 151–157.

Birkinshaw, P. (2006) Freedom of information and openness: Fundamental human rights? *Law Review*, 58, 177–218.

Birnbaum, Z.W. (1969) On the importance of different components in a multicomponent system. In P.R. Krishnaiah (ed.), *Multivariate Analysis-II*. New York: Academic Press, pp. 581–592.

Bjerga, T. and Aven, T. (2015) Adaptive risk management using the new risk perspectives – an example from the oil and gas industry. *Reliability Engineering and System Safety*, 134, 75–82.

Bjerga, T. and Aven, T. (2016) Some perspectives on risk management – a security case study from the oil and gas industry. *Journal of Risk and Reliability*, 230(5), 512–520.

Blind, K. and Mangelsdorf, A. (2016) Motives to standardize: Empirical evidence from Germany. *Technovation*, 48–49, 13–24.

Bloom, D.L., Byrne, D.M. and Andresen, J.M. (1993) *Communicating Risk to Senior EPA Policy Makers: A Focus Group Study*. Washington, DC: US Environmental Protection Agency.

Böhm, G. (2003) Emotional reactions to environmental risks: Consequentialist versus ethical evaluation. *Journal of Environmental Psychology*, 23, 199–212.

Böhm, G. and Pfister, H.R. (2000) Action tendencies and characteristics of environmental risks. *Acta Psychologica*, 104, 317–337.

Borgonovo, E. and Plischke, E. (2016) Sensitivity analysis: A review of recent advances. *European Journal of Operations Research*, 248(1), 869–887.

Bostrom, A., Böhm, G. and O'Connor, R.E. (2013) Targeting and tailoring climate change communications. *Wiley Interdisciplinary Reviews: Climate Change.* doi:10.1002/wcc.234.

Bostrom, A., Böhm, G. and O'Connor, R.E. (2018) Communicating risks: Principles and challenges. Chapter 11 in M. Raue et al. (eds.), *Psychological Perspectives on Risk and Risk Analysis*. Cham, Switzerland: Springer International Publishing AG, Part of Springer Nature.

Bostrom, A., O'Connor, R.E., Böhm, G., Hanss, D., Bodi, O. Ekström, F. . . . Sælensminde, I. (2012) Causal thinking and support for climate change policies: International survey findings. *Global and Environmental Change: Human and Policy Dimensions*, 22(1), 210–222.

Bouder, F., Way, D., Löfstedt, R. and Evensen, D. (2015) Transparency in Europe: A quantitative study. *Risk Analysis*, 35(7), 1210–1229.

Bourdieu, P. and Wacquant, L.J.D. (1992) *An Invitation to Reflexive Sociology*. Chicago: University of Chicago Press.

CDC (2020) *Implementation of Mitigation Strategies for Communities with Local COVID-19 Transmission*. www.cdc.gov/coronavirus/2019-ncov/downloads/community-mitigation-strat egy.pdf. Accessed April 1, 2020.

Chauvin, B. (2018) Individual differences in the judgment of risks: Sociodemographic characteristics, cultural orientation, and level of expertise. Chapter 2 in M. Raue et al. (eds.), *Psychological Perspectives on Risk and Risk Analysis*. Cham, Switzerland: Springer International Publishing AG, Part of Springer Nature.

Chermack, T.J. (2011) *Scenario Planning in Organizations*. San Francisco: BK Publishers.

Chief, E. and Nagale, D. (2020) *NWS Hazard Simplification Project*. www.weather.gov/media/ hazardsimplification/Haz%20Simp%20Proposal%20Slides.pdf. Accessed November 12, 2020.

Choi, T.M. and Lambert, J.H. (2017) Advances in risk analysis with big data. *Risk Analysis*, 37(8), 1435–1442.

Coburn, B.J., Wagner, B.G. and Blower, S. (2009) Modeling influenza epidemics and pandemics: Insights into the future of swine flu (H1N1). *BMC Medicine*, 7(1), 30.

Coglianese, C. (2009) The transparency president? The Obama administration and open government. *Governance*, 22(4), 529–544.

Cohen, J. and Kupferschmidt, K. (2020) Countries test tactics in 'war' against COVID-19. *Science*, 367(6484), 1287.

Connor, K.M., Davidson, J.R.T. and Lee, L.C. (2003) Spirituality, resilience, and anger in survivors of violent trauma: A community survey. *Journal of Traumatic Stress*, 16(5), 487–494.

Copeland, T.E. and Weston, J.F. (1988) *Finance Theory and Corporate Policy*. 3rd ed. Los Angeles: Addison-Wesley Publishing Company.

COSO (2020) *Enterprise Risk Management*. www.coso.org/Documents/COSO-ERM-Executive-Summary.pdf; www.coso.org/Pages/erm-integratedframework.aspx. Report Produced 2004. Accessed November 24, 2020.

Cox, T. (2008) What's wrong with risk matrices? *Risk Analysis*, 28(2), 497–512.

Cox, T. (2011) Clarifying types of uncertainty: When are models accurate, and uncertainties small? *Risk Analysis*, 31, 1530–1533.

Crosby, N. (1995) Citizen juries: One solution for difficult environmental questions. In O. Renn, T. Webler and P. Wiedemann (eds.), *Fairness and Competence in Citizen Participation*,

Evaluating New Models for Environmental Discourse. Dordrecht and Boston: Kluwer, pp. 157–174.

Dekker, S.W.A. and Nyce, J.M. (2004) How can ergonomics influence design? Moving from research findings to future systems. *Ergonomics*, 47(15), 1624–39.

Deming, W.E. (2000) *The New Economics*. 2nd ed. Cambridge, MA: MIT CAES.

Diecidue, E. and Wakker, P.P. (2001) On the intuition of rank-dependent utility. *The Journal of Risk and Uncertainty*, 23(3), 281–298.

Dietz, K. (1993) The estimation of the basic reproduction number for infectious diseases. *Statistical Methods in Medical Research*, 2(1), 23–41.

Dohle, S., Keller, C. and Siegrist, M. (2012) Fear and anger: Antecedents and consequences of emotional responses to mobile communication. *Journal of Risk Research*, 15, 435–446.

Douglas, M. and Wildavsky, A. (1982) *Risk and Culture: The Selection of Technological and Environmental Dangers*. Berkeley, CA: University of California Press.

Dubois, D. (2010) Representation, propagation and decision issues in risk analysis under incomplete probabilistic information. *Risk Analysis*, 30, 361–368.

Duhigg, C. (2012) *The Power of Habit: Why We Do What We Do in Life and Business*. New York: Random House.

Earle, T.C. and Cvetkovich, G. (1995) *Social Trust: Toward a Cosmopolitan Society*. Westport, CT: Praeger.

Earle, T.C., Siegrist, M. and Gutscher, H. (2012) Trust, risk perception and the TCC model of cooperation. In M. Siegrist, T.C. Earle and N.F. Pidgeon (eds.), *Trust in Cooperative Risk Management*. New York: Routledge, pp. 19–68.

Easterling, R.G. (1972) A personal view of the Bayesian controversy in reliability and statistics. *IEEE Transactions on Reliability*, R-21(3), 186–194.

Efron, B. (2005) Bayesians, frequentists, and scientists. *Journal of the American Statistical Association*, 100(469), 1–5.

Ellsworth, P.C. and Scherer, K.R. (2003) Appraisal processes in emotion. In R.J. Davidson, H. Goldsmith and K.R. Scherer (eds.), *Handbook of the Affective Sciences*. New York: Oxford University Press, pp. 572–595.

Epstein, S. (1994) Integration of the cognitive and the psychodynamic unconscious. *American Psychologist*, 49, 709–724.

Ersdal, G. and Aven, T. (2008) Risk management and its ethical basis. *Reliability Engineering and Systems Safety*, 93, 197–205.

Ethik-Kommission (2011) *Deutschlands Energiewende. Ein Gemeinschaftswerk für die Zukunft*. Berlin: Endbericht.

EU (2002) Consolidated version of the treaty establishing the European community. *Official Journal of the European Commission*, C325, 33–184, December 24, Brussels.

EU (2017) *Science for Environment Policy: The Precautionary Principle: Decision-Making Under Uncertainty*. Future Brief 18. Bristol: Produced for the European Commission DG Environment by the Science Communication Unit, UWE. http://ec.europa.eu/science-environment-policy.

FAA (2018, July) *Federal Aviation Administration*. Fact sheet. https://www.faa.gov/news/fact_sheets/news_story.cfm?newsId=21274. Accessed April 30, 2021.

Fenton, M. and Neil, M. (2018) *Risk Assessment and Decision Analysis with Bayesian Networks*. 2nd ed. Boca Raton: CRC Press.

Ferguson, N.M., Laydon, D., Nedjati-Gilani, G., Imai, N., Ainslie, K., Baguelin, M. . . . Dighe, A. (2020) *Impact of Non-Pharmaceutical Interventions (NPIs) to Reduce COVID-19 Mortality and Healthcare Demand*. doi:10.25561/77482; www.imperial.ac.uk/mrc-global-infectious-disease-analysis/covid-19/report-9-impact-of-npis-on-covid-19/. Accessed November 12, 2020.

Fine, P., Eames, K. and Heymann, D.L. (2011) 'Herd immunity': A rough guide. *Clinical Infectious Diseases*, 52(7), 911–916.

Finucane, M.L., Alhakami, A., Slovic, P. and Johnson, S.M. (2000) The affect heuristic in judgments of risks and benefits. *Journal of Behavioral Decision-making*, 13, 1–17.

Fischhoff, B. (1995) Risk perception and communication unplugged: Twenty years of process. *Risk Analysis*, 15(2), 137–145.

Fischhoff, B., Brewer, N. and Downs, J. (2011) *Communicating Risks and Benefits: An Evidence Based User's Guide*. Washington, DC: Government Printing Office.

Fischhoff, B. and Kadvany, J. (2011) *Risk: A Very Short Introduction*. Oxford: Oxford University Press.

Fischhoff, B., Slovic, P., Lichtenstein, S., Read, S. and Combs, B. (1978). How safe is safe enough? A psychometric study of attitudes towards technological risks and benefits. *Policy Sciences*, 9(2), 127–152.

Fisher, M. (2020). Coronavirus 'hits all the hot buttons' for how we misjudge risk. *The New York Times*, February 13. www.nytimes.com/2020/02/13/world/asia/coronavirus-risk-interpreter.html. Accessed November 12, 2020.

Fjæran, L. and Aven, T. (2019) Making visible the less visible – how the use of an uncertainty-based risk perspective affects risk attenuation and risk amplification. *Journal of Risk Research*. doi:10.1080/13669877.2019.1687579.

Fjæran, L. and Aven, T. (2020a) Creating conditions for critical trust – how an uncertainty-based risk perspective relates to dimensions and types of trust. *Safety Science*, 105008.

Fjæran, L. and Aven, T. (2020b) *Incorporating the Stakeholder Battles and Dynamics of Amplification and Attenuation Processes into the SARF*. PhD thesis. University of Stavanger, Stavanger.

Flage, R., Aven, T., Baraldi, P. and Zio, E. (2014) Concerns, challenges and directions of development for the issue of representing uncertainty in risk assessment. *Risk Analysis*, 34(7), 1196–1207.

Flage, R. and Røed, W. (2012) A reflection on some practices in the use of risk matrices. In *Proceedings of the 11th International Probabilistic Safety Assessment and Management Conference and the Annual European Safety and Reliability Conference 2012*. Helsinki, Finland: Curran Associates, Inc., pp. 881–891, June 25–29. ISBN 978-1-62276-436-5.

Flanders, W.D., Lally, C.A., Zhu, B.P., Henley, S.J. and Thun, M.J. (2003). Lung cancer mortality in relation to age, duration of smoking, and daily cigarette consumption. *Cancer Research*, 63, 6556–6562.

Florini, A. (1999) *Does the Invisible Hand Need a Transparent Glove? The Politics of Transparency*. Washington, DC: Annual World Bank Conference on Development Economics, 46.

Ford, L.R. and Fulkerson, D.R. (1956) Maximal flow through a network. *Canadian Journal of Mathematics*, 8, 399–404.

Francis, R. and Bekera, B. (2014) A metric and framework for resilience analysis of engineered and infrastructure system. *Reliability Engineering and System Safety*, 121, 90–103.

Freestone, D. and Hey, E. (1996) Origins and development of the precautionary principle. In D. Freestone and E. Hey (eds.), *The Precautionary Principle and International Law: The Challenge of Implementation*. The Hague: Kluwer Law International, pp. 3–15.

Freudenburg, W.R., Gramling, R. and Davidson, D.J. (2008) Scientific certainty argumentation methods (SCAMs): Science and the politics of doubt. *Sociological Inquiry*, 78, 2–38.

Frewer, L.J. (2017) Consumer acceptance and rejection of emerging agrifood technologies and their applications. *European Review of Agricultural Economics*, 44(4), 683–704.

Frewer, L.J., Howard, C., Hedderley, D. and Shepherd, R. (1996) What determines trust in information about food-related risks? Underlying psychological constructs. *Risk Analysis*, 16(4), 473–486.

Frewer, L.J., Hunt, S., Brennan, M., Kuznesof, S., Ness, M. and Ritson, C. (2003) The views of scientific experts on how the public conceptualize uncertainty. *Journal of Risk Research*, 6, 75–85.

Frewer, L.J. and Salter, B. (2010) Societal trust in risk analysis: Implications for the interface of risk assessment and risk management. In M. Siegrist, T.C. Earle and N.F. Pidgeon (eds.), *Trust in Cooperative Risk Management*. New York: Routledge, pp. 161–176.

Frewer, L.J., Scholderer, J. and Bredahl, L. (2003) Communicating about the risks and benefits of genetically modified foods: The mediating role of trust. *Risk Analysis*, 23(6), 1117–1133.

Friborg, O., Hjemdal, O., Rosenvinge, J.H. and Martinussen, M. (2003) A new rating scale for adult resilience: What are the central protective resources behind healthy adjustment? *International Journal of Methods in Psychiatric Research*, 12, 65–76.

Gates (2015) *The Next Outbreak? We're Not Ready: Bill Gates.* www.youtube.com/watch?v=6Af6b_wyiwI. Accessed November 12, 2020.

Gillespie-Marthaler, L., Nelson, K., Baroud, H. and Abkowitz, M. (2019) Selecting indicators for assessing community sustainable resilience. *Risk Analysis*, 39(11), 2479–2498.

Glette-Iversen, I. and Aven, T. (2021) On the meaning of and relationship between dragon-kings, black swans and related concepts. *Reliability Engineering and System Safety*, 211, 107625.

Goerlandt, F., Khakzad, N. and Reniers, G. (2017) Validity and validation of safety-related quantitative risk analysis: A review. *Safety Science*, 99, 127–139.

Goh, G. (2020) *Epidemic Calculator*. http://gabgoh.github.io/COVID/index.html. Accessed April 1, 2020.

Greenberg, M.R. (2017) *Explaining Risk Analysis*. London: Routledge.

Gregory, M.R., Failing, L., Harstone, M., Long, G., McDaniels, T. and Ohlson, D. (2012) *Structured Decision-making: A Practical Guide to Environmental Management Choices*. New York: Wiley.

Guikema, S.D. (2009) Natural disaster risk analysis for critical infrastructure systems: An approach based on statistical learning theory. *Reliability Engineering and System Safety*, 94(4), 855–860.

Guikema, S.D. (2020). Artificial intelligence for natural hazards risk analysis: Potential, challenges, and research needs. *Risk Analysis*, 40(6), 1117–1123.

Haimes, Y.Y. (2009) On the definition of resilience in systems. *Risk Analysis*, 29, 498–501.

Hanson, J.L. (2008) Shared decision-making: Have we missed the obvious? *Archives of Internal Medicine*, 168, 1368–1370.

Hansson, S.O. (2003) Ethical criteria of risk acceptance. *Erkenntnis*, 59(3), 291–309.

Hansson, S.O. (2013) Defining pseudoscience and science. In M. Pigliucci and M. Boudry (eds.), *Philosophy of Pseudoscience*. Chicago: University of Chicago Press, pp. 61–77.

Hansson, S.O. and Aven, T. (2014) Is risk analysis scientific? *Risk Analysis*, 34(7), 1173–1183.

Hashimito, T., Stedinger, J.R. and Loucks, D.P. (1982) Reliability, resilience and vulnerability criteria for water-resource system performance evaluation. *Water Resource Research*, 18, 14–20.

Hastie, T., Tibshirani, R. and Friedman, J. (2017) *The Elements of Statistical Learning – Data Mining, Inference and Prediction*. 2nd ed. New York: Springer.

Helton, J.C., Johnson, J.D., Sallaberry, C.J. and Storlie, C.B. (2006) Survey of sampling-based methods for uncertainty and sensitivity analysis. *Reliability Engineering and System Safety*, 91, 1175–1209.

Henry, D. and Ramirez-Marquez, J.E. (2012) Generic metrics and quantitative approaches for system resilience as a function of time. *Reliability Engineering and System Safety*, 99, 114–122.

Hilgartner, S. (1992). The social construction of risk objects: Or, how to pry open networks of risk. In J.F. Short and L. Clarke (eds.), *Organizations, Uncertainties, and Risk*. New York: Westview Press, pp. 39–53.

Hine, D.W., Phillips, W.J., Cooksey, R., Reser, J.P., Nunn, P., Marks, A.D., . . . Watt, S.E. (2016) Preaching to different choirs: How to motivate dismissive, uncommitted, and alarmed audiences to adapt to climate change? *Global Environmental Change*, 36, 1–11.

Hoffrage, U. and Garcia-Retamero, R. (2018) Improving understanding of health-relevant numerical information. Chapter 12 in M. Raue et al. (eds.), *Psychological Perspectives on Risk and Risk Analysis*. Cham: Switzerland: Springer International Publishing AG, Part of Springer Nature.

Hollnagel, E. (2004) *Barriers and Accident Prevention*. Aldershot: Ashgate.

Hollnagel, E., Woods, D.D. and Leveson, N. (eds.) (2006) *Resilience Engineering, Concepts and Precepts*. Burlington: Ashgate.

Hopkin, P. (2010) *Fundamentals of Risk Management*. London: The Institute of Risk Management.

Hosseini, S., Barker, K. and Ramirez-Marquez, J.E. (2016) A review of definitions and measures of system resilience. *Reliability Engineering and System Safety*, 145, 47–61.

HSE UK (2001) *Reducing Risk, Protecting People*. London: HES Books.

Hughes, J.F. and Healy, K. (2014) *Measuring the Resilience of Transport Infrastructure*. NZ Transport Agency Research Report 546. Wellington: NZ Transport Agency.

IHME (2020). *COVID-19 Projections*. https://covid19.healthdata.org/projections. Accessed April 5, 2010.

IPCC (2014) *Climate Change 2014 – Synthesis Report*. https://www.ipcc.ch/report/ar5/syr/. Accessed April 30, 2021.

IRGC (International Risk Governance Council) (2005) *Risk Governance: Towards an Integrative Approach, White Paper No. 1, O. Renn with an Annex by P. Graham*. Geneva: IRGC.

ISO (2018) *ISO 31000 Risk Management*. www.iso.org/iso-31000-risk-management.html. Accessed September 10, 2020.

ISO (2020a) *The Main Benefits of ISO Standards*. www.iso.org/benefits-of-standards.html. Accessed September 10, 2020.

ISO (2020b) *Standards in Our World*. www.iso.org/sites/ConsumersStandards/1_standards.html. Accessed November 12, 2020.

ITC (2011) *ITC Netherlands Survey: Report on Smokers' Awareness of the Health Risks of Smoking and Exposure to Second-Hand Smoke*. Ontario, Canada: University of Waterloo.

Jansen, T., Claassen, L., van Kamp, I. and Timmermans, D.R. (2019) 'It is not entirely healthy': A qualitative study into public appraisals of uncertain risks of chemical substances in food. *Public Understanding of Science*, 29(2), 139–156.

Jasanoff, S. (1999) The songlines of risk. *Environmental Values. Special Issue: Risk*, 8(2), 135–152.

Jenkins-Smith, H.C. and Silva, C.L. (1998) The role of risk perception and technical information in scientific debates over nuclear waste storage. *Reliability Engineering and System Safety*, 59(1), 107–122.

Jensen, A. and Aven, T. (2015) *Hazard/Threat Identification – Using Different Creative Methods to Support the Anticipatory Failure Determination Approach: Safety and Reliability of Complex Engineered Systems*. Boca Raton: CRC Press.

Jensen, A. and Aven, T. (2017) Hazard/threat identification: Using functional resonance analysis method in conjunction with the anticipatory failure determination method. *Journal of Risk and Reliability*, 231(4), 383–389.

Johnson, B.B. and Slovic, P. (1998) Lay views on uncertainty in environmental health risk assessment. *Journal of Risk Research*, 1, 261–279.

Jones, H.W. (2019) *NASA's Understanding of Risk in Apollo and Shuttle*. https://ntrs.nasa.gov/archive/nasa/casi.ntrs.nasa.gov/20190002249.pdf. Accessed November 12, 2020.

Jones-Lee, M. and Aven, T. (2011) ALARP – what does it really mean? *Reliability Engineering and System Safety*, 96(8), 877–882.

Josephson, M. (2002) *Making Ethical Decisions*. Los Angeles: Josephson Institute. https://store.charactercounts.org/wp-content/uploads/sites/10/2015/09/50-0450-E.pdf. Accessed August 19, 2020.

Kahan, D.M., Braman, D., Gastil, J., Slovic, P. and Mertz, C.K. (2007) Culture and identity-protective cognition: Explaining the white-male effect in risk perception. *Journal of Empirical Legal Studies*, 4, 465–505.

Kahan, D.M., Jenkins-Smith, H., Tarantola, T., Silva, C.L., and Braman, D. (2015) Geoengineering and climate change polarization: Testing a two-channel model of science communication. *The Annals of the American Academy of Political and Social Science*, 658(1), 192–222.

Kahan, D.M., Peters, E., Wittlin, M., Slovic, P., Ouellette, L.L., Braman, D. and Mandel, G. (2012) The polarizing impact of science literacy and numeracy on perceived climate change risks. *Nature Climate Change*, 2, 732–735.

Kahn, J.S. and McIntosh, K. (2005) History and recent advances in coronavirus discovery. *The Pediatric Infectious Disease Journal*, 24(11), S223–S227.

Kahneman, D. (2003) A perspective on judgment and choice. *American Psychologist*, 58(9), 697–720.

Kahneman, D. (2011) *Thinking, Fast and Slow*. New York: Farrar, Straus and Giroux.

Kaplan, S. and Garrick, B.J. (1981) On the quantitative definition of risk. *Risk Analysis*, 1, 11–27.

Kaplan, S., Visnepolschi, S., Zlotin, B. and Zusman, A. (1999) *New Tools for Failure and Risk Analysis: Anticipatory Failure Determination (AFD) and the Theory of Scenario Structuring*. Southfield, MI: Ideation International Inc.

Kasperson, R.E. (1992) The social amplification of risk: Progress in developing an integrative framework. In S. Krimsky and D. Golding (eds.), *Social Theories of Risk*. New York: Praeger.

Kasperson, R.E., Pidgeon, N.F. and Slovic, P. (2003) *The Social Amplification of Risk*. Cambridge: Cambridge University Press.

Kasperson, R.E., Renn, O., Slovic, P., Brown, H.S., Emel, J., Goble, R., . . . Ratick, S. (1988) The social amplification of risk: A conceptual framework. *Risk Analysis*, 8(2), 177–187.

Keller, C., Bostrom, A., Kuttschreuter, M., Savadori, L., Spence, A. and White, M. (2012) Bringing appraisal theory to environmental risk perception: A review of conceptual approaches of the past 40 years and suggestions for future research. *Journal of Risk Research*, 15(3), 237–256.

Keller, C., Kreuzmair, C., Leins-Hess, R. and Siegrist, M. (2014) Numeric and graphic risk information processing of high and low numerates in the intuitive and deliberative decision modes: An eye-tracker study. *Judgment and Decision-making*, 9(5), 420–432.

Keller, C., Visschers, V. and Siegrist, M. (2012) Affective imagery and acceptance of replacing nuclear power plants. *Risk Analysis*, 32, 464–477.

Knight, F.H. (1921) *Risk, Uncertainty, and Profit*. Boston: Houghton Mifflin Co.

Kraus, N., Malmfors, T. and Slovic, P. (1992) Intuitive toxicology: Expert and lay judgments of chemical risks. *Risk Analysis*, 12, 215–232.

Krueger, R.A. and Casey, M.A. (2000) *Focus Groups: A Practical Guide for Applied Research*. Thousand Oaks: Sage.

Langdalen, H., Abrahamsen, E.B. and Selvik, J.T. (2020) On the importance of systems thinking when using the ALARP principle for risk management. *Reliability Engineering and System Safety*, 107–222.

Latour, B. (2005) *Reassembling the Social: An Introduction to Actor Network Theory*. Oxford: Oxford University Press.

Le Coze, J.C. (2019) Vive la diversité! High Reliability Organisation (HRO) and Resilience Engineering (RE). *Safety Science*, 117, 469–478.

Leiss, W. (1996) Three phases in risk communication practice. In H. Kunreuther and P. Slovic (eds.), *Annals of the American Academy of Political and Social Science, Special Issue: Challenges in Risk Assessment and Management*. Thousand Oaks, CA: Sage, pp. 85–94.

Lerner, J.S. and Keltner, D. (2001) Fear, anger, and risk. *Journal of Personality and Social Psychology*, 81(1), 146–159.

Leveson, N. (2004) A new accident model for engineering safer systems. *Safety Science*, 42, 237–270.

Leveson, N. (2007) *Modeling and Analyzing Risk in Complex Socio-Technical Systems*. NeT-Work Workshop, Berlin, September 27–29.

Leveson, N. (2011) Applying systems thinking to analyse and learn from events. *Safety Science*, 49, 55–64.

Lillestøl, J. (2014) *Statistical Inference: Paradigms and Controversies in Historic Perspective.* www.nhh.no/globalassets/departments/business-and-management-science/research/lillestol/ statistical_inference.pdf.

Lindell, M.K. and Brooks, H. (eds.) (2012) *Workshop on Weather Ready Nation: Science Imperatives for Severe Thunderstorm Research, Birmingham AL.* Final Report. Sponsored by National Oceanic and Atmospheric Administration and National Science Foundation. Hazard Reduction & Recovery Center, Texas A&M University, College Station TX 77843–3137, September 17.

Lindley, D.V. (1985) *Making Decisions.* New York: Wiley.

Lindley, D.V. (2000) The philosophy of statistics. *The Statistician,* 49, 293–337. With discussions.

Lindley, D.V. (2006) *Understanding Uncertainty.* Hoboken, NJ: Wiley.

Linkov, I., Bridges, T., Creutzig, F., Decker, J., Fox-Lent, C., Kröger, W., . . . Thiel-Clemen, T. (2014) Changing the resilience paradigm. *Nature Climate Change,* 4, 407–409.

Linkov, I. and Trump, B. (2019) *The Science and Practice of Resilience.* Cham, Switzerland: Springer Verlag, e-book.

Linstone, H.A. and Turoff, M. (eds.) (2002) *The Delphi Method: Techniques and Applications.* Newark: ISA Applications.

Logan, T., Aven, T., Guikema, S. and Flage, R. (2021) The role of time in risk and risk analysis: implications for resilience, sustainability, and management. *Risk Analysis.* doi: 10.1111/risa.13733.

L'Orange Seigo, S., Arvai, J., Dohle, S. and Siegrist, M. (2014) Predictors of risk and benefit perception of carbon capture and storage (CCS) in regions with different stages of deployment. *International Journal of Greenhouse Gas Control,* 25, 23–32.

Löfstedt, R. (2013) Communicating food risks in an era of growing public distrust: Three case studies. *Risk Analysis,* 33(2), 192–202.

Lurie, N. (2009) H1N1 influenza, public health preparedness, and health care reform. *New England Journal of Medicine,* 361(9), 843–845.

Machina, M.J. (2008) Non-expected utility theory. In S.N. Durlauf and L.E. Blume (eds.), *The New Palgrave Dictionary of Economics.* 2nd ed. Basingstone and New York: Palgrave Macmillan.

MacInnis, D.J. (2011) A framework for conceptual contributions in marketing. *Journal of Marketing,* 75(4), 136–154.

Malka, A., Krosnick, J.A. and Langer, G. (2009) The association of knowledge with concern about global warming: Trusted information sources shape public thinking. *Risk Analysis,* 29(5), 633–647.

Martin, R. (2009) *The Opposable Mind.* Boston: Harvard Business Press.

Masys, A.J. (2012) Black swans to grey swans: Revealing the uncertainty. *Disaster Prevention and Management,* 21(3), 320–335.

McKenzie-Mohr, D., Lee, N.R., Schultz, P.W. and Kotler, P. (2012) *Social Marketing to Protect the Environment: What Works.* Irvine, CA: Sage.

Merkelsen, H. (2011) Institutionalized ignorance as a precondition for rational risk expertise. *Risk Analysis,* 31, 1083–1094.

Merton, R.K. (1973) Science and technology in a democratic order. *Journal of Legal and Political Sociology,* 1942(1), 115–126. Reprinted as The normative structure of science. In R.K. Merton. *The Sociology of Science: Theoretical and Empirical Investigations.* Chicago: University of Chicago Press, pp. 267–278.

Mertz, C.K., Slovic, P. and Purchase, F.H. (1998) Judgments of chemical risks: Comparisons among senior managers, toxicologists, and the public. *Risk Analysis,* 18, 391–404.

Meyer, T. and Reniers, G. (2013) *Engineering Risk Management.* Berlin: De Gruyter Graduate.

Mileti, D.S. and Sorensen, J.H. (2015) *A Guide to Public Alerts and Warnings for Dam and Levee Emergencies.* Boulder: U.S. Army Corps of Engineers.

Mohaghegh, Z. and Mosleh, A. (2009) Incorporating organizational factors into probabilistic risk assessment of complex socio-technical systems: Principles and theoretical foundations. *Safety Science*, 47(8), 1139–1158.

Murphy, K.P. (2012) *Machine Learning: A Probabilistic Perspective*. Cambridge, MA: MIT Press.

Myers, T.A., Nisbet, M.C., Maibach, E.W. and Leiserowitz, A.A. (2012) A public health frame arouses hopeful emotions about climate change: A letter. *Climatic Change*, 113(3–4), 1105–1112.

NASA (2011) *Probabilistic Risk Assessment Procedures Guide for NASA Managers and Practitioners Report NASA/SP-2011-3421*. Washington, DC: NASA.

Nateghi, R. and Aven, T. (2021) Risk analysis in the age of big data: the promises and pitfalls. *Risk Analysis*. doi: 10.1111/risa.13682.

National Archives (2020) *The Deadly Virus*. www.archives.gov/exhibits/influenza-epidemic/. Accessed April 1, 2020.

Nøkland, T.E. and Aven, T. (2013) On selection of importance measures in risk and reliability analysis. *International Journal of Production Economics*, 9(2), 133–148.

Normile (2020) *Why Airport Screening Won't Stop the Spread of Coronavirus*, March 6. www.sciencemag.org/news/2020/03/why-airport-screening-wont-stop-spread-coronavirus.

North, W.D. (2011) Uncertainties, precaution, and science: Focus on the state of knowledge and how it may change. *Risk Analysis*, 31, 1526–1529.

NRC (1975) *Wash 1400. Report NUREG-75/014*. Washington, DC: U.S. Nuclear Regulatory Commission.

NRC (1989) *National Research Council (NRC), Committee on Risk Perception and Communication: Improving Risk Communication*. Washington, DC: National Academy Press.

NSC (1987) *North Sea Conferences: Second International Conference on the Protection of the North Sea (1987, November 24–25)*. London: Shelton & Roush. www.ospar.org/about/international-cooperation/north-sea-conferences.

O'Brien, M. (2000) *Making Better Environmental Decisions*. Cambridge, MA: MIT Press.

O'Hagan, A. and Oakley, J.E. (2004) Probability is perfect, but we can't elicit it perfectly. *Reliability Engineering and System Safety*, 85, 239–248.

O'Keefe, D.J. (2013) The relative persuasiveness of different message types does not vary as a function of the persuasive outcome assessed: Evidence from 29 meta-analyses of 2,062 effect sizes for 13 message variations. *Annals of the International Communication Association*, 37(1), 221–249.

O'Neill, S., Flanagan, J. and Clarke, K. (2016) Safewash! Risk attenuation and the (Mis)reporting of corporate safety performance to investors. *Safety Science*, 83, 114–130.

Osborn, A.F. (1963) *Applied Imagination: Principles and Procedures for Creative Problem Solving*. 3rd ed. New York: Charles Scribner's Sons.

Otani, A., Ostroff, C. and Yu, X. (2020) *U.S. Stocks Close Higher After Central Banks' Action*. www.wsj.com/articles/global-stocks-and-currencies-slide-as-cash-shortage-worsens-11584595586?mod=article_inline&mod=hp_lead_pos5&mod=article_inline. Accessed November 12, 2020.

Pasman, H.J., Rogers, W.J. and Mannan, M.S. (2017) Risk assessment: What is it worth? Shall we just do away with it, or can it do a better job? *Safety Science*, 99, 140–155.

Paté-Cornell, M.E. (1996) Uncertainties in risk analysis: Six levels of treatment. *Reliability Engineering and System Safety*, 54(2–3), 95–111.

Paté-Cornell, M.E. (2012) On black swans and perfect storms: Risk analysis and management when statistics are not enough. *Risk Analysis*, 32(11), 1823–1833.

Paté-Cornell, M.E. and Cox, A. (2014) Improving risk management: From lame excuses to principles practice. *Risk Analysis*, 34(7), 1228–1239.

Peters, E. and Slovic, P. (1996) The role of affect and worldviews as orienting dispositions in the perception and acceptance of nuclear power. *Journal of Applied Social Psychology*, 26, 1427–1453.

Peters, G.J.Y., Ruiter, R.A. and Kok, G. (2013) Threatening communication: A critical reanalysis and a revised meta-analytic test of fear appeal theory. *Health Psychology Review*, 7(1), 8–31.

Peterson, M. (2006) The precautionary principle is incoherent. *Risk Analysis*, 26(3), 595–601.

Peterson, M. (2007) Should the precautionary principle guide our actions or our beliefs? *Journal of Medical Ethics*, 33(1), 5–10.

Peterson, M. (2017) Yes, the precautionary principle is incoherent. *Risk Analysis*, 37(11), 2035–2038.

Pidgeon, N., Poortinga, W. and Walls, J. (2010) Scepticism, reliance and risk managing institutions: Towards a conceptual model of 'critical trust'. In M. Siegrist, T.C. Earle and N.F. Pidgeon (eds.), *Trust in Cooperative Risk Management*. New York: Routledge, pp. 131–156.

Polasek, W. (2000) The Bernoullis and the origin of probability theory: Looking back after 300 years. *Resonance*, 5, 26–42.

Poortinga, W. and Pidgeon, N.F. (2003) Exploring the dimensionality of trust in risk regulation. *Risk Analysis*, 23(5), 961–972.

Popper, K. (1962) *Conjectures and Refutations: The Growth of Scientific Knowledge*. New York: Basic Books.

Poumadere, M. and Mays, C. (2003) The dynamics of risk amplification and attenuation in context: A French case study. In N. Pidgeon, R.E. Kasperson and P. Slovic (eds.), *The Social Amplification of Risk*. New York: Cambridge University Press, pp. 89–106.

Proctor, R.N. (2011) The history of the discovery of the cigarette–lung cancer link: Evidentiary traditions, corporate denial, global toll. *Tobacco Control*, 21, 87–91.

Rabinovich, A. and Morton, T.A. (2012) Unquestioned answers or unanswered questions: Beliefs about science guide responses to uncertainty in climate change risk communication. *Risk Analysis*, 32(6), 992–1002.

Rae, A., Alexander, R. and McDermid, J. (2014) Fixing the cracks in the crystal ball: A maturity model for quantitative risk assessment. *Reliability Engineering and System Safety*, 125, 67–81.

Rausand, M. and Haugen, S. (2020) *Risk Assessment: Theory, Methods, and Applications*. 2nd ed. New York: Wiley.

Rechard, R.P. (1999) Historical relationship between performance assessment for radioactive waste disposal and other types of risk assessment. *Risk Analysis*, 19(5), 763–807.

Rechard, R.P. (2000) Historical background on performance assessment for the waste isolation pilot plant. *Reliability Engineering and System Safety*, 69(3), 5–46.

Renn, O. (1998) Three decades of risk research: Accomplishments and new challenges. *Journal of Risk Research*, 1(1), 49–71.

Renn, O. (2008) *Risk Governance: Coping with Uncertainty in a Complex World*. London: Earthscan.

Renn, O. (2014) Four questions for risk communication: A response to Roger Kasperson. *Journal of Risk Research*, 17(10), 1277–1281.

Renn, O. (2015) Ethikkommission: Wie legitim ist die Legitimation der Politik durch Wissenschaft? In P. Weingart and G.G. Wagner (eds.), *Wissenschaftliche Politikberatung im Praxistest*. Weilerswist: Velbrück, pp. 17–34.

Renn, O. and Levine, D. (1991) Credibility and trust in risk communication. In R.E. Kasperson and P.J.M. Stallen (eds.), *Communicating Risks to the Public*. Dordrecht: Kluwer, pp. 175–218.

Robinson, L.A. and Hammitt, J.K. (2015) Research synthesis and the value per statistical life. *Risk Analysis*, 35(6), 1086–1100.

Rosa, E.A. (1998) Metatheoretical foundations for post-normal risk. *Journal of Risk Research*, 1, 15–44.

Rosa, E.A. (2003) The logical structure of the social amplification of risk framework (SARF): Metatheoretical foundation and policy implications. In N. Pidgeon, R.E. Kaspersen and P. Slovic (eds.), *The Social Amplification of Risk*. Cambridge: Cambridge University Press.

Ross, S.M. (2009) *Introduction to Probability Models*. 9th ed. Amsterdam: Academic Press. www.mobt3ath.com/uplode/book/book-41265.pdf.

Russell, J.A. and Barrett, L.F. (1999) Core affect, prototypical emotional episodes, and other things called emotion: Dissecting the elephant. *Journal of Personality and Social Psychology*, 76(5), 805.

Salignac, F., Marjolin, A., Reeve, R. and Muir, K. (2019) Conceptualizing and measuring financial resilience: A multidimensional framework. *Social Indicators Research*, 145, 17–38.

Saltelli, A. (2002) Sensitivity analysis for importance assessment. *Risk Analysis*, 22, 579–590.

Saltelli, A., Ratto, M., Andres, T., Campolongo, F., Cariboni, J., Gatelli, D., Saisana, M. and Tarantola, S. (2008) *Global Sensitivity Analysis: The Primer*. New York: Wiley.

Sandin, P. (1999) Dimensions of the precautionary principle. *Human and Ecological Risk Assessment*, 5, 889–907.

Sandin, P., Peterson, M., Hansson, S.O., Rudén, C. and Juthe, A. (2002) Five charges against the precautionary principle. *Journal of Risk Research*, 5, 287–299.

Santos, J.R., May, L. and Haimar, A.E. (2013) Risk-based input-output analysis of influenza epidemic consequences on interdependent workforce sectors. *Risk Analysis*, 33(9), 1620–1635.

Savadori, L., Savio, S., Nicotra, E., Rumiati, R., Finucane, M. and Slovic, P. (2004) Expert and public perception of risk from biotechnology. *Risk Analysis*, 24, 1289–1299.

Scherer, K.R. (1999) Appraisal theory. In T. Dalgleish and M. Power (eds.), *Handbook of Cognition and Emotion*. New York: Wiley, pp. 637–663.

Senge, P. (1990) *The Fifth Discipline: The Art and Practice of the Learning Organization*. New York: Doubleday Currency.

Shi, J., Visschers, V.H.M. and Siegrist, M. (2015) Public perception of climate change: The importance of knowledge and cultural worldviews. *Risk Analysis*, 35, 2183–2201.

Siegrist, M. (2020) Trust and risk perception: A critical review of the literature. *Risk Analysis*. doi:10.1111/risa.13325.

Siegrist, M. and Árvai, J. (2020) Risk perception: Reflections on 40 years of research. *Risk Analysis*, 40, 2191–2206.

Siegrist, M. and Cvetkovich, G. (2000) Perception of hazards: The role of social trust and knowledge. *Risk Analysis*, 20, 713–720.

Siegrist, M., Keller, C. and Kiers, H.A.L. (2005) A new look at the psychometric paradigm of perception of hazards. *Risk Analysis*, 25(1), 211–222.

Siegrist, M. and Sütterlin, B. (2014) Human and nature-caused hazards: The affect heuristic causes biased decisions. *Risk Analysis*, 34, 1482–1494.

Singpurwalla, N.D. (1988) Foundational issues in reliability and risk analysis. *SIAM Review*, 30(2), 264–282.

Sjöberg, L. (2000) Factors in risk perception. *Risk Analysis*, 220(1), 1–11.

Sjöberg, L. (2003) Risk perception is not what it seems: The psychometric paradigm revisited. In K. Andersson (ed.), *VALDOR Conference 2003*. Stockholm: VALDOR, pp. 14–29.

Slovic, P. (1987) Perception of risk. *Science*, 236(4799), 280–285.

Slovic, P. (1999) Trust, emotion, sex, politics, and science: Surveying the risk-assessment battlefield. *Risk Analysis*, 19(4), 689–701.

Slovic, P. (2000) *The Perception of Risk*. London: Earthscan.

Slovic, P. (2016) Understanding perceived risk: 1978–2015. *Environment*, 58, 25–29.

Slovic, P., Finucane, M.L., Peters, E. and MacGregor, D.G. (2004) Risk as analysis and risk as feelings: Some thoughts about affect, reason, risk and rationality. *Risk Analysis*, 24(2), 31–322.

Slovic, P., Fischhoff, B. and Lichtenstein, S. (1979) Rating the risks. *Environment*, 21, 14–20, 36–39.

SRA (2015) *Glossary Society for Risk Analysis*. www.sra.org/resources. Accessed November 12, 2020.

SRA (2017a) *Core Subjects of Risk Analysis.* www.sra.org/resources. Accessed November 12, 2020.

SRA (2017b) *Risk Analysis: Fundamental Principles.* www.sra.org/resources. Accessed November 12, 2020.

Stefánsson, O. (2019) On the limits of the precautionary principle. *Risk Analysis*, 39(6), 1204–1222. With Replay: 1227–1228.

Sternberg, R.J. (1999) *Handbook of Creative Thinking.* Cambridge: Cambridge University Press.

Stirling, A. (1998) Risk at a turning point? *Journal of Risk Research*, 1, 97–109.

Stirling, A. (2007) Science, precaution and risk assessment: Towards more measured and constructive policy debate. *European Molecular Biology Organisation Reports*, 8, 309–315.

Strack, F. and Deutsch, R. (2004) Reflective and impulsive determinants of social behavior. *Personality and Social Psychology Review*, 8, 220–247.

Sunstein, C.R. (2005) *Laws of Fear: Beyond the Precautionary Principle.* Cambridge: Cambridge University Press.

Szalay, L.B. and Deese, J. (1978) *Subjective Meaning and Culture: An Assessment Through Word Associations.* Hillsdale, NJ: Lawrence Erlbaum.

Taleb, N.N. (2007) *The Black Swan: The Impact of the Highly Improbable.* London: Penguin.

Taleb, N.N. (2012) *Anti Fragile.* London: Penguin.

Tannenbaum, M.B., Hepler, J., Zimmerman, R.S., Saul, L., Jacobs, S., Wilson, K. and Albarracín, D. (2015) Appealing to fear: A meta-analysis of fear appeal effectiveness and theories. *Psychological Bulletin*, 141(6), 1178–1204.

Taubenberger, J.K. and Morens, D.M. (2006). 1918 influenza: The mother of all pandemics. *Emerging Infectious Diseases*, 12(1), 15.

Teigen, K.H. (2012) Risk communication in words and numbers. In K.P. Knutsen, S. Kvam, P. Langemeyer, A. Parianou and K. Solfjeld (eds.), *Narratives of Risk: Interdisciplinary Studies.* Münster: Waxmann Verlag, pp. 240–254.

Thaler, R.H. and Sunstein, C.R. (2008) *Nudge: Improving Decisions About Health, Wealth, and Happiness.* New Haven, CT: Yale University Press.

Thekdi, S. and Aven, T. (2016) An enhanced data-analytic framework for integrating risk management and performance management. *Reliability Engineering and System Safety*, 156, 277–287.

Thekdi, S. and Aven, T. (2019) An integrated perspective for balancing performance and risk. *Reliability Engineering and System Safety*, 190, 106525.

Thekdi, S. and Aven, T. (2021a) A risk-science approach to vulnerability classification. *Risk Analysis.* doi: 10.1111/risa.13637.

Thekdi, S. and Aven, T. (2021b) Risk science in higher education: the current and future role of risk science in the university curriculum. *Risk Analysis.* doi: 10.1111/risa.13748.

Tijms, H. (2007) *Understanding Probability: Chance Rules in Everyday Life.* 2nd ed. Cambridge: Cambridge University Press.

Timmermans, S. and Esptein, S. (2010) A world of standards but not a standard world: Toward a sociology of standards and standardization. *Annual Review of Sociology*, 36, 69–89.

Trouwborst, A. (2016) Precautionary rights and duties of states. *Nova et Vetera Iuris Gentium*, 25. https://brill.com/view/title/12925.

Tuler, S., Kasperson, R.E., Golding, D. and Downs, T.J. (2017) How do we move forward when there is no trust? In R.E. Kasperson (ed.), *Risk Conundrums.* New York: Routledge, pp. 33–46.

Turner, B. and Pidgeon, N. (1997) *Man-Made Disasters.* 2nd ed. London: Butterworth-Heinemann.

Tversky, A. and Kahneman, D. (1974) Judgment under uncertainty: Heuristics and biases. *Science*, 185(4157), 1124–1131.

Tversky, A. and Kahneman, D. (1992) Advances in prospect theory: Cumulative representation of uncertainty. *Journal of Risk and Uncertainty*, 5, 297–323.

Tyler, T.R. (2000) Social justice: Outcome and procedure. *International Journal of Psychology*, 35, 117–125.

van Asselt, M.B., Vos, E. and Rooijackers, B. (2009) Science, knowledge and uncertainty in EU risk regulation. In M. Everson and E. Vos (eds.), *Uncertain Risks Regulated*. New York: Routledge, pp. 359–388.

van den Bos, K. (2005) What is responsible for the fair process effect. In J. Greenberg and J.A. Colquitt (eds.), *Handbook of Organizational Justice*. Mahwah, NJ: Lawrence Erlbaum, pp. 273–300.

van de Walle, S., van Roosbroek, S. and Bouckaert, G. (2008) Trust in the public sector: Is there any evidence for a long-term decline? *International Review of Administrative Sciences*, 74(1), 47–64.

Vanem, E. (2012) Ethics and fundamental principles of risk acceptance criteria. *Safety Science*, 50, 958–967.

Vapnik, V. (2013) *The Nature of Statistical Learning Theory*. New York: Springer.

Veland, H. and Aven, T. (2013) Risk communication in the light of different risk perspectives. *Reliability Engineering and System Safety*, 110, 34–40.

Viscusi, W.K. (2018) *Pricing Lives: Guideposts for a Safer Society*. Princeton, NJ: Princeton University Press.

Visschers, V.H.M., Keller, C. and Siegrist, M. (2011) Climate change benefits and energy supply benefits as determinants of acceptance of nuclear power stations: Investigating an explanatory model. *Energy Policy*, 39, 3621–3629.

Visschers, V.H.M. and Siegrist, M. (2013) How a nuclear power plant accident influences acceptance of nuclear power: Results of a longitudinal study before and after the Fukushima disaster. *Risk Analysis*, 33, 333–347.

Visschers, V.H.M. and Siegrist, M. (2014) Find the differences and the similarities: Relating perceived benefits, perceived costs and protected values to acceptance of five energy technologies. *Journal of Environmental Psychology*, 40, 117–130.

Visschers, V.H.M. and Siegrist, M. (2018) Differences in risk perception between hazards and between individuals. Chapter 3 in M. Raue et al. (eds.), *Psychological Perspectives on Risk and Risk Analysis*. Cham, Switzerland: Springer International Publishing AG, Part of Springer Nature.

Vlek, C. (2011) Straightening out the grounds for precaution: A commentary and some suggestions about Terje Aven's "on different types of uncertainties. . .". *Risk Analysis*, 31, 1534–1537.

Vose, D. (2008) *Risk Analysis*. 3rd ed. Chichester: Wiley.

Walls, J., Pidgeon, N., Weyman, A. and Horlick-Jones, T. (2004) Critical trust: Understanding lay perceptions of health and safety risk regulation. *Health, Risk & Society*, 6(2), 133–150.

WBGU (2000) *German Advisory Council on Global Change (Wissenschaftlicher Beirat der Bundesregierung Globale Umweltveränderungen). World in Transition: Strategies for Managing Global Environmental Risk*. Heidelberg: Springer Verlag.

Weber, P., Medina-Oliva, G., Simon, C. and Iung, S.B. (2012) Overview on Bayesian networks applications for dependability, risk analysis and maintenance areas. *Engineering Applications of Artificial Intelligence*, 25(4), 671–682.

Webler, T., Levine, D., Rakel, H. and Renn, O. (1991) The group Delphi: A novel attempt at reducing uncertainty. *Technological Forecasting and Social Change*, 39(3), 253–263.

Weick, K.E. and Sutcliffe, K.M. (2007) *Managing the Expected*. San Francisco: Wiley.

Weick, K.E., Sutcliffe, K.M. and Obstfeld, D. (1999) Organizing for high reliability: Processes of collective mindfulness. *Research in Organizational Behavior*, 2, 13–81.

WHO (2015) *Anticipating Emerging Infectious Disease Epidemics*. https://apps.who.int/iris/bitstream/handle/10665/252646/WHO-OHE-PED-2016.2-eng.pdf. Accessed May 6, 2021.

WHO (2020a) *Novel Coronavirus – China*. www.who.int/csr/don/12-january-2020-novel-coronavirus-china/en/. Accessed March 18, 2020.

WHO (2020b) *WHO Director-General's Opening Remarks at the Media Briefing on COVID-19,* February 27, 2020. www.who.int/dg/speeches/detail/who-director-general-s-opening-remarks-at-the-media-briefing-on-covid-19-27-february-2020. Accessed March 18, 2020.

WHO (2020c) *Coronavirus Disease 2019 (COVID-19) Situation Report – 58.* www.who.int/docs/default-source/coronaviruse/situation-reports/20200318-sitrep-58-covid-19.pdf?sfvrsn=20876712_2. Accessed April 1, 2020.

WHO (2020d) *Situation by Country, Territory & Area.* https://covid19.who.int/table. Accessed August 6, 2020.

Wiener, J.B. and Rogers, M.D. (2002) Comparing precaution in the United States and Europe. *Journal of Risk Research,* 5, 317–349.

Wilson, K., Leonard, B., Wright, R., Graham, I., Moffet, J., Phuscauskas, M. and Wilson, M. (2006) Application of the precautionary principle by senior policy officials: Results of a Canadian survey. *Risk Analysis,* 26, 981–988.

Windle, G., Bennett, K.M. and Noyes, J. (2011) A methodological review of resilience measurement scales. *Health and Quality of Life Outcomes,* 9(8).

Wolf, A. (2018) *Essentials of Scientific Method.* New York: Routledge.

Wolfs, F. (1996) *Introduction to the Scientific Method: Physics Laboratory Experiments, Appendix E, Department of Physics and Astronomy.* New York: University of Rochester.

Wu, J.T., Leung, K. and Leung, G.M. (2020) Nowcasting and forecasting the potential domestic and international spread of the 2019-nCoV outbreak originating in Wuhan, China: A modelling study. *The Lancet,* 395(10225), 689–697.

Wynne, B. (1992) Misunderstood misunderstanding: Social identities and public uptake of science. *Public Understanding of Science,* 1(3), 281–304.

Wynne, B. (2006) Public engagement as a means of restoring public trust in science – hitting the notes, but missing the music? *Public Health Genomics,* 9(3), 211–220.

Yadav, M. (2010) The decline of conceptual articles and implications for knowledge development. *Journal of Marketing,* 74, 1, January.

Yamaguchi, N., Kobayashi, Y.M. and Utsunomiya, O. (2000) Quantitative relationship between cumulative cigarette consumption and lung cancer mortality in Japan. *International Journal of Epidemiology,* 29(6), 963–968.

Yoe, C. (2012) *Principles of Risk Analysis.* Boca Raton, FL: CRC Press.

Zhen, X., Vinnem, J.E., Yang, X. and Huang, Y. (2020) Quantitative risk modelling in the offshore petroleum industry: Integration of human and organizational factors. *Ships and Offshore Structures,* 15(1), 1–18.

Zio, E. (2007) *An Introduction to the Basics of Reliability and Risk Analysis.* Singapore: World Scientific Publishing.

Zio, E. (2009) Reliability engineering: Old problems and new challenges. *Reliability Engineering and System Safety,* 94, 125–141.

Zio, E. (2013) *The Monte Carlo Simulation Method for System Reliability and Risk Analysis.* London: Springer Verlag.

Zio, E. (2016) Critical infrastructures vulnerability and risk analysis. *European Journal for Security Research,* 1(2), 97–114.

Zio, E. (2018) The future of risk assessment. *Reliability Engineering and System Safety,* 177, 176–190.

Index

Note: page numbers for figures are given in *italics*. For tables they are given in **bold**.

Printed in the United States
by Baker & Taylor Publisher Services